Darwin, Then and Now

iUniverse, Inc.
New York   Bloomington

Darwin, Then and Now
The Most Amazing Story in the History of Science

iUniverse books may be ordered through booksellers or by contacting:

iUniverse
1663 Liberty Drive
Bloomington, IN 47403
www.iuniverse.com
1-800-Authors (1-800-288-4677)

ISBN: 978-0-595-51375-8 (sc)
ISBN: 978-0-595-51575-2 (hc)
ISBN: 978-0-595-61871-2 (ebook)

Printed in the United States of America

iUniverse rev. date: 07/17/2009

Darwin, Then and Now

The Most Amazing Story in the History of Science

Richard William Nelson

I am quite conscious that my speculations run quite beyond the bounds of true science.

—Charles Darwin, 1857

Contents

Preface

Of the revolutionary thinkers who have shaped the history of the past century, Charles Darwin certainly stands as one of the most provocative and influential. *USA Today,* in a leading story in January 4, 1999, recognized Darwin as one of the top ten most influential persons of the twentieth century.[1]

Armed with the prestige of the Darwin family legacy, Charles Darwin was positioned for fame long before the HMS *Beagle* voyage even set sail in 1832. What has unfolded since has certainly become one of the most amazing stories in the history of science. The *Wall Street Journal* stated in an editorial in May 1999, "Whatever the controversies that surround him, Charles Darwin was certainly the most important natural scientist of the past century; he may become the most important social scientist of the next."[2]

The publication of *The Origin of Species* established Darwin as a cornerstone in emerging modern thought, which has clearly extended beyond the realm of natural sciences. In 1883, Friedrich Engels wrote, "As Darwin discovered the law of evolution in organic nature, so [Karl] Marx discovered the law of evolution in human history."[3]

Weighing in on the impact of Darwin, the eminent American philosopher John Dewey wrote in 1909, "The greatest dissolvent in contemporary thought of old questions, the greatest precipitant of new methods, new intentions, new problems, is the one affected by the scientific revolution that found its climax in *The Origin of Species*."[4]

At the turn of the century, the leading steel tycoon Andrew Carnegie commented on the validity of Darwin's theory, declaring, "There is no more possibility of defeating the operation of these laws (natural selection) than there is of thwarting the laws of nature which determine the humidity of the atmosphere or the revolution of the Earth upon its axis." Embracing Darwin's theory even changed Carnegie's perspective on life confiding, "Not only had I gotten rid of theology and the supernatural, but I had found the truth of evolution. 'All is well since all grows better,' became my motto, my source of comfort."[5]

Darwin's influence extends beyond the academic intellectual and industrial elite circles, even into the church. In speaking to his Sunday school class, American petroleum industrialist John D. Rockefeller said, "The growth of a large business is merely a survival of the fittest."[6] What is not widely recognized is that evolution had been widely popular even in the Church of England during the nineteenth century. After British Parliamentary deliberations, Darwin was eventually buried adjacent to Isaac Newton in the Westminster Abbey Nave.

Darwin has essentially become synonymous with the theory of evolution. Ironically, while Darwin's influence and popularity continues, Darwin's life and words have become largely a distant enigma. As pivotal as *The Origin of Species* has been, it is rarely studied, and almost never quoted. The question is why?

Evolution is intuitively intriguing. Ever since earning a bachelor of science degree from the University of California, followed by a doctor of pharmacy degree from the University of Southern California, studying the fascinating life and writings of Darwin has been a continued passion of mine. Since 2000, as a professional clinical pharmacologist, I have had the privilege to present excerpts of this material to thousands of people in the settings of junior and senior high schools, colleges and universities, and community centers.

Through a biographical and historical approach, this most amazing story in the history of science unfolds. This book highlights Darwin's life, the origins of evolutionary thought, Darwin's writings, and what scientists have discovered during the past 150 years.

To this end, the book includes over one thousand quotations and encompasses evidence from the fossil record, molecular biology, embryology, and genetics in the context of Darwin's theory of evolution. Most of the references attributable to Darwin are from *The Voyage of the Beagle, The Autobiography of Charles Darwin, 1809–1882,* Darwin's letters, and especially *The Origin of Species*, which is the focus of this book.

The first three chapters cover the motivational events of Darwin's life, followed by chapter four, which demonstrates how the stage was set for Darwin to gain an audience for *The Origin of Species*. Abandoning the scientific method for a subjective point of view is presented in chapter five. Chapter six addresses how Darwin handled the "species problem." Chapter seven is an exposé on natural selection following the VISTA format developed by Niles Eldredge for the touring *Darwin* exhibit of the American Museum of Natural History. Eldredge was a key player in the theory of natural selection, but even Darwin explained the theory is inconsistent. The chapter concludes with Darwin's top fifteen contradictions.

How the initial excitement over *The Origin of Species* culminated in the commission of the HMS *Challenger* by the British Parliament is incorporated in chapter eight. The last five chapters take a scientific method evidence approach to the history of evidence discovered in the fossil record, molecular biology, embryology, and genetics.

Though Darwin's theory was challenged and certainly abandoned by the mid-twentieth century, leading biologist Theodosius Dobzhansky of the California Institute of Technology captured the new emerging role of neo-Darwinian evolution, stating, "Nothing in biology makes sense except in the light of evolution."[7]

Two-time Nobel Prize winner Linus Pauling stated, "Science is the search for truth."[8] As the 1998 booklet published by the U.S. Academy of Science explains, "It is the nature of science to test and retest explanations against the natural world." The booklet continues, "All scientific knowledge is, in principle, subject to change as new evidence becomes available."[9]

In 2002, Nobel Foundation Board Chairman Bengt Samuelsson, quoting Israeli statesman Shimon Peres at the Nobel Prize Award Ceremony, stated, "Science and lies cannot coexist."[10] Truth reigns sovereign, even if unpopular. The history of science is replete with examples. Copernicus and Galileo debunked the idea that the Earth was the center of the universe—and paid the price.

Eighteenth century British medical doctor Edward Jenner was scorned for suggesting that an attenuated form of live smallpox should be injected into healthy people to ward off the deadly disease. It was not until 1980 that the World Health Organization finally announced the eradication of smallpox worldwide.

Nineteenth century Austrian physician Ignaz Semmelweis was ridiculed for suggesting that deaths among surgical patients resulted from the surgeon's hands. Today, infection control by washing hands is an essential component in every surgery. The history of biology follows successive waves of knowledge.

In 1880, Darwin's nineteenth century bulldog, Thomas Henry Huxley wrote, "History warns us ... that it is the customary fate of new truths to begin as heresies and to end as superstitions."[11] Darwin's goal was to find the evolutionary laws of nature as Isaac Newton had previously discovered the laws of gravity. This is an exposé on the life and works of Darwin and scientific discoveries during the 150 years of investigation.

As a challenge to embracing change, Darwin wrote, "Ignorance more frequently begets confidence than does knowledge." In arising to Darwin's challenge, now is the time to take the journey through the most amazing story in the whole history of science.

Chapter One
The Early Years

Upon the whole the three years which I spent at Cambridge were the most joyful in my happy life; for I was then in excellent health, and almost always in high spirits.

—Charles Darwin[1]

Charles Darwin and Abraham Lincoln were born on the very same day, February 12, 1809. Today, while both are honored on their countries' paper currency, Lincoln on the U.S. five-dollar bill and Darwin on the English ten-pound note, they were born into two different worlds, with two different destinies.

America was bracing for a civil war. England was on the verge of entering the Victorian era and the height of the Industrial Revolution with an unprecedented prosperity.

Abraham Lincoln was born in a one-room Kentucky log cabin. Charles Robert Darwin was born in a legendary estate. Lincoln was destined to free the American slaves; Darwin was destined to intellectually free minds from a divine creation. Lincoln sought the emancipation of men from men, and Darwin sought the emancipation of men from God. Lincoln died a martyr, and Darwin died in misery.

The Mount

Charles Robert Darwin was the son of Robert and Susannah Darwin. Darwin was born at the grand family estate known as "The Mount" in the beautiful "town of flowers," Shrewsbury, England. The Mount was built by Darwin's father in 1797 on two and a half acres, now called the "Darwin Gardens."

Things were run efficiently and orderly at The Mount. Susannah Darwin maintained a "perennial garden diary" to record the details of flowerings and fruiting in the kitchen garden in their pleasure gardens and glasshouses. Darwin was the fifth of six children: three older sisters Marianne, Caroline, and Susanne; one younger sister, Emily Catherine; and an older brother, Erasmus.

As a young child, Darwin was given the nickname of "Babba," taken from his middle name Robert. As a young teenager, his brother Erasmus just called him "Bobby."

The Darwin home was loving, caring, and cultured. Susannah Darwin skillfully used family teaching moments. When Darwin brought a flower to her, he remembers her saying, "by looking at the inside of the blossom, the name of the plant [can] be discovered;"[2] a lifetime lesson for a budding naturalist.

But in July 1817, when Darwin was only eight years old, his mother abruptly died at the age of fifty-two, leaving five children. Darwin recorded little other remembrances about her, writing, "It is odd that I can remember hardly anything about her except her death-bed, her black velvet gown, and her curiously constructed work-table."[3] Darwin attributed lack of remembering to his sisters. He wrote, "I believe my forgetfulness is partly due to my sisters, owing to their great grief, never able to speak about her or mention her name."[4]

School Days

Darwin along with his younger sister, Emily Catherine, was educated at home by their older sister Caroline until 1818, when their father enrolled them in Dr. Butler's boarding school in Shrewsbury, one mile from home. Darwin wrote that collecting insects was his greatest interest: "By the time I went to this day-school my taste for natural history was well developed."[5]

The family remained cohesive, however. Along with his brother Erasmus, Darwin grew to become a sportsman, riding horses and shooting game, especially birds, nearly to an extreme.

Collecting was soon to become a passion that Darwin would eventually weave into the history of Western civilization. Even as a young boy, Darwin was engaging. As is typical of healthy young boys, Darwin had a measure of mischievousness: "I may here also confess that as a little boy I was much given to inventing deliberate falsehoods, and this was always done for the sake of causing excitement. For instance, I once gathered much valuable fruit from my father's trees and hid it in the shrubbery, and then ran in breathless haste to spread the news that I had discovered a hoard of stolen fruit."[6]

As a boy, Darwin was a runner and racer, and often successful. In explaining the reason for success, Darwin wrote, "When in doubt I prayed earnestly to God to help me, and I well remember that I attributed my success to my prayers and not to my quick running, and marveled how generally I was aided."[7]

The active side of Darwin was balanced with long, solitary walks, hours of reading, and exploration: "I was fond of reading various books, and I used to sit for hours reading the historical plays of Shakespeare, generally in an old window in the thick walls of the school."[8]

Reading opened the world. The works of Milton, Gray, Byron, Wordsworth, Coleridge, and Shelley "gave me great pleasure," especially historical plays: "Early in my school days a boy had a copy of the *Wonders of the World*, which I often read and disputed with other boys about the veracity of some of the statements; and I believe this book first gave me a wish to travel to remote countries, which was ultimately fulfilled by the voyage of the *Beagle*."[9]

Walking developed into a favorite pastime, providing time for reflection. After walking, Darwin wrote, "what I thought about I know not."

In his autobiography, Darwin recalls that he once became so absorbed that "whilst returning to school on the summit of the old fortifications round Shrewsbury, which had been converted into a public foot-path with no parapet on one side, I walked off and fell to the ground, but the height was only seven or eight feet."[10]

School days were filled with boyhood activities. Darwin's greatest passion was shooting. Darwin wrote, "In the later part of my school life I became passionately fond of shooting, and I do not believe that anyone could have shown more zeal for the most holy cause than I did for shooting birds. How well I remember killing my first snipe, and my excitement was so great that I had difficulty in reloading my gun from the trembling of my hands."[11]

At Butler's school, Darwin pursued an extracurricular interest in chemistry with his older brother Erasmus, later writing that he "was allowed to aid him as a servant in most of his experiments."[12]

Encouraged by reading Henry and Parker's *Chemical Catechism,* on campus Erasmus and Darwin eventually became known for producing gases; Darwin picked up the nickname "Gas." Apparently, though, the school headmaster was not impressed. Darwin recalls, "I was also publicly rebuked by the headmaster, Dr. Butler, for thus wasting my time over such useless subjects; and he called me 'poco curante,' and as I did not understand what he meant it seemed to me a fearful reproach."[13]

Even though Darwin attended church, by the age of thirteen Darwin "swore like a trooper." In the end, Darwin remembers that the boarding school was a waste of time: "Nothing could have been worse for the development of my mind than Dr. Butler's school, as it was strictly classical, nothing else being taught, except a little ancient geography and history. The school as a means of education to me was simply a blank. During my whole life I have been singularly incapable of mastering any language."[14]

Apparently, the feeling was mutual. Neither did Darwin's outlook on education impress his teachers. Darwin recalls, "I believe that I was considered by all my masters and by my father as a very ordinary boy, rather below the common standard in intellect."[15]

By the age of sixteen, Darwin's father eventually took him out of the school because he was not paying attention, getting poor grades, and demonstrated excessive laziness. His father declared to him that he "cared for nothing but shooting, dogs, and rat-catching, and you will be a disgrace to yourself and all your family."[16]

None of this caused any anxiety for Darwin; he was destined to be heir to the Darwin family fortune. While leaving Dr. Butler's school, Darwin reflected later in his autobiography, "Soon after this period I became convinced from various small circumstances that my father would leave me property enough to subsist on with some comfort, though I never imagined that I should be so rich a man as I am; but my belief was sufficient to check any strenuous effort to learn medicine."[17]

Upon leaving Butler's school, even though taking on the name of Gas, Darwin was essentially an unremarkable student. Not one of his instructors considered him noteworthy. Taking on a positive approach to a negative experience, Darwin expressed his perspective on how to achieve success: "I am inclined to agree … that education and environment produce only a small effect on the mind of any one, and that most of our qualities are innate."[18]

The Darwin Family

Darwin's father, Robert Darwin, was a prosperous and prominent physician in Shrewsbury. He had the distinction of being a large man, some six feet and two inches in height, eventually weighing over 360 pounds. When Darwin inquired why he did not get out and exercise, he replied, "every road out of Shrewsbury is associated in my mind with some painful event."

Darwin had a strained relationship with his father. But, one of his father's golden rules, which Darwin remembered and attempted to follow was, "Never become the friend of anyone whom you cannot respect."

In the realm of science, Darwin claimed he had a different approach: "My father's mind was not scientific… yet he formed a theory for almost everything which occurred," much like his own father's father, Erasmus Darwin. Yet, Darwin was destined to follow in his father's footstep, too.

Darwin's grandfather, Erasmus Darwin, was also a prominent and wealthy English physician. As a physician in Lichfield from 1756 to 1781, he acquired a reputation for being a great healer. He was so successful that King George III asked him to be his doctor, but Erasmus Darwin refused the appointment.

Becoming a noted naturalist, writer, poet, and inventor during his own time, Erasmus' intellectual curiosity eventually led him to be one of the founding members of the Lunar Society. Members of this society were of influence, largely becoming the engine-driving force of the British Industrial Revolution.

Some things run in families, and this is particularly true in the Darwin family. Erasmus was the grandfather of Sir Francis Galton, one of the founders of eugenics. Eugenics uses a process of selective breeding to improve a species over generations.

As a writer, Erasmus authored several important works of poetry and of science. His most important published work was a book entitled *Zoönomia,* Latin for "law of life," published in 1794. In *Zoönomia,* Erasmus entertains the basic tenets of evolution and asks the question: "Would it be too bold to imagine that all warm-blooded animals have arisen from one living filament, which the great First Cause endued with animality... possessing the faculty of continuing to improve by its own inherent activity, and of delivering down these improvements by generation to its posterity, world without end?"[19]

The spontaneous origin of life and evolution formed the central theme. Erasmus was a pantheist, believing that God is everything and everything is God. Concepts of creation and evolution ran parallel in varying measures in the Darwin family. Erasmus was a Unitarian but maintained a public connection with the Church of England, the only officially established government church in England. In the Victorian era, professional Englishmen typically maintained their reputations and respectability by associating with the Church of England.

Erasmus engaged in marital and extramarital relationships. Erasmus married Mary Howard in 1757, and together they had fourteen children. Additionally, at least two and possibly three illegitimate children existed. She died at the age of thirty-one in 1770 from alcohol-induced liver failure. Erasmus then hired Mary Parker to look after the children. By late 1771, Darwin and Miss Parker had become involved, eventually having two daughters, Susanna Parker and Mary Parker. They never married.

In 1775, Erasmus met and developed an attraction to Elizabeth Pole, the wife of Edward Pole. Since she was married, he could only make his feelings known for her through poetry. Five years later, in 1780, Edward Pole died, and Darwin married Elizabeth Pole in 1781. They eventually had four sons, one of whom died in infancy, and three daughters. Later, when Erasmus died, he was buried in All Saints Church in Breadsall.

Like his grandfather, while Charles Darwin was baptized as a young boy in the Church of England, he regularly attended a Unitarian church with his mother. As a young boy, little did Darwin know that he would soon be studying his grandfather's work in college.

Edinburgh University

During the summer of 1825, Darwin was introduced to the practice of medicine by assisting his father as an apprentice, treating the poor of Shropshire. That autumn, at the age of sixteen, Darwin was sent by his father to Edinburgh University in Scotland to study medicine with his brother. Edinburgh was the leading European medical school of the day. Attendance at Edinburgh fulfilled a long Darwin alumni tradition.

At his father's insistence, Darwin was to study medicine and become a third-generation physician, continuing the Darwin physician mystique legacy. Since his father insisted that the practice of medicine was certain to "run in the family," Darwin was expected to follow suit.

That was the plan, but contrary to the insistence of his father, it became apparent that Darwin had little interest in studying medicine, or even in attending school. Darwin wrote that even the sight of surgery being performed was certainly beyond the scope of his interests. It even "haunted" him: "I also attended on two occasions the operating theatre in the hospital at Edinburgh, and saw two very bad operations, one on a child, but I rushed away before they were completed. Nor did I ever attend again, for hardly any inducement would have been strong enough to make me do so; this being long before the blessed days of chloroform. The two cases fairly haunted me for many a long year."[20]

Darwin's real passion, the study of nature, came to light during the second year at Edinburgh. On campus, the naturalist activities drew Darwin's attention. In these activities, Darwin became acquainted with

Professor Robert Edmund Grant, a proponent of evolution and student of Erasmus Darwin.

In his doctoral thesis, Grant quoted from Darwin's grandfather's book, *Zoönomia*. Evolution even at that time was strongly rooted in academic circles. Grant espoused the Lamarckian theory: evolution through acquired characteristics. In his autobiography, Darwin recalls an early conversion with Grant: "He one day, when we were walking together he burst forth in high admiration of Lamarck and his views on evolution. I listened without any effect on my mind. Nevertheless it is probable that the hearing rather early in life such views maintained and praised may have favoured my upholding them under a different form in my *Origin of Species*."[21]

Darwin developed a relationship with Grant through activities of the Plinian Society, a student forum for naturalists, even though Darwin was not fond of his grandfather's perspective on nature. Studying nature was Darwin's passion. Activities of the society included trawling for oysters with Newhaven anglers. In examining the oysters, Darwin discovered the differentiation between the ova and larva forms of the oyster. Darwin presented these findings to the Plinian Society in early 1826, joining the society later in the autumn of 1826. In time, Darwin became one of Grant's keenest students and assisted him with collecting specimens. Grant introduced Darwin to the academic elite of the day, connections that were to become invaluable for his future.

At Edinburgh, Darwin was taught taxidermy by John Edmonstone, a freed black slave from Guyana, South America. They met together often. John's vivid pictures of the South American tropical rain forests, along with the horrors of the slave trade, revealed a completely new realm to Darwin.

Passion for collecting, analyzing, and presenting specimens continued at Edinburgh to the point that Darwin's collecting expeditions went to extremes. Following one such episode, Darwin recorded: "One day, on tearing off some old bark, I saw two rare beetles, and seized one in each hand; then I saw a third and new kind, which I could not bear to lose, so that I popped the one which I held in my right hand into my mouth. Alas! It ejected some intensely acrid fluid, which burnt my tongue so that I was forced to spit the beetle out, which was lost, as was the third one."[22]

Darwin continued to assist Grant in collecting evidence for a "unity of plan" theory. One project culminated with Grant's announcement to the Wernerian Society that Darwin had identified the pancreas in shellfish, demonstrating similarity between animals. This similarity, referred to as homology, between animals was logically thought to be evidence for the "unity of life," and concepts of a Tree of Life theory. Darwin later applied these concepts of homology in the development of his theory.

Though sent to study medicine at Edinburgh, Darwin enrolled in Professor Robert Jameson's natural history course to learn about geology and assist with the collections at the Edinburgh University museum. For Darwin though, Jameson's lectures "were incredibly dull."

Actually, Darwin found the lecturing format, as a means of learning, to be a waste of time and found that Jameson's lectures "completely sickened me of that method of learning." At the time, Darwin resolved to never read a geology book again.

Opinions can change over time though. Eventually, Darwin took Sir Charles Lyell's geology book on his voyage around the world. Darwin studied Lyell thoroughly. The long geological ages envisioned by Lyell eventually persuaded Darwin to change his worldview. Ironically, geology played a foundational role in developing Darwin's theory.

Studying medicine was clearly not in the picture for Darwin. This became a great disappointment to his father. For Darwin, listening to lectures on any subject was too passive, boring, and dull. Being action-oriented, Darwin found hunting and collecting were certainly more interesting than sitting through a lecture. Darwin preferred to learn by reading and doing.

Even on his own admission, Darwin considered himself to be academically "rather below average." The disconnection with medicine was likely related to the learn-by-lecture format. Darwin's 1876 autobiography records his impression of Edinburgh, that "the instruction at Edinburgh was altogether by lectures, and these were intolerably dull, with the exception of those on Chemistry by Hope; but to my mind there are no advantages and many disadvantages in lectures compared with reading. Dr. Duncan's lectures on Materia Medica at 8 o'clock on a winter's morning are something fearful to remember." [23] Darwin continues, "Dr. Munro made his lectures on human anatomy as dull as he was himself, and the subject disgusted me."[23]

In the autumn, Darwin spent time at "Uncle Jos's" at Maer and at "Mr. Owen's" Woodhouse. At Woodhouse, Darwin was introduced to Fanny Owen by his sisters. They soon discovered they had common interests. Darwin began courting Fanny Victorian style: riding horses, shooting birds, and playing billiards. During this time, shooting and riding played the trump card. Darwin remembers that the "autumns were devoted to shooting, chiefly at Mr. Owen's Woodhouse, and at my Uncle Jos's, at Maer. My zeal was so great that I used to place my shooting boots open by my bed-side when I went to bed, so as not to lose half-a-minute in putting them on in the morning."[24]

The picture was getting clearer: Darwin's interests were in exploring nature and not in practicing medicine. But he could not get the courage to tell his father, especially since his older brother Erasmus had already given up studying medicine. Eventually his sisters broke the news to his father.

Christ's College, University of Cambridge

Fearing that Darwin would "ne'er do well," his father enrolled him at Christ's College, University of Cambridge, in 1827 to obtain a bachelor of arts degree in theology. A theology degree would qualify Darwin to become a clergyman in the Church of England—a guaranteed government professional.

For Darwin's father, this was seen as a sensible career move. A "living" as an English clergyman would at least provide a comfortable income. And, clergymen in the Victorian era were trained as naturalists. Studying nature and exploring the wonders of creation were essential for clergymen to gain an understanding of God's creative handiwork.

Studying nature was perfect for Darwin, but the aspect of becoming a clergyman was something new, but eventually Darwin "liked the thought." On signing the required paper that infers acceptance of the Thirty-nine Articles of the Church of England to enter Christ's College in 1828 at the age of nineteen, Darwin wrote in his autobiography, "I asked for some time to consider, as from what little I had heard or thought on the subject I had scruples about declaring my belief in all the dogmas of the Church of England; though otherwise I liked the thought of being a country clergyman."[25] Fully embracing the Bible, Darwin continues, "I did not then in the least doubt the strict and literal truth of every word in the Bible, I soon persuaded myself that our Creed must be fully accepted."[25]

For Darwin, interest in the Bible was more than a passing intellectual pursuit. In the characteristic free-spirit legacy, Darwin recalls, "inventing day-dreams of old letters between distinguished Romans and manuscripts being discovered at Pompeii or elsewhere which confirmed in the most striking manner of all that was written in the Gospels."[26]

While interests can be motivating, passion is life's driving force. At Cambridge, Darwin continued to be passionate about riding and shooting with his cousin William Darwin Fox. Darwin eventually became engrossed in the craze for the competitive collecting of beetles, writing, "No pursuit at Cambridge was followed with nearly so much eagerness, or gave me so much pleasure as collecting beetles."[27]

In fueling the intellectual side of his passions, Darwin became deeply rooted in developing a relentless thirst for reading. Reflecting on his time at Cambridge, Darwin wrote, "I read with care and profound interest Humboldt's 'Personal Narrative.' This work, and Sir J. Herschel's 'Introduction to the Study of Natural Philosophy,' stirred up in me a burning zeal to add even the most humble contribution to the noble structure of Natural Science." [28] These became the most influential books in Darwin's life. Darwin continued: "No one or a dozen other books influenced me nearly so much as these two."[28]

Reading Alexander von Humboldt's book, *Personal Narrative*, introduced Darwin to the area known as Tenerife, the largest of the seven Canary Islands in the Atlantic Ocean off the coast of Africa, and sowed more seed for his emerging global destiny. Searching beyond the shores of the isles was the continuing quest of the emerging generation; the British Empire was still expanding. Writing in his autobiography, Darwin recalls, "I had talked about the glories of Tenerife, and some of the party declared they would endeavour to go there; but I think that they were only half in earnest. I was, however, quite in earnest, and got an introduction to a merchant in London to enquire about ships; but the scheme was, of course, knocked on the head by the voyage of the *Beagle*."[29]

In reflection later in life, Darwin concluded that Humboldt's *Personal Narrative* was the single most influential book in his life, followed by Sir John Herschel's *Introduction to the Study of Natural Philosophy.*

Cambridge, like other elite university campuses, had a reputation for catering to young men just like Darwin who had probably too much money and too little discipline. A publication of the day described "in lurid detail the 'corrupt state' of the university: habitual drunkenness, gambling, and falling into debt; a profligacy so common that one could hardly find a female servant in a university lodging house who had managed to preserve her virtue; and a condition of moral laxity in which the highest aspiration was to be recognized as an authority of food and drink."[30]

Not immune to the range of available opportunities, Darwin became active to the point of excess, even at the Gourmet Club. Writing in his autobiography Darwin recalls, "Although as we shall presently see there were some redeeming features of my life at Cambridge, my time was sadly wasted there and worse than wasted. From my passion for shooting and for hunting and when this failed, for riding across the country I got into the sporting set, including some dissipated low-minded young men."[31]

Dining at Cambridge was often accompanied by liquid spirits. Darwin continues, "We used often to dine together in the evening, though these dinners often included men of a higher stamp, and we sometimes drank too much, with jolly singing and playing at cards afterwards. I know that I ought to feel ashamed of days and evenings thus spent, but as some of my friends were very pleasant and we were all in the highest spirits, I cannot help but looking back to these times with much pleasure."[31]

On this subject, Darwin eventually did admit to his son, Francis Darwin, that once he did drink too much while at Cambridge; but he wrote to Sir Joseph Hooker that he was drunk only "three times in early life." Biographer Peter Brent in 1981 encapsulated Darwin's undergraduate years: "The fact is that Charles Darwin was in almost all respects a fairly standard example of the nineteenth century student, well off, active in field sports, working hard enough to avoid academic failure, but a long way from academic success."[32]

At Cambridge, Darwin even developed a taste for pictures and engravings with his friend Whitley, frequenting the Fitzwilliam Museum together. With his friend John Herbert, Darwin "also got into a music set," frequently visiting the anthems in King's College Chapel, even employing the choir boys for entertainment. Darwin recalls, "I sometimes hired the chorister boys to sing in my rooms."[33]

After his first year at Cambridge, Darwin started the summer at home in Shrewsbury. Later in June, he went to the Welsh coast at Cardigan Bay, taking a math tutor to improve his algebra, a subject he found very difficult to grasp.

The tutoring only lasted a few weeks, and then Darwin got back to fly fishing and collecting beetles. He also went on a reading tour at Barmouth with his Cambridge friends, Herbert and Thomas Butler. During the tour, Darwin confided with Herbert that he had serious doubts about entering the clergy. As the summer was ending, Darwin returned to court Fanny Owen at the Owen's estate.

Like the study of medicine at Edinburgh, in time it became obvious that the study of theology at Cambridge was certainly not one of his interests either. Charles summarized his experience: "During the three years which I spent at Cambridge my time was wasted, as far as academic studies were concerned, as completely as Edinburgh."[34]

What Darwin did have at Cambridge, however, was a group of like-minded friends with whom he could relax and enjoy life as a college student. As far as nature was concerned, though, Darwin was fortunate to have exciting and enthusiastic instructors who encouraged Darwin to continue studies in natural history.

During the winter break of 1828, Darwin visited London, where his brother Erasmus showed him around to the Royal Institution, Linnean Society, and Zoological Gardens, further igniting his interest in natural history. Afterward, Darwin returned to visit the Woodhouse to see his girlfriend, Fanny Owen.

It became obvious, though, that Darwin's passion was in collecting, riding, and hunting, not courting. During the winter break of 1829, Darwin stayed in Cambridge to hunt beetles without even visiting Fanny. By February 1830, Darwin's relationship with Fanny was well on the way to dissolution, and Fanny was being pursued by suitors that were more attentive. Just after a "little go" exam, similar to a mid-term test, at Cambridge, Darwin and Fanny broke up.

Final exams were looming, and Darwin was focusing on studying. It was during this time that Darwin became particularly enthused by William Paley's *Evidences of Christianity,* which espoused a divine design in nature. Paley wrote, "The marks of design are too strong to be gotten over. Design must have a designer. That designer must have been a person. That person is God." On commenting on Paley's work, Darwin wrote, "I did not at that time trouble myself about Paley's premises, and taking these on trust I was charmed and convinced by the long line of argumentation."[35]

At Cambridge, Darwin's interest in Euclid's mathematics, and geometry equaled that of his interest in Paley's *Evidences of Christianity.* Darwin aligned with Paley's classic design perspective of creation. Darwin writing, "I am convinced that I could have written out the whole of the *Evidences* with perfect correctness... The logic of this book as I may add of his *Natural Theology* gave me as much delight as did Euclid."[36]

In retrospect, the years at Cambridge were the best times of his life. Writing in his autobiography, Darwin reflected, "Upon the whole the three years which I spent at Cambridge were the most joyful in my happy life; for I was then in excellent health, and almost always in high spirits."[37]

Professors Henslow and Sedgwick

The time at Cambridge allowed Darwin to begin a lifelong friendship with Professor John Stevens Henslow, professor of mineralogy and of botany. Darwin had heard of Henslow earlier from his brother, Erasmus, who revered the professor as "a man who knew every branch of science."

Henslow held open house once every week, becoming a campus nucleus for the science intellectuals. After finally getting an invitation to the open house, Darwin was hooked. The teaching style of Henslow matched Darwin's learning style perfectly. Henslow used progressive teaching techniques, relying heavily on field and garden work and encouraging students to make their own observations. Darwin attended

Henslow's field trips assiduously, and was soon taking long, almost daily walks with Henslow.

Darwin became such a regular companion that he earned the title Henslow's "favourite pupil," and "the man who walks with Henslow." In time, it was Henslow that recommended Darwin for the *Beagle* expedition.

Henslow was "deeply religious, and so orthodox, that he told me one day, he should be grieved if a single word of the Thirty-nine Articles were altered. He was free from every tinge of vanity or other petty feeling; and I never saw a man who thought so little of himself or his own concerns."[38] Reflecting later in life, Darwin considered Henslow the single most influential person in his life and "influenced my career more than any other."[39]

From Edinburgh University, Darwin developed long-lasting friend-ships with leading men of science from a range of different disciplines. One of these was Professor Adam Sedgwick. Henslow introduced Darwin to Sedgwick.

As a professor of geology, Sedgwick eventually became recognized as one of the founders of modern geology. Sedgwick is noted for intro-ducing the geological term "Cambrian." The Sedgwick Museum of Earth Sciences, founded in his honor at the University of Cambridge, opened in 1904. It was while on an expedition with Sedgwick that Darwin received an invitation from Henslow that ultimately changed his life and the world.

Darwin continued to correspond with Sedgwick while aboard the *Beagle*. Their lifelong relationship, though, eventually turned rocky. After receiving a copy of *The Origin of Species* from Darwin, Adam Sedgwick wrote a letter to him in December 1859, "I laughed till my sides were almost sore."[40]

As an exploratory thinker, curiosity drove Darwin to discover things in a hands-on manner, writing, "I consider that all I have learnt of any value has been self-taught."[41] *Hmmm . . .*

15

Chapter Two
The Voyage

If you can find any man of common sense who advises you to go I will give my consent.

—Robert Darwin[1]

After a flurry of studying over Christmas break, in January of 1831, at the age of twenty-one, Darwin passed his examination for the bachelor of arts in theology, Euclid, and the classics from the University of Cambridge, finishing tenth out of a field of 178.

Darwin had a passion for reading. While studying at the university, Darwin was especially drawn to one of the most popular texts studied: Paley's *Evidences of Christianity*, which emphasizes divine natural designs.

Remaining at Cambridge for two more terms after passing the final examination, Darwin became obsessed with the desire to travel and began planning a trip to the Canary Islands while reading Humboldt's *Personal Narrative.* Darwin wrote his sister Caroline, "My head is running about the Tropics.… My enthusiasm is so great that I cannot hardly sit still.… I have written myself into a tropical glow."[2]

After graduation, Henslow persuaded Darwin to pursue studies in geology. Henslow arranged for Darwin to accompany Sedgwick on field trips through Wales during the summer of 1831. Little did they realize just how soon Darwin would take his newfound geological training around the world. Darwin recalled, "This tour was of decided use in teaching me a little how to make out the geology of a country."[3]

Continuing to be inspired by Alexander von Humboldt's book, *Personal Narrative,* which Darwin said was "the parent of a grand progeny of scientific travelers," he wanted to study natural history in the tropics, and planned to visit Madeira with one of his classmates, Marmaduke Ramsay, after graduation. While tentatively approving the trip, his father left it to Darwin to work out the logistics and expenses.

As a stroke of fate, while Darwin was on a geological surveying tour in Wales with Sedgwick, Darwin received a message that Marmaduke had died, completely dashing the Madeira plans. But awaiting Darwin upon returning home was a letter from Professor John Henslow, with the offer of a lifetime. Darwin wrote, "On returning home from my short geological tour in N. Wales, I found a letter from Henslow, informing me that Captain Fitz-Roy was willing to give up part of his own cabin to any young man who would volunteer to go with him without pay as naturalist to the Voyage of the *Beagle*."[4]

Offer of a Lifetime

Henslow had previously been offered the post of a gentleman's naturalist companion to Robert FitzRoy, the captain of the HMS *Beagle,* for a planned two-year expedition to chart the coastline of South America in December. But Henslow had been dissuaded from accepting the offer by his wife.

Perceiving the perfect opportunity for his protégé, Henslow immediately dashed off a letter to FitzRoy recommending Charles Darwin. On August 24, 1831, Henslow wrote a letter to Charles Darwin stating that he had been recommended as being the most qualified candidate for the expedition.

On sharing the letter with his father, his father said, "If you can find any man of common sense who advises you to go I will give my consent."[5] Not knowing who to ask, on August 31, 1831, Darwin wrote to Henslow reluctantly turning down the offer.

By pure coincidence on the next day, Josiah Wedgwood II, Darwin's uncle, arrived to visit Darwin's father. Since Josiah was considered "one of the most sensible men in the world" by his father, Darwin discussed the situation with Josiah, who immediately made the case for Darwin to join Henslow on his expedition.[6]

Reluctantly, Darwin's father approved the invitation. Sealing the deal, Josiah offered to pay Darwin's cost for the two-year expedition; an expedition that would eventually stretch to nearly five years. The next day Darwin left for Cambridge to meet with Henslow to intercept the letter he had just sent. On September 5, 1831, Henslow introduced Darwin to FitzRoy in London.

Earlier, FitzRoy had asked a friend to accompany him on the *Beagle*, but just moments before Darwin's appointment, FitzRoy's friend informed him that he was no longer able to leave his job for the lengthy voyage.

FitzRoy was a wealthy nobleman, a descendant of the Duke of Grafton, and the Marquis of Londonderry. He was widely admired for his tight reign on his men, but as Darwin was soon to discover, his commanding was accompanied by a fiery temper.

At the age of twenty-six, FitzRoy was not much older than Darwin. At first, FitzRoy was not impressed with Darwin. FitzRoy thought the shape of Darwin's nose was too weak to take a lengthy sea voyage.

Eventually, Captain FitzRoy accepted Henslow's recommendation. Darwin was appointed to be a "gentleman's naturalist" and assist the "official" naturalist, surgeon Robert McKormick. As a paying passenger, Darwin was granted full use all the onboard facilities to perform research as a naturalist.

FitzRoy outlined the details of the voyage, including the impending sail date, October 10. Not wasting any time, Darwin took up residence at 17 Spring Gardens in London and began shopping and discussing the details of the voyage with FitzRoy; a dynamic relationship had just been launched.

Convinced "that he would find scientific proof that Genesis was literally true," FitzRoy wanted a like-minded naturalist on board the *Beagle* to find the evidence. Darwin's interest in Paley's perspective on nature made Darwin the perfect applicant. Ironically, prior to leaving England, FitzRoy gave Darwin a copy of the just-released first volume of Charles Lyell's *Principles of Geology,* which argues in favor of a long geologic Earth history, something FitzRoy would later regret, forever. For Darwin, the voyage was the chance of a lifetime.

HMS *Beagle* and Crew

Originally, as a ninety-foot-long, twenty-four-foot-wide, ten-gun, two-square–rigged-mast vessel, the HMS *Beagle* was launched from the Woolwich dockyard on the River Thames in May 1820. Since there was no immediate need for the brig, the *Beagle* was placed in reserve.

Then five years later, in September 1825, the *Beagle* was pulled out of reserve and dry-docked at Woolwich to be refitted for her new duties in the navy's new surveying program. Her guns were reduced from ten cannons to six, and her main deck was raised by eighteen inches to increase the space below the main deck.

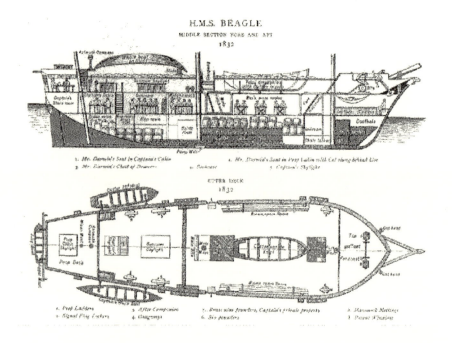

A mizzenmast was added to improve her maneuverability, changing her from a brig to a barque, making her a three-masted vessel. The guns were retained only for emergencies, since the twenty-two chronometers on board for surveying and navigation easily could become unbalanced.

In May 1826, the *Beagle* set sail for her first voyage to survey South America under the command of Captain Pringle Stokes, with Lieutenant Robert FitzRoy on board as an expert meteorologist. The command changed while surveying Tierra del Fuego, an archipelago at the southernmost tip of South America. After becoming despondent, Captain Pringle Stokes intentionally shot himself, and he died eleven days later. FitzRoy took command of the *Beagle*.

On board returning to Plymouth, England, in October 1830 were four young natives from Tierra del Fuego. The Fuegians were taken hostage after stealing one of FitzRoy's boats. FitzRoy, seizing the opportunity, planned for them "to become useful as interpreters, and be the means of establishing a friendly disposition towards Englishmen on the part of their countrymen."[7]

Captain FitzRoy, who had a missionary zeal, reasoned, "If these 'Indians' resided some years among 'the civilized,' they would, upon their return, transfer to their relatives the rudiments of civilization." Once the Fuegians were on board, FitzRoy treated them as first class passengers; they were scheduled to eat before the officers, and the crew assigned nicknames to the four of them: York Minster, Jemmy Button, Fuegia Basket, and Boat Memory. Back in England, FitzRoy paid for all of their expenses, including educational expenses.

The intrigue of the Fuegian's arrival made *Beagle* well-known, even from her first maiden voyage. Newspapers throughout England soon started publishing details of these exotic guests; they became immediate celebrities. In London, they met King William IV and Queen Adelaide.

Queen Adelaide personally gave young Fuegia Basket, the only girl, a bonnet. In Tierra del Fuego, Jemmy Button was "paid for" with a mother of pearl button, hence his name. Two of the Fuegians married while in England. Boat Memory died of smallpox shortly after arrival in England.

Captain FitzRoy had championed the Fuegian project without the approval of the Royal Navy and without a government contract. Eventually, FitzRoy was granted permission for another surveying expedition, but at his own expense, to return the three remaining Fuegians. This now-famous second expedition became known as the *Voyage of the Beagle*.

Success of leaders is measured by their skill in commanding the respect and allegiance of those under their command. To this end, Captain FitzRoy was highly respected; two-thirds of the crew on the previous voyage "signed-up" for the next voyage.

Darwin's first night on board the *Beagle* was on December 3, 1831. He was given quarters in the Chart Room just one deck above Captain FitzRoy's quarters at the stern of the ship. Darwin hung on a hammock over a table, looking up at the stars through a skylight that Captain FitzRoy had installed for sleeping.

Living was compact. The Chart Room was just nine by eleven feet, with only five feet of headroom. The walls were lined with bookshelves, cabinets, an oven, and a wash stand leaving only about six feet by eight feet of open space. The Chart Room even included a mizzenmast extending through the quarters, and a large four-foot-by-six-foot chart table located in the middle of the room. Little did Darwin realize that these would be his living quarters at sea for the next five years.

Plymouth, December 1831

After having been twice driven back by heavy southwestern gales, the *Beagle* finally successfully set sail just two days after Christmas near noon on December 27, 1831, with a crew of seventy-four under clear skies and with finally a good wind. The twenty-two-year-old crew member Charles Darwin held the position of a "gentlemen's" naturalist.

Shortly after setting sail, seasickness took hold of Darwin almost immediately. Darwin wrote, "The misery I endured from seasickness is far beyond what I ever guessed at."[8]

The *Beagle* was originally chartered for a two-year expedition. From England, the *Beagle* was charted to sail down the east coast of South America across the South Pacific to the Atlantic, following northward along the west coast of Africa back to England. Along with returning the Fuegians, the expedition was to survey the coastlines, chart the harbors, and collect information on the natural resources.

With the expanding British Empire, the surveying was to produce charts and drawings of the terrain and hills as seen from the sea, with height measurements to aid navigation. In particular, the exact longitude of Rio de Janeiro was to be determined using calibrated chronometers. The measurements were then checked through repeated astronomical observations to correct previous discrepancies, and the tides and meteorological conditions were recorded.

Lesser priority was given to surveying the approaches to harbors on the Falkland Islands and, season permitting, the Galápagos Islands. Then the *Beagle* was to proceed to Tahiti and on to Port Jackson, Australia, both of which were points known to verify the chronometers. An additional requirement was for a geological survey of a circular coral atoll in the Pacific Ocean, including investigation of its profile and of tidal flows.

For the voyage, the ship was supplied with over six thousand cases of vegetables, tinned meats, and barrels of lime juice, medicines, and preservatives for specimens. Eventually, after displacing more than five hundred tons, the *Beagle* finally set sail.

Canary Islands, January 1832

The first scheduled stop was Darwin's long-dreamed-of island, the Tenerife Island in the Canary Islands. However, because of an outbreak of cholera in England, the *Beagle* was prevented from landing on the islands. Waiting through twelve days of quarantine was an option, but Captain FitzRoy gave orders for the ship to set sail. Darwin was devastated, missing the chance to explore the island, and he watched the silhouette of Tenerife vanish beyond the horizon.

After twenty-one days at sea and just off the west coast of Africa, Darwin finally had the chance to depart and explore the volcanic island of St. Jago (now known as Santiago) in the Cape Verde Islands. On the island, Darwin made his first "discovery:" a horizontal white band of shells within a cliff face along the shoreline about forty-five feet above sea level. It is here that Darwin actually started the *Journal*, exploring the island for twenty-three days.

How to approach an investigation on a remote island was a new experience for Darwin. Being acutely aware of his inexperience, in a letter to Henslow, Darwin confided, "One great source of perplexity to me is an utter ignorance whether I note the right facts, and whether they are of sufficient importance to interest others."[9]

Exploring East South America, 1832–1834

Sailing west, the *Beagle* arrived at Salvador, Brazil, on February 28 at All Saints Bay. Finally ashore, Darwin once again pursued his favorite pastime—long walks. Darwin was living his dream, but this time in the tropics. In a letter to Henslow, Darwin wrote, "Here I first saw a tropical forest in all its sublime grandeur.... I never experienced such intense delight."[10]

Looming in striking contrast to the splendor of the landscape, though, Darwin found the horrors of slavery everywhere, and was long haunted by it: "To this day, if I hear a distant scream, it recalls with painful vividness my feelings ... Near Rio de Janeiro, I lived opposite an old lady, who kept screws to crush the fingers of female slaves. I have stayed in a house where a young household mulatto, daily and hourly, was reviled, beaten and persecuted enough to break the spirit of the lowest animal."[11]

Darwin's aversion to slavery quickly set him on the wrong side of Captain FitzRoy's irascible, hot temper. Darwin challenged FitzRoy and tempers clashed. Darwin wrote, "We had several quarrels; for instance, early in the voyage at Bahia, in Brazil, he defended and praised slavery, which I abominated, and told me that he had just visited a great slave-owner, who had called up many of his slaves and asked them whether they were happy, and whether they wished to be free, and all answered 'No.'"[12] But Darwin could not just leave the issue there: "I then asked him, perhaps with a sneer, whether he thought that the answer of slaves in the presence of their master was worth anything? This made him excessively angry, and he said that as I doubted his word we could not live any longer together."[12]

The other officers had nicknamed Captain FitzRoy "hot coffee" for such outbursts, but within hours, FitzRoy apologized and asked Darwin to remain. Darwin often bore the brunt of a good deal of laughter "from several of the officers for quoting the Bible as final authority on some moral point."[13]

Being acquainted to life on board the ship took some time. But gradually, Darwin discovered that life at sea was at least comfortable and convenient. Darwin wrote in a letter to his father, "I find to my great surprise that a ship is singularly comfortable for all sorts of work. Everything is so close at hand.... If it was not for seasickness the whole world would be sailors."[14]

The *Beagle* sailed further down the eastern seaboard of South America and finally dropped anchor at Rio de Janeiro, Brazil, on April 3, where the crew received its first mail from England. It was in Rio de Janeiro Darwin learned that his former girlfriend, Fanny Owen, had married a wealthy politician named Robert Biddulph.

Heartbroken, Darwin turned ever more passionately to collecting insects and plants in the tropical forest, this time with Patrick Lennon, a local English merchant. The inland trek to Rio Macao eventually extended to 150 miles and lasted eighteen days. By May 1832, thoughts of the new world were becoming an ever-increasing consuming fervor. In a letter to his second-cousin Fox, Darwin wrote, "My mind has been since leaving England, in a perfect hurricane of delight and astonishment, and to this hour scarcely a minute has passed in idleness."[15]

While Darwin was ashore collecting specimens, preserving them, taking notes, and writing letters, the *Beagle* sailed back to Salvador to perform survey readings. After returning from Salvador in early June, four crew members were gone; three crew members had died from a fever; and the ship's surgeon, Robert McKormick, after being upstaged by Darwin, was in the process of resigning and heading back to England on the HMS *Tyne*.

By British custom, the ship's surgeon traditionally took the position of the official "naturalist." Darwin's role was to be a "gentleman's naturalist" and assist McKormick and Captain FitzRoy. It was Darwin though, and not McKormick, who began receiving all the notoriety and the invitations from dignitaries on shore. Reasonably, McKormick felt upstaged by Darwin. The fame of the Darwin name was widespread; the Darwin's were the Kennedy's of the nineteenth century.

Being sufficiently disgruntled, McKormick left the *Beagle* at Rio de Janeiro. McKormick's status was "invalided out" back to Britain. Darwin assumed the quasi-official duties of the naturalist, and McKormick's assistant, Benjamin Bynoe, assumed the role of surgeon. Bynoe later made several voyages to Australia as a surgeon on convict ships.

Without question, Darwin was becoming a crew member to be reckoned with. As a symbol of his status, Darwin started accumulating a variety of new nicknames from the crew, like "Dear Old Philosopher," "Philos," and "Flycatcher."

From Brazil, having doubts about the quality of his collections and writings and wondering what Henslow might think, Darwin sent off his first load of specimens and notes to Henslow back at Cambridge. He included a personal note for Sedgwick: "Tell Prof: Sedgwick he does not know how much I am indebted to him for the Welch expedition.

It has given me an interest in geology, which I would not give up for any consideration."[16]

The shipment, while small, included several rocks, tropical plants, four bottles of animals in spirits, beetles, and various marine animals, all numbered, cataloged, and described.

By late August 1832, the *Beagle* reached the Patagonia coastline on the southern extreme of South America. Once ashore, Darwin joined in with the gauchos, using bolas, bringing down "ostriches," and eating roast armadillo. As an avid rider himself, and impressed by their equestrian finesse, Darwin had to comment, "The Gauchos are well known to be perfect riders. The idea of being thrown, let the horse do what it likes; never enters their head."[17]

Darwin spent weeks collecting fossils, of which he knew very little, but hoped they may be of some interest to Henslow back in England. Captain FitzRoy had a difficult time understanding why Darwin was bringing all sorts of apparently "useless junk" aboard the ship. The compact quarters of the ship lead to lively debates on all matters, which Darwin later described as quarrels sometimes "bordering on insanity."

In September, Darwin found fossilized bones of extinct giant mammals at Punta Alta. At first Darwin thought, they may be related to the rhinoceros, but he soon discovered that the bones were from a *Megatherium*, a giant ground sloth with large claws on its feet and weighing almost as much as an African bull elephant. For Darwin the excitement was mounting: "I have been wonderfully lucky with fossil bones. Some of the animals must have been of great dimensions: I am almost sure that many of them are quite new."[18]

By later in October 1832, Darwin sent his second shipment of notes, journals, and collections back to Henslow in England from Montevideo, Uruguay, and received a copy of the new second volume of Charles Lyell's book, *Principles of Geology.*

In *Principles of Geology,* Lyell departs from traditional geology and proposes that geological features developed as gradual processes over longer periods of time. Lyell's concept of long geological periods of time eventually became foundational to Darwin's theory of evolution.

While reading *Principles of Geology* on the voyage, Darwin wrote that he was "seeing" landforms as if through the eyes of Lyell.

After sailing through the straight of Le Maire and finally arriving at Tierra del Fuego later in December, the *Beagle* anchored at Good Success Bay. Landing was a challenge, and Darwin noted, "Whilst in Tierra del Fuego it is impossible to find an acre of land not covered by the densest forest."[19]

Falling under the ether of the unfolding landscape, the thoughts of Darwin turned toward the meaning and complexity of life—life and death. Observing life in Brazil and death in Tierra del Fuego, Darwin reflects, "Among the scenes which are deeply impressed on my mind, none exceed in sublimity the primeval forests undefaced by the hand of man; whether those of Brazil, where the powers of Life are predominant, or those of Tierra del Fuego, where Death and Decay prevail."[20] Darwin was exploring a way to unite life and death: "Both are temples filled with the varied productions of the God of Nature: no one can stand in these solitudes unmoved, and not feel that there is more in man than the mere breath of his body."[20]

Now it was time for the three Fuegians, York Minster, Jemmy Button, and Fuegia Basket, to depart the ship and return to living in their native homeland. For many of the crew, including Darwin, this was his first encounter with native Fuegians.

Initially the crew's efforts to communicate with the native Fuegians failed. But, when the natives were given gifts of bright red clothes, they began making friendly expressions by patting on the crew members' chests. The crew soon discovered the Fuegians had an amazing ability to mimic the crew's gestures, even speaking using complete English sentences.

By the beginning of 1833, work on founding a mission in Tierra del Fuego was well underway at Woolya Cove just off the Beagle Channel. Huts were built and gardens were planted. York Minster, Jemmy Button, and Fuegia Basket stayed to run the mission, along with the missionary Richard Matthews, to civilize the natives. After leaving generous supplies, the *Beagle* set sail.

When the *Beagle* returned just nine days later to check in, they were dumbfounded to discover that the mission had been completely looted by the natives. Richard Matthews, overwhelmed by the turn of events, returned to the *Beagle,* leaving York Minster, Jemmy Button, and Fuegia Basket on their own to run what was left of the mission.

In early March 1833, the *Beagle* landed in the Falkland Islands at Port Louis. The British Navy had just seized control of the islands from Argentina earlier in January. The British claim was contested. Darwin records: "After the possession of these miserable islands had been contested by France, Spain, and England, they were left uninhabited. The government of Buenos Aires then sold them to a private individual, but likewise used them, as old Spain had done before, for a penal settlement. England claimed her right and seized them. The Englishman who was left in charge of the flag was consequently murdered."[21] The British were not to concede, however. Darwin continues: "A British officer was next sent, unsupported by any power: and when we arrived, we found him in charge of a population, of which rather more than half were runaway rebels and murderers."[21]

While in the Falklands, Captain FitzRoy purchased a schooner to aid in the surveying work, but without permission from the admiralty. He named the schooner the HMS *Adventure* after a supply ship he used on the previous *Beagle* voyage.

After leaving the Falkland Islands, Darwin was let ashore at Maldonado, Uruguay, while the *Beagle* continued on to the larger city of Montevideo, Uruguay, arriving in early May. From Maldonado, Darwin commanded a twelve-day interior expedition, with two hired gauchos and a team of horses.

In a letter to his sister Catherine, Darwin asked his father to support the hiring of a servant for sixty pounds a year. His father approved the request, and Darwin hired Syms Covington, who had been the *Beagle's* odd-job man.

On July 18, more mail arrived, and the third shipment of specimens was sent back to Henslow. This third shipment consisted of about eighty species of birds, twenty quadrupeds, four barrels of skins and plants, geological specimens, and some fish. Completely intrigued by what he had already discovered, Darwin was even more ready to venture to the western side of South America and explore the unknown, vast ranges of the Andes Mountains.

After setting sail south, the *Beagle* made progress, arriving at the Rio Negro River in Argentina. Darwin coordinated an inland horseback expedition with Gauchos upstream to the town of Patagones, then overland to General Juan Rosas' camp on the Rio Colorado and received permission to proceed overland to Bahia Blanca. Commenting on the South American lifestyle, Darwin wrote, "There is high enjoyment in the independence of the Gaucho life—to be able at any moment to pull up your horse, and say, 'Here we will pass the night.'"[22]

At Bahia Blanca, Darwin discovered another very large fossil that was complete. Geological location of the fossil find was problematic. The location of the fossil was below a layer of white seashells, similar to the layer he found on the island of Santiago. This puzzled Darwin. How could the large fossil be located below an ocean deposit, not above? Darwin knew this observation contradicted what Lyell had proposed in his *Principles of Geology.*

This period of life was good for Darwin; the days were filled with riding on the plains, while the nights were spent drinking, smoking cigars, and singing songs with the gauchos. Darwin immensely enjoyed life on the open plains.

Darwin was developing a fascination with fossil collecting. This was fueled further by the discovery of the head of a fossilized *Toxodon* while exploring the Mercedes region of Uruguay. At the time, there was widespread flooding in the area delaying travel.

The adventure, though, did become a bittersweet experience. Darwin was bitten by an insect called the "Great Black Bug of the Pampas" or *Trypanosoma cruzi.* The bite was soon to become an immense influence on his health, lasting a lifetime. In his own words: "At night I experienced an attack (for it deserves no less a name) of the *Vinchuca,* a species of *Reduvius,* the great black bug of the Pampas. It is most disgusting to feel soft wingless insects, about an inch long, crawling over one's body."[23]

Civil unrest, though, was beginning to spread through Argentina. Feeling the pressure, Darwin moved as quickly as possible to board a cargo ship in Buenos Aires on November 2, bound to join the *Beagle* back at Montevideo, Uruguay. From Montevideo, Darwin shipped a fourth group of specimens back to Henslow in England. The shipment this time consisted of approximately two hundred animal skins, mice, a jar of fish, insects, rocks, seeds, and a growing collection of fossils and geological samples.

By late November, Darwin was back to Montevideo, wanting to get back on board the *Beagle,* even if it meant becoming seasick again.

Finding Jemmy Button, February 1834

The *Beagle* and *Adventure* returned to Woolya Cove in Tierra del Fuego on February 12, 1834, to check on the missionary camp. This time, however, they found that the mission had been completely abandoned; the planted gardens were left in ruins. Soon after landing, a few Fuegians began to arrive in canoes, including Jemmy Button. What was obvious was that the civilizing experiment was short-lived. In a letter to Catherine in April, Darwin wrote, "We could hardly recognise poor Jemmy. Instead of the clean, well-dressed stout lad we left him, we found him a naked, thin, squalid savage."[24] Jemmy was glad to see the crew, but did not want to rejoin the *Beagle* crew. Darwin continues: "Poor Jemmy was very glad to see us, and, with his usual good feeling, brought several presents (otter-skins, which are most valuable to themselves) for his old friends. The Captain offered to take him to England, but this, to our surprise, he at once refused. In the evening, his young wife came alongside and showed us the reason. He was quite contented."[24]

Certainly, the effect of the cultural exchange in England on the Fuegians had quickly vanished. Darwin records: "Nature by making habit omnipotent, and its effects hereditary, has fitted the Fuegian to the climate and productions of his miserable country."[25]

On February 12, 1834, Darwin's twenty-fifth birthday, Captain FitzRoy honored Darwin by naming the highest mountain in the region after him—Mt. Darwin, a 7,200 foot-high, snow-capped, glaciered peak located in the Andes. Captain FitzRoy had developed a sincere friendship with Darwin.

Challenged by West South America, 1834 to 1835

By this time, more than two years had elapsed since the *Beagle* left Plymouth, England. In early April 1834, Captain FitzRoy set sail for the Pacific Ocean. As the *Beagle* rounded Cape Horn through the Strait of Magellan and entered the Pacific Ocean, Darwin received a shipment of mail from another passing freight ship.

Sailing up the western coast of South America, the crew continued surveying the Chilean coast up to the island of Chiloe.

The *Beagle* and *Adventure* arrived at Valparaiso, Chile, near the city of Santiago later in July 1834. The warmer climate was a welcome experience for Darwin, and his physical condition was certainly more agreeable in the calmer Pacific Ocean. Over the next several weeks, both ships were refitted for the impending crossing of the Pacific Ocean.

In Valparaiso, Darwin found his old Shrewsbury classmate, Richard Cornfield. Darwin was invited to stay with Cornfield in the house he owned in the town. Once ashore, however, Darwin was anxious to lead an inland expedition. Arriving back at Valparaiso later in September 1834, Darwin discovered that he could not seem to recover from his new and lingering health problem: fevers would last several weeks at a time. Fortunately, Darwin was able to return to Cornfield's house, staying until late October 1834.

In an act he would later regret, Darwin wrote a letter to his sisters back home describing his adventures and revealing his new health status. Darwin's health deteriorated to a critical point. In appreciation for the care he received from Mr. Bynoe, the *Beagle's* onboard surgeon, and realizing the seriousness of his condition, years later Darwin wrote in the first paragraph of the preface in his autobiography: "I must take this opportunity of returning my sincere thanks to Mr. Bynoe, the surgeon of the "*Beagle*," for his very kind attention to me when I was ill at Valparaiso."[26]

Darwin's life was in the balance. Captain FitzRoy prayed that Darwin's life would be spared. Later, in appreciation for FitzRoy's compassion, Darwin named a newly discovered dolphin species *Delphiniums fitzroyi* in his honor.

For Darwin, though, an emerging version of the origin of life was developing. In 1834, while in Valparaiso, Darwin wrote in his diary, "It seems not very improbable conjecture that the want of animals may be owing to none having been created since this country was raised from the sea."

From Valparaiso, Darwin sent the fifth and last shipment of the voyage to Henslow, including bird skins, insects, seeds, plants, and water and gas samples from hot springs in the Andes. As Darwin's health was improving, the *Beagle* picked him up in early November and headed south to survey the Chronos Archipelago and the waters around Chiloe Island, and then sailed on towards the town of Valdivia.

Just prior to arriving in Valdivia, a massive earthquake followed by a tsunami on February 20, 1835, had devastated the town. The damage was extensive. Darwin wrote, "The island itself plainly showed the overwhelming power of the earthquake, as the beach did that of the consequent great wave.... Shortly after the shock, a great wave was seen from the distance of three or four miles, approaching in the middle of the bay with a smooth outline; but along the shore it tore up cottages and trees, as it swept onwards with irresistible force." [27] Darwin continues: "In my opinion, we have scarcely beheld, since leaving England, any sight so deeply interesting."[27]

Movements from the earthquake seemed to Darwin to support the concept that South America was very slowly moving and rising above the ocean. This phenomenon seemed to support Lyell's geological theory that the Earth had been changing incrementally over long periods of time. This was a revolutionary concept.

The massive earthquake was novel to Darwin, and intriguing. Investigating South American geology further, Darwin journeyed into the Andes in the spring of 1835. But rather than slow and incremental movements, Darwin wrote in April 1835 that the mountains were the result of violent massive movements acting as "tossed about like the crust of a broken pie."[28]

In July 1835, the *Beagle* sailed into Lima, Peru, where the crew began storing away provisions in anticipation of the next venture: exploring the vast realms of the Pacific Ocean. In Lima, Darwin received two letters from his sisters telling him how worried they had been about his being ill for such a long time at Valparaiso.

Fearing his health may be ruined for the rest of his life, his sisters pleaded with him to return to England at once. He immediately wrote a letter home stating he was resolute to see the voyage to the end, healthy or not. And early in September, the *Beagle* set sail from Peru to the now-famous Galápagos Archipelago, one of the most active volcanic areas in the world.

Thirty-Five Days on Galápagos Islands, 1835

After nearly four years since leaving England, the *Beagle* finally reached the first of the Galápagos Islands on September 16, 1835. Early the next morning a team set out on a boat to examine the shorelines, and by noon, the second boat was launched to survey the central islands of the archipelago. Later that afternoon, *Beagle* reached Chatham Island.

For Darwin, the islands were far from fascinating: "Nothing could be less inviting than the first appearance. A broken field of black basaltic lava, thrown into the most rugged waves, and crossed by great fissures, is everywhere covered by stunted, sunburnt brushwood, which shows little signs of life."[29]

Not only was the appearance uninviting, the weather was equally unwelcoming. Darwin continues: "The dry and parched surface, being heated by the noonday sun, gave to the air a close and sultry feeling, like that from a stove: we fancied even that the bushes smelt unpleasantly."[29]

Galápagos, meaning "tortoise" in Spanish, is a series of islands forming an archipelago: landforms consisting of a chain or cluster of islands. The Galápagos archipelago, located on the equator, includes more than twenty-four islands encompassing a combined landmass of approximately 2,800 square miles. Except for on the higher volcanic mountains' slopes, the Galápagos archipelago has little vegetation or cultivable soil.

The American author Herman Melville, stopping on the Galápagos Islands on the whaler *Acushnet* just shortly after the visit of the *Beagle,* vividly supports Darwin's impressions: "Take five-and-twenty heaps of cinders dumped here and there in an outside city lot—imagine some of them magnified into mountains, and the vacant lot the sea; and you will have a fit idea of the general aspect of the Encantadas, or Enchanted Isles."[30]

While acclimating to the islands, the collecting of specimens began slowly. On the second day, Darwin and the assistant surveyor, John Stokes, started from the northeast end of Chatham, eventually collecting ten plants.

Acclimating proved to be a challenge. On October 8, the *Beagle* anchored at the northern tip of James Island and met up with a party of Spanish settlers salting fish and extracting oil from tortoises. But the hot temperatures on the island continued to be an issue. Darwin records: "During the greater part of our stay of a week the sky was cloudless, and if the trade-wind failed for an hour the heat became very oppressive." Darwin continues that the sand was almost too hot, "even in thick boots it was quite disagreeable to walk over it."[31]

Beyond the hot weather and rough topography, though, Darwin discovered an abundance of life. Indeed the islands and the encircling ocean lagoons were teeming with life unique to the Galápagos Islands. During the twenty-one-day investigation, Darwin gathered a range of specimens. The collection of specimens included tortoises, some weighing up to 500 pounds; iguanas; and what was to become one of the most well-known natural collections in natural history—finches. With time running out, Darwin had collected the finches hurriedly and, unfortunately, with little or no documentation on the location of the finches. But, at the time, the importance of the Galápagos Islands collection was not apparent, much to his later chagrin. Darwin later confided in his autobiography: "Unfortunately most of the specimens of the finch tribe were mingled together."[32]

With surveying nearing completion and FitzRoy ready to set sail, Darwin's time on the islands was running out. Lamenting the brief stay, Darwin wrote a consolatory perspective: "It is the fate of every voyager, when he has just discovered what object in any place is more particularly worthy of his attention, to be hurried from it."[33]

After leaving, Darwin tried to reconstruct the locations by relying on the notes taken by other *Beagle* crew members and Captain FitzRoy.

From this "nothing could be less inviting" cluster of islands, the seeds of the theory of evolution were beginning to sprout. Darwin recorded in his *Journal:* "Seeing this gradation and diversity of structure in one small, intimately related group of birds, one might really fancy that from an original paucity of birds in this archipelago, one species had been taken and modified for different ends."[34]

Utterly Home Sick

Just thirty-five days after arriving in the Galápagos, on the evening of October 20, the crew, under full sail and a strong wind, set for the island of Tahiti. Tahiti is a long way from England. From this point in the voyage, it will be more than a year for the *Beagle* to complete the voyage back in Plymouth.

Before returning to England in October 1836, the *Beagle* ported in Tahiti, New Zealand, Australia, Tasmania, Cocos Islands, Mauritius Island, South Africa, and again, South America. Five years at sea took a toll on the twenty-seven-year-old: Darwin had developed a growing case of homesickness. While on the Mauritius Islands, Darwin confided to Caroline, "there is no country which has now any attractions for us, without it is seen right astern, and the more distant and indistinct the better. We are all utterly home sick."[35]

By August 1836, just two months from home and exhausted from seasickness and homesickness, Darwin became utterly dismayed when Captain FitzRoy decided to make an unscheduled detour to South America to gather more surveying measurements. Understandably, Darwin had now lost his love for life on the sea and wanted to get home, writing, "This zig-zag manner of proceeding is very grievous.... I loathe, I abhor the sea, and all ships which sail on it."[36]

The rigors of the voyage, though, paled in comparison to the passion for collecting and developing a unifying theory to explain the origin of species. It appeared that the origins of different animals were similar and connected, rather than being created to be unique, separate, and stable.

In September 1835, just nine months after leaving the Galápagos, reflecting on the "mingled" finches, Darwin wrote the central concept of his emerging theory in his *Journal*: "When I see these islands in sight of each other and possessed of but a scanty stock of animals, tenanted by these birds but slightly differing in structure and filling the same place in nature, I must suspect they are varieties ... if there is the slightest foundation for these remarks, the zoology of the archipelagoes will be well worth examining: for such facts would undermine the stability of species."[37] ?

Nearly a decade later in 1845, Darwin wrote in the second edition of the *Voyage of the Beagle,* and hinting at his emerging theory of evolution based on observing the Galápagos finches, "Seeing the gradations and the diversity of structure in one small, intimately related, group of birds, one might really fancy that from an original paucity of birds in this archipelago, one species had been taken and modified for different ends."[38] This is the essence of Darwin's theory that was to stay a guarded secret until challenged by Russel Wallace later in 1865.

Throughout the nearly five-year voyage, Darwin filled twenty-four notebooks with daily entries and sent thirty-nine letters back to England. Reflecting in the experience, Darwin wrote that the "voyage of the Beagle has been by far the most important event of my life and has determined my whole career."[42]

The greatest captivating interest during the voyage and driving force of the emerging theory was geology, however. For Darwin, geology took precedence over biology. Darwin wrote that the "investigation of the geology of all the places visited was far more important."[39] Darwin's biological perspective was emerging from Charles Lyell's new theory of geology.

Collecting biological specimens had become secondary to geology. By Darwin's own admission, the collection of specimens proved to be "almost useless." In his autobiography, Darwin acknowledges: "Another of my occupations was collecting animals of all classes, briefly describing and roughly dissecting many of the marine ones; but from not being able to draw and from not having sufficient anatomical knowledge, [it was] a great pile of MS [manuscripts] which I made during the voyage has proved almost useless. I thus lost much time."[40]

The expedition challenged Darwin. Earlier experiences on the British Isles were fading in comparison to new findings in the expansive new world; old concepts were in the process of being replaced by the new. Toward the end of the voyage on the *Beagle*, while still at sea, Darwin was well on the way to questioning the "stability of species" and ultimately the origin of species, following in the footsteps of his grandfather, Erasmus.

The Andes and Galápagos archipelago in the Pacific appeared to support Lyell's view on the longer age of the Earth. By 1844, Darwin wrote to Joseph Hooker, "At last gleams of light have come, and I am almost convinced (quite contrary to the opinion I started with) that species are not (it is like confessing a murder) immutable."[41]

Falmouth, England, October 1836

After more than five years, the *Beagle* finally pulled into the southern seaport town of Falmouth on October 2, 1836. With the sighting of England, and with the memories of exploring the inexhaustibly beautiful, diverse, and dangerous world, Darwin recalls his first thoughts: "My head is quite confused with so much delight."[42]

During the five years at sea, Darwin's interests began changing. While interests in shooting gradually waned, they were replaced by intellectual pursuits. Darwin wrote, "Gradually I gave up my gun more and more, and finally altogether to my servant, as shooting interfered with my work, more especially with making out the geological structures of a country."[43] Now, Darwin had a new passion: "I discovered, though unconsciously and insensibly, that the pleasure of observing and reasoning was a much higher one than that of skill and sport."[43]

Not only did Darwin have a new set of interests, but also apparently a new look. Speaking of seeing his father, Darwin wrote, "For on first seeing me after the voyage, he turned round to my sisters and exclaimed, 'Why, the shape of his nose is quite altered.'" On that night, although "confused," Darwin was ever more convinced who he was: "I was born a naturalist."[44, 45]

Darwin never left the shoreline of England again. Darwin took the enlightenment and unrelenting and grueling hardships of the voyage to construct a new purpose for life. Later, writing in his autobiography, Darwin explains: "As far as I can judge of myself, I worked to the utmost during the voyage from the mere pleasure of investigation, and from my strong desire to add a few facts to the great mass of facts in Natural Science."[46] Darwin continues to explain that his purpose was to achieve a noticeable place: "But I was also ambitious to take fair place among scientific men – whether more ambitious or less so then most of my fellow-workers, I can form no opinion."[46]

Without question, Darwin achieved reaching the goal. The name Darwin will now stand forever archived in the halls of the history of science as one of the world's most influential scientists.

Captain FitzRoy's Legacy

Captain FitzRoy was a legend in his own right. After entering Royal Naval College at the age of twelve, he promoted the term "port," replacing "larboard" since "larboard" was too easily confused with "starboard," and developed a new system of weather forecasting and storm warnings.

FitzRoy was the first to publish daily weather forecasts. The "Weather Book" that he published in 1863 was far in advance of the scientific opinion of the time, and remnants of the system still exist.

After returning from the voyage on the *Beagle*, FitzRoy entered public life by being elected a member of parliament for Durham in

1841; he later was appointed the Governor of New Zealand in the South Pacific, serving from 1843 to 1846.

In the summer of 1854, the British government decided to set up a meteorological office to look for new ways to understand the weather and improve the safety of shipping. Robert FitzRoy was chosen to be its first director. Eventually, FitzRoy was promoted to rear admiral in 1857 and vice admiral in 1865.

Later in his career, FitzRoy came to regret his decision to allow Darwin to take part in the voyage of *Beagle*. Both FitzRoy and Darwin were passionate about their work, but they had different perspectives. FitzRoy was devoted to the Church of England and a member of the conservative Tory party. While Darwin had a bachelor of arts degree in theology from Cambridge University, he was a member of the liberal Whig party.

In the ensuing years, FitzRoy progressively personalized the guilt and betrayal over the publication of *The Origin of Species*. Years later, following a very successful career, in a culmination of other personal and professional failures, FitzRoy took his own life.

Chapter Three
Sketching

When I am dead, know that many times, I have kissed and cried over this.

—Charles Darwin[1]

Just after returning from the HMS *Beagle* voyage in 1836, Darwin embarked on what were some of the most active years of his life. Darwin was motivated, wanting to "take a fair place among scientific men."[47] In his autobiography, Darwin states his merits: "I think that I am superior to the common run of men in noticing things which easily escape attention, and in observing them carefully. My industry as been nearly as great as it could have been in the observation and collection of facts."[2]

And indeed, Darwin sent a massive collection of specimens during the voyage to his once professor and now colleague, Henslow. During the voyage, Henslow became Darwin's liaison, distributing specimens for analysis and publishing extracts of Darwin's letters. By the time Darwin returned home on October 2, 1836, his credentials and future were up and running.

Wasting No Time

Darwin was driven; every minute counted. The use of time was a strict discipline, with every usable hour allotted. In Darwin's words, "A man who dares to waste one hour has not discovered the value of life."[3]

Darwin did vacation, but only for preparing to work on his projects. Darwin lived to work. While on a vacation, Darwin wrote in a letter, "We have come here for rest for me, which I have much needed; and shall remain here for about ten days, and then home to work, which is my sole pleasure in life."[4]

After visiting home in Shrewsbury, Darwin hurried back to Cambridge. Taking a full circle and realizing his son's emerging fame, Darwin's father reorganized investments, enabling Darwin to become a self-funded gentleman scientist.

The young and highly charged Darwin searched through London institutions seeking experts to analyze the specimens that he had collected and shipped back to England. By then, even Charles Lyell was eager to meet Darwin. They finally met for the first time on October 29, and Lyell wasted no time introducing Darwin to one of Britain's leading biologists, Richard Owen, who agreed to work on fossil bones collected in South America. Owen had access to the facilities at the Royal College of Surgeons to analyze specimens. Darwin was now working with the inner scientific circle, taking a "fair place among scientific men."

By mid-December 1836, Darwin completed his first paper, on the movement of South American landmass, and read it to the Geological Society of London on January 4, 1837. On the same day, Darwin presented his mammal and bird specimens to the Zoological Society.

In February 1837, Darwin was elected to the Council of the Geographic Society. In the presidential address to the society, Lyell presented Owen's findings from Darwin's fossils, associating geographical continuity with uniformitarian concepts. To be closer to the action, Darwin moved in March 1837 to an apartment on Great Marlborough Street in London, where Darwin would stay until 1839, when he married Emma Wedgwood.

"I Think"

By early 1837, while continuing write in the *Red Notebook,* which he started on the *Beagle,* Darwin started writing *First Notebook on the Transmutation of Species,* which eventually launched his greatest work, *The Origin of Species.* Darwin writes: "In July [1837] I opened my first note-book for facts in relation to the *Origin of Species,* about which I had long reflected, and never ceased working for the next twenty years."[5]

Darwin wrote in *First Notebook* on the now-famous sketch the words "I think." Darwin did not write, "based on the evidence, therefore."

While beginning to write in London, not only was Darwin closer to the ongoing specimen work, but he also joined the social scene.

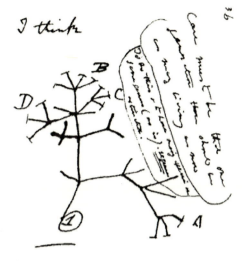

His brother, Erasmus, was associated with the freethinking Whig party and was a close friend of Harriet Matineau, who aligned with the concepts of Thomas Malthus. *Boo* Malthus, an economist, was an activist who promoted limiting the poor from breeding to prevent exhausting food resources.

Like today, London was the intellectual and social center of England. There Darwin "saw a great deal of Lyell" and began acting as an honorary secretary for the Geological Society. Darwin frequently took excursions from London, especially on the roads parallel to Glen Roy. Darwin also had some of his work published in *Philosophical Transactions*, which he later considered a "great failure." This lead Darwin to conclude: "My error has been a good lesson to me never to trust in science to the principle of exclusion."[6] For Darwin, the evidence was not the trump card.

Darwin continued his early childhood passion and remained an avid reader until later in life. While Darwin did extend his reading selection to include "some metaphysical books," Darwin explains, "but I was not at [all] well fitted for such studies."[7] Poetry and adventure books were his most favorite: "About this time I took much delight in Wordsworth's and Coleridge's poetry, and can boast that I read the *Excursion* twice through.... Milton's *Paradise Lost* had been my chief favorite."[7]

FitzRoy was planning to publish the accounts of the voyage. He invited Darwin to contribute to the captain's account of the voyage. Using the field notes he had sent home for his family to read, Darwin completed his part of the *Journal* by the summer of 1837. On Darwin's behalf, Henslow obtained a Treasury grant of 1,000 pounds to sponsor the publication of the *Voyage of HMS Beagle*.

Shortly afterward, earlier health problems returned. By September 1837, the "palpitations of the heart" returned. Taking the doctor's advice, Darwin went to Shrewsbury to spend time with the Wedgwood's at Maer Hall, but found them "too eager" for travel tales.

At the time, Darwin's charming, intelligent, and rather messy cousin Emma Wedgwood, nine months older than Darwin, was nursing his invalid aunt. Emma Wedgwood was the granddaughter of Josiah Wedgwood, who established the world-renowned pottery empire and had sponsored Darwin's voyage on the *Beagle*.

By February 1838, Darwin was beginning a new pocketbook, the maroon *C notebook*, and was investigating and documenting the breeding of domestic animals. During this time, Darwin was formulating concepts for *The Origin of Species* and the *Descent of Man,* hinting at ideas of adaptation to climate. In support of an emerging survival of the fittest theory, Darwin found the ideas in a pamphlet by Sir John Sebright intriguing: "A severe winter, or a scarcity of food, by destroying the weak and the unhealthy, has all the good effects of the most skilful selection. In cold or barren countries no animals can live to the age of maturity, but those who have strong constitutions; the weak and the unhealthy do not live to propagate their infirmities."[8]After reading the pamphlet, Darwin realized that he was aligned with Sebright's concept and commented: "excellent observations of sickly offspring being cut off."[9]

In time, Darwin extrapolated the emerging theory to include man. Darwin reflected on the sequence of emerging events in his autobiography: "As soon as I had become, in the year 1837 or 1838, convinced that species are mutable productions, I could not avoid the belief that man must come under the same law."[10] But Darwin judiciously withheld any discussion of the developing theory, stating, "Accordingly I collected notes on the subject for my own satisfaction, and nor for a long time with any intention of publishing."[10]

Networking

After the stint on the *Beagle,* Darwin never held another paying job, but he did hold a number of positions in several different professional societies. In March 1838, persuaded by William Whewell, Darwin accepted the position of Secretary of the Geological Society of London, a position he held until his health was too poor. His last geology excursion was in 1842, when he went to Wales to observe the evidence of glacial action.

Darwin's "mixed-up finches" were eventually handed over to John Gould, an ornithologist who identified the Galápagos finches as three separate species because of the "perfect gradation in the size of the beaks in the different species." Later, Darwin alluded to his emerging theory in 1839 in the first published account about the *Voyage of the Beagle*, stating: "Seeing this gradation and diversity of structure in one small intimately related group of birds, one might really fancy that from an original paucity of birds in this archipelago, one species had been taken and modified for different ends."[11]

The account was published in May 1839, in four volumes, as the *Narrative of the Surveying Voyages of HMS Adventure and Beagle*, with Darwin's *Journal and Remarks, 1832–1836* as the third volume. Darwin's contribution proved remarkably popular, and the publisher, Henry Colburn, took it upon himself to reissue the same text in August, with a new title page, as *Journal of Researches into the Geology and Natural History of the Various Countries Visited by HMS Beagle,* apparently without seeking Darwin's permission or paying him a fee. The book went through several editions, the best known being the second edition in 1845, and was published with several different titles. In 1842, Darwin published the *Sketch*, including an explanation for the life cycle of coral atolls in the South Pacific.

By now, Darwin had radically departed from his earlier position: "I did not then in the least doubt the strict and literal truth of every word in the Bible." The prospect of species changing over time continued to lurk in Darwin's mind. Species, "kind after kind," seemed to be more like "kind after another kind" that is a contradiction to the Genesis account and Paley's work, *Evidences of Christianity.*

By the spring of 1837, Darwin was becoming increasingly convinced that not only were all species descended from a previously existing species, but that the authority of the Bible was certainly in question. Darwin wrote a detailed explanation why he gradually turned against the authority of the Bible and Christianity:

During these two years (March 1837–January 1839) I
was led to think much about religion. Whilst on board
the Beagle I was quite orthodox, and I remember being
heartily laughed at by several officers (though themselves
orthodox) for quoting the Bible as an unanswerable
authority on some point of morality. I suppose it was
the novelty of the argument that amused them. But I
had gradually come by this time (i.e., 1836 to 1839) to
see the Old Testament, from its manifestly false history
of the world, with the Tower of Babel, the rain-bow as
a sign ... and from its attributing to God the feelings
of a revengeful tyrant, was no more to be trusted
than the sacred books of the Hindus, or the beliefs
of any barbarian.... I gradually came to disbelieve in
Christianity... beautiful as is the morality of the New
Testament, it can hardly be denied that its perfection
depends in part on the interpretation.[12]

While rejecting Christianity, Darwin was not willing to depart
from a belief of inventions. Darwin amalgamated a belief through
"daydreams." Darwin explains: "But I was very unwilling to give up
my belief; I feel sure of this for I can well remember often and often
inventing daydreams of old letters between distinguished Romans and
manuscripts being discovered at Pompeii or elsewhere, which confirmed
in the most striking manner all that was written in the Gospels."[13]

By June 1838, during the same period when Darwin was developing
his theory and rejecting the tenets of the Bible, he was laid up for days
with stomach problems, headaches, and heart symptoms. The crucible of
these events allowed Darwin to clarify his discordant issues. Reflecting
on these events, Darwin later wrote in his autobiography:

Thus disbelief crept over me at a very slow rate but at
least it was complete. The rate was so slow that I felt no
distress, and have never since doubted even for a single
second that my conclusion was correct. I can hardly see
how anyone ought to wish Christianity to be true; for
if so the plain language of the text seems to show that

the men who do not believe, and this would include my Father, Brother and almost all my best friends, will be everlasting punishment. This is a damnable doctrine.[14]

Even though Darwin claimed to have personal peace, episodes of stomach pains, vomiting, boils, and palpitations, often becoming more acute during times of stress, continued throughout his life.

Marry—Not to Marry

After recuperating from his June bout of illness, Darwin returned to Shrewsbury in July 1838, following a "geologizing" tour in Scotland. Jotting down notes on animal breeding, Darwin recorded rambling thoughts. On scraps of paper were written two columns headed "Marry" and "Not Marry." Advantages included "constant companion and a friend in old age ... better than a dog anyhow"; disadvantages listed were "less money for books" and "terrible loss of time."

Leaning toward marriage after consulting with his father, Darwin ventured on to Maer Hall, the Wedgwood's seventeenth century country estate, to find favor with Emma Wedgwood in July 1838. But rather than proposing, and against his father's advice, Darwin only discussed his theory, never broaching the big question.

In November 1838, Darwin returned to Maer Hall and finally proposed to Emma—along with discussing the theory. Darwin wrote that Emma had "grey eyes, a firm, humorous mouth, and rich chestnut hair."

Emma accepted, and they began exchanging love letters, showing how she valued his openness, but as an Anglican, expressing fears that his lapses of faith might endanger prospects of meeting in the afterlife. Emma wrote of her fiancée's many virtues to her favorite aunt, Jesse Sisimondi, in November 1838: "He is the most open, transparent man I ever saw, and every word expresses his real thoughts. He is particularly affectionate . . . and possesses some minor qualities that add particularly to one's happiness, such as not being fastidious, and being humane to animals."[15]

While confessing that the solitary time on the *Beagle* was "the commencement of my real life," the Sunday night before the wedding Darwin was looking forward to an even better life with Emma. Darwin writes, "I think you will humanize me, and soon teach me there is greater happiness, than building theories, and accumulating facts in silence and solitude. My own dearest Emma, I earnestly pray, you may never regret the great, and I will add very good deed, you are to perform on the Tuesday: my own dear future wife. God bless you."[16]

Within two months of proposing, Charles Darwin and Emma Wedgwood were married on Tuesday, January 30, 1839, at Maer Hall in an Anglican ceremony that had been arranged to accommodate Unitarians. Leaving on a train, the newly married couple went directly to their newly acquired "McCaw Cottage" on Upper Gower Street in London after the wedding. Just five days before the wedding, on January 24, Darwin was elected as a Fellow of the Royal Society.

The wedding continues a Darwin–Wedgwood legacy dating back several generations. It began with Charles and Emma's grandparents, Josiah Wedgwood and Erasmus Darwin. They were best of friends and associates in the pottery business dating back to 1780.

The families were united when Erasmus's son, Robert Darwin, married one of Josiah's daughters. Charles Darwin, Erasmus Darwin's grandson, lengthens the Darwin–Wedgwood legacy by marrying Josiah's granddaughter. Darwin and Emma were first cousins.

Josiah had been a leader in the English industrial revolution, transforming artisan pottery works into the first true pottery factory. In 1763, he patented beautiful cream-colored pottery. As this was very popular with Queen Charlotte, the wife of George III, the pottery became known as the "Queen's Ware."

Josiah Wedgwood was a Unitarian. As a Unitarian, Wedgwood worked for reform: political, social, economic, and biblical. In the town of his manufacturing plant, Josiah started a school in Etruria, appointing a Unitarian minister to run the school.

In this school Erasmus's son, Robert Darwin, and Charles's mother, Susannah Wedgwood, were educated. The Unitarian theology in the Darwin–Wedgwood family proved to be foundational to the free-thinking nature of the future generations. Darwin had long since affectionately called Emma's father "Uncle Jos." It was Uncle Jos's support that overrode Darwin's father's opposition to the voyage, and even paid Darwin's financial expenses during the five-year *Beagle* expedition.

Now, it was this double-barreled inheritance of the Darwin–Wedgwood legacy that eventually underwrote Darwin's life as a country gentleman and his development of the theory of evolution. Upon his father's death in 1848, Darwin inherited approximately 45,000 pounds. He had also received 13,000 pounds from his father upon his marriage in 1839 and a five-thousand-pound dowry that Emma Wedgwood brought into the marriage. Today, Charles Darwin stands as the most notable member of the Darwin–Wedgwood family legacy.

After the wedding, Darwin and Emma made London their home. Living close to the British Museum gave Darwin the opportunity to attend professional meetings and engage in research. The following December, as Emma's first pregnancy progressed, Darwin fell ill and accomplished little during the following year.

Darwin did accept a position on the Council of the Geographical Society in May 1840. By 1841, he could work only for short periods, a couple of days a week. Darwin produced a paper on stones and debris carried by ice floes, but his condition did not improve.

During this time, Richard Owen was one of the few scientific friends to visit Darwin. Owen's opposition to any hint of transmutation, however, was certainly a point of contention, so Darwin kept quiet about his theory.

Down

Darwin was in search of a better and healthier life. The city life in London was convenient but far from pristine. Darwin began looking to the country life: "I miss a walk in the country very much; this London is a vile smoky place, where a man loses a great part of the best enjoyments in life."[17]

After consulting his father, Darwin began looking for a house in the country to escape from the city, which was now suffering from poor air, economic depression, and civil unrest. Darwin pined that London "suited my health so badly that we resolved to live in the country, which we both preferred and have never repented."[18]

So the Darwin family moved to the village of Down in Kent on September 14, 1842, sixteen miles southeast of London. According to the census of 1841, Down had just 444 residents.

Emma was strong and spirited. When Darwin had gone looking for a house, she wrote, "It is as well that I am coming to look after you, my poor old man, for it is quite evident that you are on the verge of insanity." In looking for Darwin, Emma had some fun and published the following advertisement: "Lost in the vicinity of Bloomsbury, a tall thin gentleman quite harmless."

The newly acquired home in Down was spacious, and while under furnished, a comfortable place to raise a family. Their furniture was once described by their granddaughter as "ugly in a way, but dignified and plain." At first, the house had no running hot water, but it had two serviceable outhouses, as well as a study and a dining room; the Darwin's were set to raise their family. The Darwin home is now named the "Down House."

Darwin's study became an intellectual center, where questions and answers were continuously processed via the mail. From Down, Darwin orchestrated the ever-expanding intellectual revolution. Most of the letters were exchanged with his closest scientific friends—Charles Lyell, Joseph Hooker, Asa Gray, and Thomas Henry Huxley.

Darwin wrote thousands of letters, more than fifteen thousand which are still found in collections and libraries all over the world. Through the letters, Darwin gained access to the intellectual leaders of the day. Asa Gray was a professor at Harvard University in Cambridge, Massachusetts.

But the largest international audience was in Germany. Darwin corresponded with more than one hundred different scientists in Germany. Reflecting the worldwide British influence, many of Darwin's letters went to naturalists in India, Jamaica, New Zealand, Canada, Australia, China Borneo, and the Hawaiian Islands.

Over time, the Down House was gradually expanded to include two bathrooms. Situated on fifteen acres, with cherry and walnut trees as well as Scotch pine and Silver Fir trees, Emma and Charles provided their growing family with a loving and caring environment typical of an aristocratic English family living at the height of the Victorian age. Emma and Charles Darwin had ten children, seven reaching the age of maturity.

In the Down House, while Darwin launched his passion, "a considerable revolution in natural history," Emma became the family spiritual leader: regularly reading the Bible, teaching the Unitarian Creed, and baptizing all the children in the Church of England.

Darwin stood about five feet, eleven and a half inches and weighed, after his famous *Beagle* voyage, 148 pounds, gaining weight in his later years. He had dark brown hair, with a receding hairline, and thick curly sideburns. His piercing blue-gray eyes were made even larger by his bushy eyebrows.

As the years progressed, the hairline receded further, his face turned a healthy ruddy color, and his forehead developed deep horizontal wrinkles. By 1866, Darwin had grown a large, flowing, scruffy gray-white beard. Attending a meeting of the Royal Society, his closest friends hardly recognized him, since they had not seen him for some time.

Industrious Library

Writing was a way of life for Darwin, who was a compulsive note-taker and list-maker. As a husband, he recorded every household expense, and despite being independently wealthy, made year-end resolutions for the smallest of savings.

Darwin noted how the children cried and when they blushed. Later in life, suffering from chronic ill health, Darwin tracked and tallied his woes: each headache or bout of flatulence was noted, each morning and evening was ranked, and each month was proclaimed as one in which it either felt well or poorly.

In his lifetime, Darwin published twenty-five books, in addition to writing nearly fifteen thousand letters. Today, over nine thousand of these letters written by Darwin are at Cambridge University, and more than six thousand are known to be in private collections worldwide.

In developing the theory of natural selection and encouraged by his brother, Erasmus, Darwin found the key in 1838 by reading the sixth edition of *An Essay on the Principle of Population* by Thomas Robert Malthus, an English political economist. Darwin recalls in his autobiography the sentinel moments as he read the book:

> In October 1838 … I happened to read for amusement
> Malthus On Population, and being well prepared to
> appreciate the struggle for existence which everywhere

goes on from long-continued observation of the habits of animal and plants, it at once struck me that under these circumstances favorable variations would tend to be preserved, and unfavorable ones to be destroyed. The result of this would be the formation of new species. Here, then, I had at last got a theory by which to work; but I was so anxious to avoid prejudice, that I determined not for some time to write even the briefest sketch of it. In June 1842 I first allowed myself the satisfaction of writing a very brief abstract of my theory in pencil in thirty-five pages.[19]

Based on Malthus' premise, Darwin reasoned that as species breed beyond available resources, those with favorable variations will survive, and those with less favorable variations will become extinct. The preservation of favorable variations would eventually lead to the formation of new species. Darwin now had a logical theory—natural selection.

To test the waters, in January 1842 Darwin sent a tentative description of his ideas in a letter to Lyell, who was then touring America. Lyell wrote that he was dismayed that Darwin was becoming a "Transmutationist." Lyell was not persuaded by the concept of natural selection.

The anonymous best seller publication in October, *Vestiges of the Natural History of Creation,* dismissing Lamarckian evolution, also challenged Darwin's confidence. But the popular interest in the quest to answer the question of all questions, "Where did we come from?" was present. Darwin wrote that the topic was "in the air." The time for Darwin's entrance was emerging.

By July 1844, Darwin had expanded his sketch into a 230-page "essay," expanding his early ideas on natural selection by giving the analogy of overpopulation and competition leading to "natural selection" through the "war of nature" supplying the mechanism of common descent. Darwin wrote to Emma on July 5, 1844, "I have just finished my sketch of my species theory. If, as I believe, my theory in time be accepted by even one competent judge, it will be a considerable step in science."[20]

So confident in the theory, Darwin sent a letter sent to Emma that included a note with instructions to publish his essay in the event of his death, specifying a sum of 400 pounds to support the publication.

Darwin thought the concept "derogatory" that God would lower himself and create the world of parasites and worms. He believed that everything resulted from grand laws, stating that should "exalt our notion of the power of the omniscient Creator."[21] Inventing a new paradigm, Darwin envisioned that from "death, famine, rapine and the concealed war of nature we can see that the highest good, which we can conceive, the creation of the higher animals has directly come."[22]

Eventually, Darwin became a close friend of the botanist Joseph Hooker, and in January 1844 wrote to him of transmutation, describing it as being like confessing "a murder," hinting at transmutation's association with radicals. Hooker' replied cautiously, there "might have been a gradual change of species. I shall be delighted to hear how you think that this change may have taken place, as no presently conceived opinions satisfy me on this subject."[23]

During the 1840s, Darwin published three geological books, including *The Structure and Distribution of Coral Reefs* (1842), *Geological Observations on the Volcanic Islands* (1844), and *Geological Observations on South America* (1846).

Between Two Worlds

Darwin's health continued to be a lifelong issue. In an attempt to improve his chronic ill health, Darwin went to a spa in Malvern in 1849. To his surprise, he found two months of the water treatment to be of benefit.

Work on *The Origin of Species* was frequently a topic of discussion between Emma and Darwin. As a Christian, Emma supported Darwin, but was not comfortable in accepting the concept of evolution. In a letter to Darwin, Emma wrote: "The state of mind that I wish to preserve with respect to you, is to feel that while you are acting conscientiously and sincerely wishing and trying to learn the truth, you cannot be wrong, but there are sine reasons that force themselves upon me, and prevent myself from being always able to give myself this comfort."[24]

Emma did not share the same faith with Darwin, which distressed her greatly. Sensing Emma's hesitation and perhaps his own, on April

22, 1881, exactly thirty years after the burial of his daughter Annie', he reread Emma's letter of that time and the passages about Annie, and added a note: "When I am dead, know that many times, I have kissed and cried over this."[25]

When Annie, his treasured daughter, fell ill, it reawakened the fear that his own illness might be hereditary. After a long series of crises, Annie died on April 24, 1851; she was just over ten years old. In experiencing her death, Darwin struggled to reconcile how a beneficent God could allow death and destruction.

Determined not to allow ill health to undermine his goal to take a place among the men of science, Darwin struggled to continue his work. Charles Lyell continued to encourage Darwin to write out his theory. Darwin recalls, "Early in 1856 Lyell advised me to write out my views pretty fully, and I began at once to do so on a scale three or four times as extensive as that which … followed in my *Origin of Species*."[26]

Thus encouraged, Darwin continued to work on his theory. Even the friendship of Captain FitzRoy continued. FitzRoy continued to visit Darwin at his home right up until 1857. Darwin is thought to have used FitzRoy as a sounding board to test the theory.

The Wallace Letter

Darwin's work eventually led him into a corresponding relationship with the naturalist Alfred Russel Wallace, who was working in the islands of the South Pacific and Indonesia. At the time, Wallace had become one of Darwin's providers of natural history specimens.

On the morning of June 18, 1858, the arrival of one letter launched Darwin into the realm of no return. The work was progressing, but this letter caused an explosion in Darwin's life. Darwin records in his autobiography: "But my plans were overthrown, for in the early summer of 1858 Mr. Wallace, who was then in the Malay archipelago, sent me an essay *On the Tendency of Varieties to depart indefinitely from the Original Type*; and this essay contained exactly the same theory as mine. Mr. Wallace expressed the wish that if I thought well of the essay, I should send it to Lyell for perusal."[27]

Wallace had essentially composed the very same theory as Darwin. To the best of his knowledge, Darwin had never disclosed to Wallace any aspects of the theory he was developing. Darwin was shaken to the core. In his *Tendency of Varieties,* Wallace arrived at the same conclusion as Darwin "that there is a general principle in nature which will cause many varieties to survive the parent species, and to give rise to successive variations departing further and further from the original type."[28]

Darwin was stunned. Wallace had actually sent the letter to Darwin for passing on the paper to Lyell for presentation at the next Linnean Society meeting. Wallace, acting in good faith, was hoping that Darwin would personally lend support by introducing the letter to Lyell. Darwin could not let his friend down, but now Darwin's theory was in serious jeopardy of being superseded.

On the very same day, June 18, 1858, Darwin quickly sent a letter to Lyell realizing that even though his own originality would be lost, Darwin recommended that Wallace's paper should be accepted. Darwin wrote:

> My dear Lyell, Some year or so ago you recommended me to read a paper by Wallace in the Annals, which had interested you, and, as I was writing to him, I knew this would please him much, so I told him. He has to-day sent me the enclosed and asked me to forward it to you. It seems to me well worth reading. Your words have come through with a vengeance – that I should be forestalled. You said this, when I explained to you here very briefly my views of Natural Selection depending on the struggle for existence. I never saw a more striking coincidence; if Wallace had my MS, sketch written in 1842, he could have made a better short abstract. Even his terms now stand as heads of my chapters. Please return me the MS [manuscript], which he does not say he wishes me to publish, but I shall of course, at once write and offer to send to any journal. So all my originality, whatever it may amount to, will be smashed, though my book, if it will ever have any value, will not be deteriorated; as all the labour consists in the application of the theory. I hope you will approve of Wallace's sketch, that I may tell him what you say. My dear Lyell, yours most truly, C. Darwin.[29]

Now Darwin even doubted whether he should make his views public at all, for fear of being accused of copying Wallace. In a demonstration of his character, Darwin confides, "I would far rather burn my whole book, than that he or any other man should think that I have behaved in a paltry spirit."[30]

Concepts of evolution were "in the air." The intellectual avant-garde was in full swing. Speculative theories were enjoying wide currency and extending the enlightenment of intellectuals of Europe, including Jean-Baptiste Lamarck, and Darwin's own grandfather, Erasmus Darwin.

Through a series of fateful events, Lyell and Hooker intervened. They sent to the Linnean Society not only Wallace's manuscript but also abstracts from Darwin's 1844 *Essay* and an excerpt from a September 5, 1857, letter Darwin had recently written to Asa Gray, which were published in the *Journal of the Proceedings of the Linnean Society* in 1858.

Wallace and Darwin's papers were jointly read by Lyell and Hooker to the Linnean Society, in London, on August 20, 1858. Neither Darwin nor Wallace was present for the reading. Darwin was at home with his son, who was dying of scarlet fever, and was too distraught to attend. Wallace was in the Far East. The paper was entitled *On the Tendency of Species to form Varieties; and on the Perpetuation of Varieties and Species by Natural Means of Selection.*

The reading brought only a brief mention in a small review. Darwin recalls only one review written by Professor Haughton of Dublin: "Our joint productions excited very little attention, and the only published notice of them which I can remember was by Professor Haughton of Dublin, whose verdict was that all that was new in them was false, and what was true was old."[31]

Even though evolution was "in the air," neither evolution nor the theory of natural selection was popular even among Darwin's closest colleagues at the time. Darwin recalls that even "Lyell and Hooker, though they would listen with interest to me, never seemed to agree. I tried once or twice to explain to able men what I meant by Natural Selection, but signally failed."[32]

Over time, the collegial comrade between Darwin and Wallace has come to stand as one of the finest examples of collaboration in the history of science. In a letter to Wallace in 1870, Darwin confided, "I hope it is a satisfaction to you to reflect—and very few things in my life have been more satisfactory to me—that we have never felt any jealousy towards each other, though in one sense rivals. I believe that I can say of myself with truth, and I am absolutely sure that it is true of you."[33] Eventually, Wallace was one of Darwin's pallbearers.

Years later in 1903, the editor of *Black and White* asked Wallace to write an essay on his relationship with Darwin. Wallace gave Darwin credit for the theory of natural selection. Wallace considers his major contribution was to compel Darwin to publish. Wallace explains, "In conclusion, I would only wish to add, that my connection with Darwin and his great work has helped to secure for my own writings on the same questions a full recognition by the press and the public; while my share in the origination and establishment of the theory of Natural Selection has usually been exaggerated. The one great result which I claim for my paper of 1858 is that it compelled Darwin to write and publish his *Origin of Species* without further delay."[34]

For the next thirteen months, Darwin worked more intensely than ever to publish what was originally intended to be an abstract of his "big book on species." Never had Darwin worked with such intensity, and the work began to take a toll on Darwin's health, again. Just less than two weeks before publication of *The Origin of Species,* Darwin described his condition to his cousin Fox in a letter, stating, "I have had a series of calamities; first a sprained ankle, and then badly swollen whole leg and face; much rash and a frightful succession of Boils—4 or 5 at once. I have felt quite ill—and have little faith in this 'unique crisis' as the Doctor calls it, doing me much good. I cannot now walk a step from bad boil on knee."[35]

Emma was concerned about the effect of the stress on Darwin. Emma wrote to Darwin, "I am sure you know I love you well enough to believe that I mind your sufferings, nearly as much as I should my own, and I find the only relief to my own mind is to take it as from God's hand, and to try to believe that all suffering and illness is meant to help us to exalt our minds and to look forward with hope to a future state. When I see your patience, deep compassion for others, and above all for the smallest thing done to help you, I cannot help longing that these precious feelings should be offered to Heaven for the sake of your daily happiness."[36]

The Origin of Species

Along with receiving encouragement from his colleagues, the first edition of *On the Origin of Species by Means of Natural Selection and the Preservation of Favoured Races* finally went on sale November 24, 1859. Darwin called the book "one long argument." The controversial term "evolution" was not included in the book, only that "endless forms most beautiful and most wonderful have been, and are being, evolved." Reflecting later in his life, Darwin considered *The Origin of Species* his chief accomplishment.

The book was an immediate success, the *Harry Potter* of the nineteenth century. All 1,250 copies of the book were sold on the very first day. During the first year, 3,800 copies were sold. Unlike the reading of Darwin and Wallace's paper in 1858, book reviews on *The Origin of Species* were published in prominent periodicals. To avoid involvement in public controversy to save his health, Darwin skillfully let his friends defend him, his book, and his theory. Darwin never publicly discussed, presented, or defended *The Origin of Species,* nor any of his other works.

Darwin's good friend Captain FitzRoy, though, was so indignant about the publication that when the first public debate concerning the publication was held at Oxford in 1860, FitzRoy, then an admiral in the Royal Navy, appeared at the meeting waving the Bible and shouting that he had warned Darwin "against holding views contrary to the word of God."

Just as popularity of *The Origin of Species* was gaining momentum, Darwin's health was likewise gaining momentum, but in the wrong direction. In 1861, as Darwin was incapacitated for weeks at a time. Emma challenged Darwin to look to God for help, "I cannot tell you the compassion I have felt for all your suffering for these weeks past that you have had so many drawbacks … 'Thou shalt keep him in perfect peace whose mind is stayed on thee' … I feel presumptuous in writing to you. I feel in my heart your admirable qualities and feelings and all I would hope is that you would direct them upwards, as well as to one who values them above everything in the world."[37]

And Darwin did bounce back, continuing to work. In time, Darwin developed a closer personal friendship with Lyell. Lyell became one of the first prominent scientists to support *The Origin of Species*, while ironically never fully accepting natural selection as the driving engine behind evolution.

Reaction

Attention and controversy gathered. Today, *The Origin of Species* continues to be one of the most controversial and discussed books ever written.

Publication of *The Origin of Species* became the catalyst that popularized the revolution in theology, sociology, philosophy, and science. Standing in Darwin's way were leading biologists, including Louis Aggasiz and Richard Owen, who opposed Darwin and continued to believe that every form of life was created uniquely.

Owen declared that *The Origin of Species* symbolized an "abuse of science." While Owen supported natural selection as a form of adaptation, he opposed any hint of transmutation—the development of a new species. Owen held that natural selection allows for adaptation of a species to the conditions of life, but could never give rise to an entirely new and distinct species, as described by Darwin.

In 1860, a new fossil was discovered in Solnhofen, Bavaria. The fossil was named the *Archaeopteryx*. This one piece of evidence had two interpretations, and the debate developed along two lines: firstly, that the *Archaeopteryx* was a bird, and secondly, that the *Archaeopteryx* was a reptile that evolved into a bird—evidence of evolution predicted by Darwin.

Owen described the *Archaeopteryx* unequivocally as a bird and in January 1863 bought the fossil for the British Museum. Thomas Henry Huxley, who eventually became Darwin's bulldog of the nineteenth century, claimed the fossilized bird fulfilled Darwin's prediction that a protobird that was developing into a new species would be found. Debate over *The Origin of Species* was tumultuous from the beginning, never experiencing a honeymoon.

Initially, while Darwin did not explicitly apply evolution to man, the subject was intriguing. In a letter to Lyell in 1860, Darwin wrote, "You ask whether I shall discuss man. I think I shall avoid the whole subject, as so surrounded with prejudices; though I fully admit that it is the highest and most interesting problem for the naturalist."[38] Darwin was playing the role of a diplomat, since he had previously revealed his position on the evolution of man. In a letter to Hooker in 1857, Darwin confided, "I cannot swallow Man [being that] distinct from a Chimpanzee."[39]

The feuding between Owen and Darwin continued. In 1872, Owen recommended ending the government's funding of Hooker's botanical collection at Kew and to bring it under the control of the British Museum. Darwin, who enjoyed Hooker's advocacy, responded: "I used to be ashamed of hating him so much, but now I will carefully cherish my hatred and contempt to the last days of my life."[40]

In several of his later biological books, *The Variation of Animals and Plants Under Domestication* (1868), *The Descent of Man and Selection in Relation to Sex* (1871), and *The Expression of Emotions in Animals and Man* (1872), Darwin expanded on many topics first introduced in *The Origin of Species. The Descent of Man,* in particular, aroused even greater argument since it theorized that humanity descended from apes. Darwin had held this position since 1837. In his autobiography, Darwin records his position on the evolution of man:

> *My Descent of Man* was published in Feb. 1871. As soon as I had become, in the year 1837 or 1838, convinced that species were mutable products, I could not avoid the belief that man must come under the same law.... Although in The Origin of Species, the derivation of any particular species is never discussed, yet I thought it best, in order that no honorable man should accuse me of concealing my views, to add that by the work in question 'light would be thrown on the origin of man and his history.' It would have been useless and injurious to the success of the book to have paraded without giving any evidence of my conviction with respect to his origin.[41]

Over time, Darwin's resentment toward the Bible became more deeply rooted. Eventually Darwin was determined to develop natural selection to "overthrowing the dogma of ... creation." By 1871, in the publication of *Descent of Man,* Darwin reveals his emerging ultimate motivation: "If I have erred in giving to natural selection great power, which I am very far from admitting, or in having exaggerated its power, which is in itself probable, I have at least, as I hope, done good service in aiding to overthrow the dogma of separate creation."[42]

Darwin corresponded extensively with the thought leaders of the day. In a letter to Karl Marx in October 1873, Darwin wrote: "Dear Sir: I thank you for the honour which you have done me by sending me your great work on *Capital*; and I heartily wish that I was more worthy to receive it, by understanding more of the deep and important subject of political Economy. Though our studies have been so different, I believe that we both earnestly desire the extension of Knowledge, and that this is in the long run sure to add to the happiness of Mankind. I remain, Dear Sir, Yours faithfully, Charles Darwin."[43]

Despite the controversies of being the "monkey man," Darwin's work received widespread recognition. Across the Atlantic, Darwin was awarded honorary memberships in the United States' American Philosophical Society (1869), California Academy of Sciences (1872), and California State Geological Society (1877). Darwin was elected a member of the French Academy of Sciences (1878) and the New York Academy of Sciences (1879).

In 1877, Darwin was awarded an honorary doctorate in law from Cambridge University, his alma mater, as well as a doctor of medicine from Leyden. The continued popularity of Darwin today measures the importance of his contributions and stands in sharp contrast to one categorized as a "family disgrace" by his father.

During these years of theorizing and writing, Darwin emerged as a very successful financial investor. He had a strong business sense. During his adulthood, Darwin nearly quadrupled his inheritance and his estate to an estimated 282,000 pounds by skillfully investing in the railroad systems that were rapidly developing throughout the British Isles.

Theory and Theology

When challenged by a sermon given by the popular theologian E. B. Pusey in 1878, Darwin responded. In a letter Darwin wrote to N. H. Ridley while working on *The Origin of Species*, he said he definitely believed in a "personal God": "Many years ago, when I was collecting facts for the "Origin," my belief in what is called a personal God was as firm as that of Dr. Pusey himself."[44]

What Darwin believes a "personal God" means is further clarified in his autobiography. Darwin believed in the involvement of an intelligence in the beginning, but considers himself a "theist"—one who believes in the existence of God, but uninvolved with the events of the world since the beginning. Darwin wrote, "I feel compelled to look to a First Cause having an intelligent mind in some degree analogous to that of a man; and I deserve to be called a Theist.... This conclusion was strong in my mind about the time, as far as I can remember, when I wrote *The Origin of Species*; and it is since that time that it has very gradually with many fluctuations become weaker."[45]

"Fluctuations" became a pattern for Darwin, not only in theology, but also in the theory of natural selection. After writing *The Origin of Species*, his theology continued to change. In a letter to Hooker in 1870, Darwin wrote that the universe has no creator: "My theology is a simple muddle; I cannot look at the universe as the result of blind chance, yet I can see no evidence of beneficent design, or indeed of design of any kind in details."[46]

Continuing with the fluctuations in the autobiography, Darwin considers himself to have become an agnostic: "The mystery of the beginning of all things is insoluble by us; and I for one must be content to remain an agnostic."[47] During his lifetime, Darwin considered himself a believer in a personal God, a theist and an agnostic.

Life Issues

Despite professional successes, life took a toll on Darwin. Things that Darwin once found pleasurable as a young man turned on him. By 1865, at the age of fifty-six, Darwin summed up his problems in writing to a new medical adviser by writing that for twenty-five years he had experienced extreme flatulence, preceded by ringing ears and visual black dots, and vomiting preceded by shivering and crying. In a letter to Wallace in 1871, Darwin wrote that at "present I feel sick of everything, and if I could occupy time and forget my daily discomforts, or rather miseries, I would never publish another word."[49]

One of Darwin's greatest legacies may be his persistence. After 1871, Darwin published ten books, just after writing he "would never publish another word."

Final Years

Perhaps sensing the brevity of life, Darwin began writing his autobiography on May 31, 1876, at the age of sixty-seven. He completed his autobiography two months later, in August. Darwin completed the work by stating that this "sketch of my life was begun about May 28th at Hopedene, and since then I have written for nearly an hour on most afternoons."[50]

After writing his autobiography, Darwin, in very ill health, continued to conduct research and write for four more years, right up until the age of seventy-four, when he died. Time took a toll on Darwin's mind:

> I have said that in one respect my mind has changed during the last twenty or thirty years. Up to the age of thirty, or beyond it, poetry of many kinds, such as the works of Milton, Gray, Byron, Wordsworth, Coleridge, and Shelley, gave me great pleasure, and even as a schoolboy I took intense delight in Shakespeare, especially in historical plays. But now after may years I cannot endure to read a line of poetry: I have tried lately to read Shakespeare, and found it so intolerably dull that it nauseated me. I have also lost my taste for pictures or music.[51]

By the winter of 1881, his heart began to give him problems. Earlier Darwin had written a letter to Hooker: "I am rather despondent about myself, and my troubles are of an exactly opposite nature to yours, for idleness is downright misery to me ... So I must look forward to Down graveyard as the sweetest place on Earth."[52]

In a letter to the family, with no prospect of recovery or change of heart, Darwin wrote, "I am not the least afraid to die."[53] Having the highest esteem for Emma, Darwin wrote to his children, "You all know well that your Mother, and what a good Mother she has been to all of you. She has been my greatest blessing, and I can declare that in my whole life I have never heard her utter one word which I had rather not been unsaid. She has never failed in the kindest sympathy towards me, and has borne with the utmost patience my frequent complaints from ill-health and discomfort."[54]

Finally, while visiting a friend in London in December 1881, Darwin experienced a mild heart attack, eventually leading to a fatal heart attack on April 19, 1882.

In appreciation of Darwin and to celebrate his wide popularity, twenty members of the British Parliament immediately asked the dean, Reverend George Granville Bradley, whether Darwin could be laid to rest in Westminster Abbey, London, the burial place of dignitaries, kings, and queens.

On April 26, 1882, a four-horse funeral carriage, accompanied by three of his children, Francis, Leonard, and Horace, made the sixteen-mile journey to London. Darwin's wife, Emma, a "stronger-minded, tougher person than Charles," did not attend the formal service in London at Westminster Abbey, preferring to remain at home and reflect in private.

The pallbearers for Darwin were intellectual leaders of the day and Darwin's personal colleagues: the president of the Royal Society; the United States Ambassador to the British Isles, Russell Lowell; the churchman Cannon Farrar; an earl; two dukes; Thomas Huxley; Sir Joseph Hooker; and Alfred Russel Wallace, who had pushed Darwin into publishing *The Origin of Species*.

The famous British philosopher, Herbert Spencer, thought the occasion of Darwin's internment at the Abbey "worthy enough to suspend his objections to religious ceremonies." Spencer attempted to apply the theory of evolution to philosophy and ethics in his series *A System of Synthetic Philosophy*.

In the area of the Abbey known as Scientists' Corner, Darwin lies a few feet from the burial place of Sir Isaac Newton and next to that of the astronomer Sir John Herschel. It was Herschel that Darwin referred to in the introduction of *The Origin of Species* as the great philosopher who coined the phrase "mystery of mysteries" to describe the change of Earth's species through time.

In writing Darwin's obituary for the April 27, 1882, issue of *Nature* (London), Thomas Huxley ended by writing that the words applied to

Socrates' "Apology" were appropriate for Charles Darwin: "The hour of departure has arrived, and we go our ways—I to die and you to live. Which is the better, God only knows."[55]

By the end of the day on April 25, 1882, the ashes of Charles Darwin were placed in St. Faith at Westminster Abbey only a few paces apart from other modern-day Western history legends, including, Sir Isaac Newton and Charles Lyell. The simple and remaining inscription on the tombstone reads,

CHARLES ROBERT DARWIN BORN 12 FEBRUARY 1809.
DIED 19 APRIL 1882.

End-of-Life Myths

Myths have circulated that Darwin recanted the theory of evolution while he was dying. Some of the stories read like this: "Shortly after Darwin's death at seventy-four on April 19, 1882, the evangelistic widow of Admiral of the Fleet Sir James Hope, told a gathering of students at Northfield Seminary in Massachusetts that she had visited Darwin in his last hours and found him reading the Epistle to the Hebrews. Darwin, she said, announced that he wished he 'had not expressed my theory of evolution as I have done,' and he also asked her to get some people together so he could speak to them of Jesus Christ and His salvation, being in a state where he was eagerly savoring the heavenly anticipation of bliss."[56]

Darwin's family all denied the story and campaigned against it. Darwin's son Francis wrote in a letter in May 1918: "Lady Hope's account of my father's views on religion is quite untrue. I have publicly accused her of falsehood, but have not seen any reply. My father's agnostic point of view is given in my *Life and Letters of Charles Darwin*, Vol. I., pp. 304–317. You are at liberty to publish the above statement. Indeed, I shall be glad if you will do so."[57]

Darwin's daughter Henrietta Litchfield also refuted the story, stating in the 1922 publication of *The Christian*: "I was present at his deathbed, Lady Hope was not present during his last illness, or any illness. I believe he never even saw her, but in any case she had no influence over him in any department of thought or belief. He never recanted any of his scientific views, either then or earlier. We think the story of his conversion was fabricated in the U.S.A.... The whole story has no foundation what-so-ever."[58]

As an agnostic, Darwin was respected by his contemporaries, and even the Church of England. The Bishop of Carlisle, Harvey Goodwin, in a memorial sermon preached in the Abbey on the Sunday following the funeral, launched to bridge the agnostic-belief gap by stating, "I think that the interment of the remains of Mr. Darwin in Westminster Abbey is in accordance with the judgment of the wisest of his countrymen ... It would have been unfortunate if anything had occurred to give weight and currency to the foolish notion which some have diligently propagated, but for which Mr. Darwin was not responsible, that there is a necessary conflict between a knowledge of Nature and a belief in God."[59]

Chapter Four
The Stage

Until recently the great majority of naturalists believed that species were immutable productions, and had been separately created. This view has been ably maintained by many authors. Some few naturalists, on the other hand, have believed that species undergo modification, and that the existing forms of life are the descendants by true generation of preexisting forms.

—Charles Darwin, 1859[1]

Evolution was ready for center stage. Darwin launched the theory of evolution, or at least that is a popular notion today, starting with *The Origin of Species.* Published during the Victorian era, the *Origin* has since become the mascot of evolution. Certainly, no other publication has been as pivotal. The question is, though, was Darwin really the originator of the theory, or was he a part of a larger movement?

Victorian Era

The Victorian era was history's theater for the drama ready to unfold. Ironically, this era is commonly viewed as setting strict moral standards—"Victorian." The Victorian era gave rise to a unique style of clothing and architecture, gaining global popularity. "Victorian" signaled a measure of sophistication and status.

Movements founded in the Victorian era have continued through the twentieth century. In 1844, George Williams founded the Young Men's Christian Association (YMCA) in London. William Booth founded the Salvation Army in 1865. Florence Nightingale, "The Lady with the Lamp," single-handedly created the modern nursing profession. Now hospitals originating from Christian church organizations comprise the largest segment of nonprofit hospitals.

What has been so stunning about *The Origin of Species* was the nearly immediate and widespread acceptance. The phenomenon speaks an underlying intellectual shift; evolution was part of a larger movement. Now, 150 years after the entrance of the *Origin* to center stage, evolution stands as the ultimate litmus test for acceptance in intellectual, academic, social, and political circles.

Gaining insight into the nature of the movement can be seen in a letter sent by one of England's leading clergymen, Charles Kingsley, written just a year after the release of *The Origin of Species* in 1859. In the letter to his wife about the tragedy, he had seen traveling through the devastation of the great Irish potato famine, Kingsley wrote, "I am haunted by the human chimpanzees I saw along that hundred miles of horrible country. ... To see white chimpanzees is dreadful; if they were black, one would not feel it so much, but their skins, except where tanned by exposure, are as white as ours."[2]

Kingsley envisions a poor man as a "human chimpanzee." Even within the halls of the traditional church, a break from the Bible was well under way. The idea that man is "created in the image of God" was waning. Elitism was leading a revolution, even in the church.

The Victorian era followed the popular philosophical movement that had earlier swept across Western Europe, known as the Age of Enlightenment, and part of the longer Age of Reason period. The driving force of the Age of Enlightenment was law, order, rationality, individualism, and prosperity. This set the stage for Victorian England to emerge as the quintessence of the movement.

The Age of Enlightenment gave rise to the Scientific Revolution that began unveiling the long-standing mysteries of nature by using the scientific method. The unexplainable forces of nature were becoming explainable. Using the scientific method, Newton, in the eighteenth century, discovered the natural laws behind the forces of gravity.

Through this gradual progression of knowledge through the discovery of natural laws, the stage was set for a culture of change. The use of newly founded natural laws sophisticated Western society into a period of expanding materialism. Newly acquired industrial tools tamed the land and spearheaded the building of sewer systems, canals and dams, roads and trains, and worldwide ocean travel. An unprecedented period of material progress was emerging.

Along with material progress was a growing interest in evolution, even within the halls of traditional clergy. What were the events that eventually set the stage for *The Origin of Species*? How did a theology graduate student from Cambridge University come to reject Moses' creation account in Genesis? The answer starts with the scripting of philosophy dating back to at least as early as the Greeks.

Evolution Origins in Greek Philosophy

The Greeks embraced a materialistic worldview; life flowed from logic and reason. For the Greek mind, meaning and purpose was to be found through intelligence and wisdom within the world, rather than from a transcendent essence.

A snapshoot of the Greek mind is revealed through their philosophers. The rudimentary elements of evolution are neatly woven into the fabric of philosophy from Thales to Aristotle.

Thales of Miletus (640–545 BC) started as a merchantman, bringing goods from afar into Greece. Actually, the works of Thales is known only from the writings of Aristotle. None of Thales' original works has lasted.

Thales reasoned water is the first principle of all things, "all things are water," and that the Earth rests on water and life originates from water. Thales reasoned that life originated by natural rather than supernatural means.[3]

The prediction of the eclipse in 585 BC by Thales is thought to have been based on information obtained from the Babylonians. It is thought that Thales was responsible for importing elements of Babylonians' philosophy into Greece.

Thales' student, Anaximander of Miletus (610–546 BC), is the first person credited with offering a detailed explanation for all aspects of nature and the first tenets of evolution. Anaximander taught that the Earth is at the center of the universe; that man achieved his physical state by adaptation to environment, "life had evolved from moisture"; and that "man developed from fish," anticipating the theory of evolution.[4] Anaximander is the first person known to produce a map.

Heraclitus of Ephesus (535–475 BC) taught, "struggle is the father of everything,"[5] a concept Darwin would later coin "natural selection." The Greek mind envisioned reason as the foundation of order and intelligence. Heraclitus taught that all things are composed of opposites, that opposites are constantly at strife with one another, and that all things are in perpetual change. Change, though, is ultimately governed by "logos" reasoning. This concept was further developed by the Frenchman Jean-Baptiste Lamarck in the eighteenth century.

Democritus (460–370 BC) is thought to have been by far the most learned Greek philosopher of his time, writing on the subjects of physics, mathematics, ethics, and music. Democritus proposed the "atomic system."

This "system" is composed of an infinite number of everlasting atoms, from whose random combinations springs an infinite number of successive world orders in which there is law but not design, stating, "everything existing in the Universe is the fruit of chance and necessity."

Democritus concepts were further developed by Epicurus of Sámos (341–270 BC). Epicurus taught that life is continuous and progressively adapts through the selection of random changes. *really?*

Epicurus held that pleasure is the chief good, which developed into movements known as Epicureanism. Life arises from "inner-forces" or "vital drives" to ever more complex and perfect ends. This concept was further developed by the Frenchman Lamarck. Darwin initially rejected but later embraced this concept in part.

Aristotle (384–322 BC) took a different approach, arguing that we can only make sense of nature if it has a purpose, since "nature does nothing in vain."[6] Aristotle became one of the most influential of the ancient Greek philosophers, along with Socrates and Plato. Aristotle founded one of the most important schools of ancient philosophy, now referred to as Aristotelian logic, or Aristotelianism. Aristotle promoted a geocentric worldview, where the "Earth is the center of the universe."[7]

Aristotle used the logic process known as natural philosophy. This logic process is more commonly known as deductive reasoning. Deductive reasoning builds on what is known or thought to be known as fact. Unfortunately, using facts and logic may not lead to a truth, otherwise known as a scientific discovery. Copernicus demonstrated that the Earth is not the center of the universe.

The concept of spontaneous generation was promoted by Aristotle, and became known as Aristotelian abiogenesis. Spontaneous generation means that life arises directly from nonlife matter. This concept of life arising spontaneously was woven into Darwin's theory of pangenesis. Long before Darwin, Aristotle wrote, "variety in animal life may be produced by variety of locality."[8] With different environments, which Aristotle refers to as "variety of locality," Aristotle reasoned that differentiation in animals would produce different results: "locality will differentiate habits … rugged highlands will not produce the same results as the soft lowlands."[9]

While Darwin aligns with the essential elements of Aristotle's theory, in the first paragraph of the preface in *The Origin of Species,* Darwin acknowledges Aristotle's limits: "We here see the principle of natural selection shadowed forth, but how little Aristotle fully comprehended the principle."[10]

Western Origins of Evolution

The origins of evolution, rooted in Greek philosophy, sprouted in Western civilization. At least a century before Darwin published *The Origin of Species,* French political philosopher, Charles De Secondat Montesquieu (1689–1755), envisioned that "in the beginning there were very few [kinds of] species, and they have multiplied since."[11]

Benoit de Maillet (1656–1738) published the book entitled *Telliamed: Of Discourses Between an Indian Philosopher and a French Missionary on the Diminution of the Sea, the Formation of the Earth, the Origin of Men and Animals,* in 1748. Maillet proposed that fish were the precursors of birds, mammals, and men.

German naturalist Johann Friedrich Gmelin (1748–1804) and German geologist and paleontologist Christian Leopold Freiherr von Buch (1774–1853) independently promoted the concept that species evolved based on their geographic environment. Buch is most remembered for defining the Jurassic System.

Some historians of science point to French mathematician and astronomer Pierre Louis Moreau de Maupertuis (1698–1759) as signaling centennial precursors in the development of evolutionary concepts. In the book *Venus Physique,* published in 1745, Maupertuis encompasses Darwin's major tenets. Maupertuis wrote, "Chance, one would say, produced an innumerable multitude of individuals" in which the "species we see today are but the smallest part of what blind destiny has produced." [12, 13] This eighteenth century Frenchman formulated the essence of Darwin's theory of natural selection over 100 years earlier. Maupertuis explains further: "Could one not say that, in the fortuitous combinations of the productions of nature, as there must be some characterized by a certain relation of fitness which are able to subsist, it is not to be wondered at that this fitness is present in all the species that are currently in existence?" [14]

In the preface of *The Origin of Species*, Darwin notes Maupertuis' contributions: "It has been maintained by several authors that it is as easy to believe in the creation of a million beings as of one; but Maupertuis' philosophical axiom 'of least action' leads the mind more willingly to admit the smaller number; and certainly we ought not to believe that innumerable beings within each great class have been created with plain, but deceptive, marks of descent from a single parent."[15]

A contemporary of Maupertuis was Denis Diderot (1713–1784), the philosopher and leading figure of the Enlightenment in France. Diderot suggested that animals evolved from one primeval organism by natural selection. For expressing his opinions, Diderot was imprisoned for three months in 1749.

Perhaps the one most responsible for the rise of European interest in natural history and evolution during the eighteenth century was the French naturalist George Louis Buffon (1707–1788). Buffon's massive book series, *Histoire naturelle*, set out to organize all that was then known about the natural world. Darwin concurs that "the first author who in modern times has treated it (natural selection) in a scientific spirit was Buffon. But as his opinions fluctuated greatly at different periods and as he does not enter on the causes or means of the trans-formation of species."[16]

Buffon's views gained influence in the next generation of naturalists, including Lamarck and Charles Darwin. Buffon wrote, "that man and ape have a common origin; that, in fact, all the families among plants as well as animals have come from a common stock."[17]

Interestingly though, after first reading Buffon, Darwin wrote to his colleague Huxley, "I have read Buffon: whole pages are laughably like mine. It is surprising how candid it makes one to see one's view in another man's words."[18]

As a member of the French Academy of Sciences, Buffon arranged for his colleague, Lamarck (1744–1829), to be appointed to the Muséum National d'Histoire Naturelle in Paris. Eventually Lamarck became one of the most influential nineteenth century French naturalists, became the first to use the term "biology" in its modern sense, and coined the term "invertebrate." According to Darwin, "Lamarck was the first man whose conclusions on the subject excited much attention. This justly celebrated naturalist first published his views in 1801. … In these works he up holds the doctrine that all species, including man, are descended from other species."[19]

Lamarck's own theory of evolution, which he referred to as "transformism," was based on the idea that individuals develop new traits during their own lifetimes and transmit them to the next generation: "Progress in complexity of organization exhibits anomalies here and there in the general series of animals, due to the influence of environment and of acquired habits."[20]

The giraffe served as his classic example of evolution, acquiring longer necks in successive generations in competition to reach the ever-scarcer leaves higher in the trees. In illustrating Lamarck's views on adaptation, Darwin wrote, "To this latter agency he seems to attribute all the beautiful adaptations in nature; such as the long neck of the giraffe for browsing on the branches of trees."[21]

Initially, Darwin did not accept Lamarck's theory: "Lamarck, who believed in an innate and inevitable tendency towards perfection in all organic beings, seems to have felt this difficulty so strongly that he was led to suppose that new and simple forms are continually being produced by spontaneous generation. Science has not as yet proved the truth of this belief."[22]

One of the most eminent pre-Darwinists was Charles Darwin's own grandfather, Erasmus Darwin (1731–1802). Erasmus discussed his ideas at length in a two-volume work, *Zoönomia*, published in 1794. Erasmus wrote that "all … have risen from one living filament."[23]

Erasmus' book was widely popular in Western Europe and was translated into German, French, and Italian. Erasmus envisioned that the driving force behind species modification was a result of "lust, hunger, and danger."[24] In line with Greek philosophy, Erasmus envisioned changes by "continuing to improve its own inherent activity."[25]

Actually how these "improvements" developed was completely unknown to Lamarck and Erasmus—evolution was a philosophy, not a science. The unknown cause of "improvements" is what drove Darwin to discover the underlying laws of nature, scientifically. Writing in the preface of *The Origin of Species*, Darwin suggests how Erasmus's work, although "erroneous," may have influenced Lamarck: "It is curious how largely my grandfather, Dr. Erasmus Darwin, anticipated the views and erroneous grounds of opinion of Lamarck in his *Zoönomia*."[26]

Growing acceptance of evolution in England can be seen in a book entitled *History of the English People,* by Élie Halévy. The book refers to a pamphlet on the Elizabethan Poor Laws, *A Dissertation on the Poor Laws by a Well-Wisher to Mankind,* written by Methodist clergyman Rev. Joseph Townsend in 1786. In the pamphlet, Townsend blames the Poor Laws for preserving the weak at the expense of the strong, which is the essence of natural selection.

While a cohesive scientific evolutionary theory had not emerged by the early 1800s, the intellectual momentum in academia toward natural reasons for the origin of life was well under way. In his autobiography, Darwin acknowledges that the "innumerable well-observed facts were stored in the minds of naturalists ready to take their proper place as soon as any theory which would receive them was sufficiently explained."[27]

Once Darwin delivered the explanation "by means of natural selection" in *The Origin of Species*, the momentum was already well under way. Evolution was reaching a new level of acceptance. Writers once tentative to sign their own work came out of the closet. In the preface of the first edition of *Origin* in 1859, Darwin chronicled only five individuals contributing to the movement. But just twelve years later, in 1872, in the sixth edition, Darwin was able to expand the list to forty-four writers. Included in the list of writers in the sixth edition were naturalists, geologists, and even clergymen, from England, France, Germany, Belgium, and the United States.

The key feature central to the acceptance of Darwin's theory of evolution was the mechanism "by means of natural selection." But how did Darwin "discover" the mechanism of natural selection? Was it from a fossil discovery, a laboratory experiment, or data analysis of field collections? The correct answer is no. Nor was the mechanism proposed by a scientist, a naturalist, or anyone working in the biological sciences of the day. Darwin based the mechanism of natural selection on a theory written by the famous English political economist Thomas Robert Malthus.

Malthus was widely read in philosophy; he studied David Hume, Adam Smith, Robert Wallace, and Rev. Joseph Townsend's *Dissertation on the Poor Laws,* published in 1786. These were philosophical, not scientific, works.

In 1789, Malthus published a paper entitled *Essay on the Principle of Population,* which predicted that the population would outgrow the food supply, leading to a decrease in food per person. Malthus predicted worldwide starvation by the middle of the nineteenth century.

In the process of searching for a theory, Darwin mused, "how selection could be applied to organisms living in a state of nature remained for some time a mystery to me."[28] Not until reading Malthus' essay on *Population* was Darwin able to arrive with the central driving force for his emerging theory. Recounting the events leading to the discovery, Darwin wrote in his autobiography:

> In October 1838, that is, fifteen months after I had begun my systematic enquiry, I happened to read for amusement Malthus on Population, and being well prepared to appreciate the struggle for existence which everywhere goes on from long-continued observation of the habits of animals and plants, it at once struck me that under these circumstances favourable variations would tend to be preserved, and unfavourable ones to be destroyed. The result of this would be the formation of new species.[29]

The influence of Malthus extended beyond Darwin. Even one of Darwin's colleagues, Alfred Russel Wallace, called Malthus' essay "the most important book I read." Just months before the publication of *The Origin of Species,* in a letter to Darwin, Wallace wrote that it was "the most interesting coincidence" that they were led independently to the theory of evolution through reading Malthus. On April 6, 1859, Darwin wrote to Wallace, "You are right, that I came to conclusion that Selection was the principle of change from study of domesticated productions; and then reading Malthus I saw at once how to apply this principle ... especially [in the] case of Galapagos Isl."[30]

Malthus had had a decisive influence on both Darwin and Wallace. Darwin became a lifelong admirer of Malthus. In a letter to J. D. Hooker in June 1860, Darwin refers to Malthus as "that great philosopher."[31]

Not only was Darwin an avid writer, he was also an ardent reader. Darwin extended his interests to include metaphysical works. In August 1838, Darwin said he read "a good deal of various amusing books paid attention to Metaphysical subjects."[32]

Darwin was known to have read books by Edmund Burke, Dugald Stewart, Gotthold Lessing, David Hartley, Thomas Reid, and David Hume, among others. In the same way that Darwin incorporated Malthus' concepts, Darwin eventually developed his theory of evolution based on a philosophy and not on any measure of scientific evidence. Deductive reasoning drove Darwin's theory. This approach to studying nature has been termed "natural philosophy."

The Scientific Revolution

Running parallel to natural philosophy was the rise of inductive reasoning to center stage. Inductive reasoning is the inverted form of deductive reasoning that relies completely on reproducible facts. Through inductive reasoning, the theory of philosophy is dependent on the facts; evidence trumps theory.

The emergence of inductive reasoning, now known as the scientific method, gave rise to the Scientific Revolution. The Scientific Revolution started with the Polish astronomer, Nicolas Copernicus (1473–1543), with the publication of *De revolutionibus orbium coelestium (On the Revolution of the Heavenly Spheres)* in 1543.

The Scientific Revolution developed along with a larger movement known as the Age of Enlightenment. In part, the movement was seeking to overthrow the Roman Catholic Church, which by the sixteenth century had even embraced the secular geocentric worldview developed by Aristotle.

Gaining momentum during in the Age of Enlightenment, Copernicus was driven to discover the truth regarding the "mechanisms of the universe," writing that the "mechanisms of the universe, wrought for us by a supremely good and orderly Creator … the system best and most orderly artist of all framed for our sake."[33]

Copernicus is thought to have been the first modern western astronomer to use inductive reasoning in studying the relationship of the Earth to the sun. The discovery that the sun, and not the Earth, was the center of our solar system placed him in opposition to the Roman Catholic Church.

Announcement of the discovery created a firestorm. Galileo Galilei (1564–1642), in support of the discovery, likewise paid a personal price inflicted by the Roman Catholic Church. But use of the scientific method set the foundation for Newton's discovery of the laws of motion. The Scientific Revolution became an unprecedented new force in the discovery of natural laws.

In the seventeenth century, English philosopher Sir Francis Bacon earned the title as the founder of the Scientific Revolution. Bacon established that the discovery of natural laws must proceed through inductive and not deductive reasoning. The early Scientific Revolution period culminated with the publication of the *Philosophiae Naturalis Principia Mathematica,* in 1687, by Isaac Newton.

Darwin and Newton were both alumni of Cambridge University. But unlike Darwin, Newton subscribed to the Genesis account of creation and the accuracy of the entire Bible. Newton was convinced that Christianity went astray in the fourth century AD, when the first Council of Nicaea propounded erroneous doctrines of the nature of Christ.

Newton further fueled the Scientific Revolution by integrating algebraic concepts acquired from the Middle East and geometric concepts from Western mathematics and synthesized calculus.

Actually, science played a secondary role in Newton's life. Convinced of God's providential role in life, during his lifetime Newton was actually more widely known for his expertise on the books of Daniel and Revelation than as a scientist. Concerning the end of times, Newton wrote, "About the time of the end, a body of men will be raised up who will turn their attention to the Prophecies, and insist upon their literal interpretation, in the midst of much clamor and opposition."[34]

Through inductive reasoning, Newton discovered the universal laws of gravity and motion of our solar system: the scientific method was now held center stage. In *Principia Mathematica,* Newton declared that this "most beautiful system of the sun, planets, and comets could only proceed from the counsel and dominion of an intelligent and powerful Being."[35]

The physical laws of nature, once a mystery, were becoming understandable using the scientific method. In the same way, Darwin wanted to discover the laws of life to answer the ultimate "mystery of mysteries," the origins of life. The answer to this question now seemed within the reach and realm of science for Darwin, just as it had been for Newton. Additionally, Darwin had the winning combination—driving passion and vast resources.

The Scientific Revolution was moving to center stage in the eighteenth century. John Herschel's book, *Preliminary Discourse on the Study of Natural Philosophy* (1830), encapsulated the concepts of the scientific method. Darwin read Herschel's *Discourse* while at Cambridge. William Whewell supported Herschel's approach to the scientific method and later published *The History of Inductive Sciences* (1837) and *The Philosophy of Inductive Sciences* (1840).

In 1865, the Austrian monk Gregor Mendel, a contemporary of Darwin's, read Darwin's paper "Experiments on Plant Hybridization," based on experiments using the scientific method, at the Natural History Society of Brünn in Moravia. Mendel reported on the discovery of two laws of inheritance after testing some twenty-nine thousand pea plants over a period of seven years. Unlike Darwin though, the paper received intense criticism, and the theory was rejected. During this time, the use of inductive reasoning was becoming varyingly accepted, even in scientific circles. It is not known whether Darwin ever read Mendel's work; Darwin does not mention Mendel in any of his publications and letters.

Mendel's laws went unrecognized until rediscovered by Hugo de Vries and Carl Correns in 1900. These laws were to become known as "Mendel's Laws of Inheritance," and served as the foundation of modern evolutionary synthesis during the twentieth century. Mendel is now recognized as the "Father of Modern Genetics."

Swinging Pendulum

Gaining and holding center stage is a dynamic deed. During this same time, the scientific method was being challenged by another school of thought popularized by John Stuart Mill. Mill captured the rising popular empiricism tide of the Victorian era, which promoted natural philosophy over the role of the scientific method.

The empiricism approach emphasizes subjective, rather than objective, evidence. And, the movement garnered an impressive following, including even Darwin. Historian David L. Hull, in 1983, wrote, "The controversy which ensued was gradually decided in Mill's favor, not because Mill's position was superior to that of Whewell, but because Whewell's version of Kantian philosophy ran contrary to the rising tide of empiricism."[36]

Natural philosophy challenged the eminent role of the scientific method. One natural philosophy adherent was English geologist Charles Lyell. Lyell interpreted geological observations without using the scientific method. Lyell envisioned a long time line for the Earth; with continuous changes occurring very slowly—a theory that has become known as uniformitarianism, which is incompatible with dramatic geological changes.

Lyell's theory became widely popular and eventually formed the cornerstone for Darwin's theory of evolution. Darwin referred to the work of Charles Lyell twenty-eight times in *The Origin of Species.*

Evidence for Lyell's theory largely came from the maps and conclusions drawn by the eighteenth century English surveyor William Smith (1769–1839), who is now considered the "Father of English Geology." Using deductive reasoning, Smith theorized that the fossil layers are successive, one layer forming over another through time. Smith reasoned that the evidence in the rock strata have archived the development of new species in fossil form over time. Smith published his first geological map of Wales and England in 1815. Now referred to as the "map that changed the world," Smith wrote in *Deductions from Established Facts of Geology,* in 1835, that as "doubts may remain in the minds of many on the Principles of Geology, I shall endeavour to exhibit the principles, long familiar to my mind, in a clear view, opened by the organized fossils, which are the medals of creation, the antiquities of nature, and records of time."[37]

The concept of successive geological layers containing fossils that are more advanced became foundational to Darwin's theory. These layers were thought to represent the chronicle of evolution over time. Swinging with the pendulum of the nineteenth century, Darwin predicted that fossil evidence in these geological layers in columns would eventually be found as evidence of evolution.

Contemporary Evolution Intellectuals

While Russel Wallace has garnered the most recognition for pioneering the concept of evolution through natural selection with Darwin, this is only part of the story. Darwin was surrounded by contemporaries working diligently to define evolution in scientific terms. One of Darwin's contemporary colleagues was Robert Chambers.

In 1844, predating *The Origin of Species* by fifteen years, a book entitled *Vestiges of Natural History of Creation* was published, arguing that life "could best be understood by appeal to natural law rather than by flight to an intervening deity."[38] The English writer, Robert Chambers (1802–1871), anticipating the storm the book would create, arranged to have the book published anonymously. These arrangements were made through Alexander Ireland of Manchester. The secret was so well kept that such different names as those of Prince Albert and Charles Lyell were often coupled with the book.

Within the powerful British intellectual circles, Chambers was eventually elected as a fellow of the Geological Society of London in 1844, and then elected as a member of the Royal Society of Edinburgh in 1840. During this time, according to his brother William, Chambers "had become occupied with speculative theories which brought him into communication with Sir Charles Bell, George Combe, his brother Dr. Andrew Combe, Dr. Neil Arnott, Professor Edward Forbes, Dr. Samuel Brown, and other thinkers on physiology and mental philosophy."[39]

Vestiges of Natural History of Creation was written to inspire interest in evolution, rather than the social and academic elites. Chambers took a distant stance from Lamarck, writing that Lamarck's ideas were "obviously so inadequate."[40]

The book was controversial, but over twenty thousand copies were sold within the decade, making it one of the best sellers of its time. Since there was no longer any reason to conceal the author's name, the preface in the twelfth edition of *Vestiges,* in 1884, identified Chambers as the author. In the concluding chapter, Chambers wrote that the "book, as far as I am aware, is the first attempt to connect the natural sciences in a history of creation."[41]

In 1851, Chambers joined publisher John Chapman to reinvigorate the journal *Westminster Review,* which became the flagship forum of philosophical radicals and developing evolutionary concepts. In the office on the famous Strand in London, Chambers was connected with John Stuart Mill, Herbert Spencer, John Tyndale, Harriet Martineau, and Thomas Huxley. Huxley, who later became known as "Darwin's Bulldog," coined the term "Darwinism" and popularized the term "agnostic."

Eventually the *Westminster Review* lent Darwin unyielding support during the ensuing furor. The term "Darwinism" was first used by Huxley in his favorable review of *The Origin of Species* in the April 1860 issue of the *Westminster Review.*

Patrick Matthew (1790–1874), a Scottish fruit grower, proposed the principle of natural selection as a mechanism of evolution more than a quarter century earlier than did Darwin and Alfred Russel Wallace. In the book *On Naval Timber and Arboriculture*, Matthew, in 1831, wrote that there "is a law universal in nature, tending to render every reproductive being the best possible suited to its condition that its kind, or organized matter, is susceptible of, which appears intended to model the physical and mental or instinctive powers to their highest perfection and to continue them so."[42]

In 1860, Matthew read a review of Darwin's *Origin of Species* in the *Gardeners' Chronicle*, which included a description of the principle of natural selection. Matthew was prompted to write a letter to the publisher to call attention to his earlier revelation of the theory. Upon reading the letter, Darwin subsequently commented in a letter to Lyell: "In last Saturday *Gardeners' Chronicle*, a Mr. Patrick Matthews publishes long extract from his work on *"Naval Timber and Arboriculture"* published in 1831, in which he briefly but completely anticipates the theory of Nat. Selection. I have ordered the Book, as some few passages are rather obscure, but it is, certainly, I think, a complete but not developed anticipation!"[43]

Recognizing Matthew's contributions, Darwin wrote to the *Gardeners' Chronicle*, "I freely acknowledge that Mr. Matthew has anticipated by many years the explanation which I have offered of *The Origin of Species*, under the name of natural selection."[44]

Starting with the third edition of *The Origin of Species*, Darwin acknowledged Matthew's earlier work, stating that Matthew "clearly saw ... the full force of the principle of natural selection."[45] Matthew claimed credit for natural selection, even having calling cards imprinted with "Discoverer of the Principle of Natural Selection." However, even with the calling cards and Darwin's statements, Matthew has continued to be a generally unknown personage in the history of evolution.

For Matthew, natural selection was a directed process with evidence of design. In a letter to Darwin in 1871, Matthew wrote, "a sentiment of beauty pervading Nature [that] affords evidence of intellect and benevolence in the scheme of Nature ... This principle of beauty is clearly from design and cannot be accounted for by natural selection."[46]

A theory of natural selection was also developed by English zoologist Edward Blyth (1810–1873). Blyth published three articles on natural selection in *The Magazine of Natural History* between 1835 and 1837. Blyth was one of the first to recognize Alfred Russel Wallace's paper, *On the Law Which has Regulated the Introduction of Species*, in 1855. In a letter to Darwin on December 8, 1855, from Calcutta, Blyth wrote, "What think you of Wallace's paper in the Ann. N. Hist.? Good! Upon the whole! Wallace has, I think, put the matter well; and according to his theory, the various domestic races of animals have been fairly developed into species. A trump of a fact for friend Wallace to have hit upon."[47]

Loren Eiseley, professor of anthropology and the history of science at the University of Pennsylvania, spent decades tracing the origin of the ideas attributed to Darwin. In a 1979 book, Eiseley concluded that in Blyth's papers published between 1835 and 1837 in *The Magazine of Natural History*, "the leading tenets of Darwin's work—the struggle for existence, variation, natural selection and sexual selection—are all fully expressed."[48]

There is no doubt that Darwin read Edward Blyth's papers. In the first chapter of just the third and fourth editions of *The Origin of Species*, Darwin writes, "Mr. Blyth, whose opinion, from his large and varied stores of knowledge, I should value more than that of almost any one."[49]

Without question, though, Darwin's earliest and closet collaborator, friend, and eventual pallbearer was the British naturalist Alfred Russel Wallace (1823–1913). Wallace was also profoundly influenced by Robert Chambers' work *Vestiges of the Natural History of Creation.* In a letter to Henry Bates in December 1845, Wallace wrote, "I have a rather more favourable opinion of the *Vestiges* than you appear to have. I do not consider it a hasty generalization, but rather as an ingenious hypothesis strongly supported by some striking facts and analogies, but which remains to be proven by more facts and the additional light which more research may throw upon the problem."[50]

In 1855, Wallace published a paper entitled *On the Law Which has Regulated the Introduction of Species.* On how species originate, Wallace wrote that every "species has come into existence coincident both in space and time with a closely allied species."[51]

Wallace had once briefly met Darwin and was one of Darwin's correspondents whose observations Darwin used to support his theories. Wallace knew that Darwin was interested in the question of how species originate, and entrusting his theory in Darwin's hands raddled Darwin's timetable. Wallace and Darwin coauthored their first paper on natural selection in 1858. Now, to accommodate a shorter time line, Darwin was rushed to finish and publish *The Origin of Species* sooner than intended.

As has long been clear, the pending Darwinism revolution was neither completely Darwinian nor completely revolutionary. Darwin's work connected with major transformations of the nineteenth century thought and came to symbolize the emerging new materialism. In 1850, poet Alfred Lord Tennyson, born the same year as Darwin, captured the essence of the struggle for existence—"nature red in tooth and claw."

In a landmark accomplishment, German chemist Friedrich Wöhler became the first to synthesize an organic compound, urea, from inorganic material in 1828. Synthesis of urea caused much excitement; a nonliving material could be made into one of life's building blocks, organic compounds. This discovery appeared to solve a lingering mystery—is there any evidence that life can emerge naturally from nonlife?

In the struggle to gain influence simply based on an argument of "imaginary instances," Darwin compares himself to Charles Lyell, stating, "I am well aware that this doctrine of natural selection, exemplified in the above imaginary instances, is open to the same objections which were first urged against Sir Charles Lyell's noble views on the modern changes of the Earth, as illustrative of geology."[52] Darwin clearly understood that his theory was not based on any new scientific evidence.

In the first edition of *The Origin of Species*, Darwin mentions in the introduction only five individuals known to support concepts of evolution: Robert Chambers, Joseph Dalton Hooker, Robert Malthus, Alfred Russel Wallace, and Charles Lyell. Clearly, the topic of evolution was "in the air." By the sixth edition, Darwin acknowledges in an added preface section forty-five individuals known to support some concept of evolution, under the heading:

AN HISTORICAL SKETCH OF THE PROGRESS OF OPINION ON THE ORIGIN OF SPECIES, PREVIOUSLY TO THE PUBLICATION OF THE FIRST EDITION OF THIS WORK.[53]

Scripts in the Marketplace

The prevailing opinion of the early nineteenth century was that species remain the same from generation to generation. This traditional view held that variations were within clearly defined limits of the species or type, "kind after its kind."

Copernicus, Galileo and Newton envisioned science and creation to be congruent, not mutually exclusive. During the nineteenth century, the study of nature played a central role in the university theology curriculum. Nature reflected the hand of God: life was created. Classic clergy education in nineteenth century England included training as a naturalist—to know the hand of God.

These concepts were influenced by Moses (1400 BC, estimated), in the Genesis account, and by the writer of Job (1500 BC, estimated). During his trial, Job reflects on the origins of life. Job suggested that even the animals know the creative hand: "But now ask the beasts, and they will tell you; and the birds of the air, and they will tell you; or speak to the Earth, and it will teach you; and the fish of the sea will explain it to you. Who among these does not know that the hand of the Lord has done this?"[54]

The Victorian era was the height of the British Industrial Revolution—the apex of the British Empire. This long period of peace, economic growth, colonialization, and industrialization was unrivaled in history.

During the long reign of Queen Victoria, from 1837 until her death in 1901, scientific advances began to rapidly transform the standard of living worldwide. The first railroad in England was built in 1825 when Victoria was a little girl. Before that, the maximum speed was that of the horse. By the time the Queen died, nearly all of Britain's extensive railroad system had been built.

In 1851, just eight years before the publication of *The Origin of Species,* the first World Fair was hosted in Hyde Park, London. The fair highlighted man's achievements and became an international success story. Prosperity ushered in the Gothic revival movement in architecture now known as Victorian architecture. With the newly acquired natural laws in hand, man was making unprecedented advances.

Industrialization was forming the foundation of the twentieth century. Leaps in engineering following the discovery of nature's physical laws were improving almost every aspect of daily life. Louis Pasteur invented the process of pasteurization in the spring of 1862. Railways and the expansion of shipping lanes have continued to lead to endless means of connecting to the world.

The Victorian era was far from monolithic: the Victorian era was divided. In many respects, worldwide British progress was certainly less than admirable. The British used opium taken from India to break a trade deadlock with China. British confrontations led to two wars with China, known as the Opium Wars, which did not end until 1860. China ceded Hong Kong to the British, another victory for the powerful British economy, supporting evidence that the strong survive.

While Copernicus had revealed the order of our solar system using the scientific method, Darwin launched evolution into the great debate using a different approach, natural philosophy. No other issue in the realms of science and philosophy has had such a staggering impact on humanity as Darwin's theory of "molecules to man."

What was at stake was a new worldview. Society was entering a philosophical crossroad. In a letter written in the summer of 1872,

Thomas H. Huxley saw the conflict "between free thought and traditional authority. One or the other will have to succumb."[55]

The pendulum was swinging. The publication of *The Origin of Species* led to the development of the "X Club" in 1864. The purpose of the X Club was to support the cause of naturalism and natural history through scientific discussions free from theological influences. In addition to Darwin, members of the X Club were the secular elite of the day and included George Busk, Edward Frankland, Thomas Hirst, Joseph Dalton Hooker, Thomas Huxley, John Lubbock, Herbert Spencer, William Spottiswoode, and John Tyndall. The members of the X Club were joined in a fight, both public and private, to unite the London scientific community, with the objective of furthering the ideas of academic liberalism.[56]

Between its inception in 1864 and its termination in 1893, the X Club and its members gained much prominence within the scientific community. Between 1870 and 1878, Hooker, Spottiswoode, and Huxley held office in the Royal Society simultaneously, and between 1873 and 1885, they consecutively held the presidency of the Royal Society.[57]

The men of the X Club continued to gain influential positions outside the Royal Society. Five members of the club held the presidency of the British Association for the Advancement of Science between 1868 and 1881. Hirst was elected president of the London Mathematical Society between 1872 and 1874, while Busk served as examiner and eventually president of the Royal College of Surgeons. Frankland served as president of the Chemical Society between 1871 and 1873.[58] Even if Darwin had not become famous for his theory of evolution, the dynamics, influence, and public relations of the X Club ensured a place for him in history in the halls of academia and far beyond.

Stage Showdown

The timing was perfect: the topic of evolution was "in the air." The stage was set, the lights were ready, and the audience was waiting with baited breath. *The Origin of Species* was ready for center stage. Up to the eighteenth century, concepts of evolution were only driven by natural philosophy. Darwin initially was determined to move evolution into the realm of the Science Revolution.

Evolution, popularized by the Greeks, followed natural-philosophy logic by envisioning life originating through a systematic sequence of events; somehow there was a molecules-to-man sequence. Natural philosophy was certainly a lingering popular player of the day, giving rise to geological theories espousing a long time line proposed by Charles Lyell.

Darwin, however, was driven to discover the natural laws of evolution, scientifically, using inductive reasoning. The scientific method, though, is an inverted form of natural philosophy. While the "hypothesis," "theory," or "logic" is natural philosophy's trump card, reproducible "evidence" is the scientific method's trump card.

These two different approaches present an irresolvable chasm between natural philosophy and the scientific method; this is deductive reasoning versus inductive reasoning. Natural philosophy uses deductive reasoning, and the scientific method uses inductive reasoning.

While natural philosophy uses the "hypothesis" to drive the evidence, by contrast the scientific method uses only "evidence" to derive the theory. The purpose of the scientific method is to discover natural laws independent of any theory: this is an inductive approach to reasoning.

Unlike natural philosophy, the scientific method is not required to follow any systematic sequence, logical process, hypothesis, or theory because the evidence is independent of any subjective influence. And unlike natural philosophy, the scientific method cannot selectively exclude any evidence.

For example, the scientific method must include evidence of the sudden explosion and the sudden extinction of species. Without a question, evidence of a discontinuous pattern of creation and flood written by Moses are irreconcilable with any concept of evolution. While natural philosophy envisions that life is the result of a systematic natural sequential process only, the scientific method is not limited to any such process, whether natural or supernatural.

With evolution on center stage, the ultimate task was to answer the question "how did we get here?" Just as Newton had discovered the natural laws of gravity, Darwin wanted to discover the natural laws of evolution. Darwin was convinced that just as the natural laws of physics were discoverable, the laws of nature governing the origination of life must likewise be discoverable.

In the center of the Victorian era of progress, Darwin was driven to find the natural laws of evolution. A material world requires a material process. Into the fray enters Darwin's book, *The Origin of Species*. By the end of the nineteenth century, hardly any field of thought remained unchallenged by evolution.

John Dewey, the great turn-of-the-century philosopher and educator, born the same year *The Origin of Species* was published, wrote that "the *Origin of Species* introduced a mode of thinking that in the end was bound to transform the logic of knowledge, and hence the treatment of morals, politics, and religion."[59]

As a skilled negotiator, Darwin brilliantly plays between these irreconcilable characters in *The Origin of Species*. In the end, questions loom. Does Darwin desert natural philosophy for the scientific method? Was Darwin's theory of evolution developed using the scientific method? Is the theory "scientific" or is it just a philosophy?

Chapter Five
The Origin of Species

Now, things are wholly changed, and almost every naturalist admits the great principle of evolution.

—Charles Darwin[1]

Popularly known as the *Origin* or *Origin of Species*, the complete title is *The Origin of Species by Means of Natural Selection, or the Preservation of Favored Races in the Struggle for Life.* The first edition was released for sale on November 24, 1859, for fifteen shillings. Darwin was fifty years of age.

Darwin first tried to collect all the reviews of the book, but that became an overwhelming task. Darwin found that conclusions drawn were as diverse then as they are now, finding even "an essay in Hebrew has appeared on it, showing that the theory is contained in the Old Testament."[2]

Success and Questions

Timing largely determines the measure of success of any endeavor, and the stage had been set. Darwin had entered center stage. Recognizing the significance of the timing, Darwin wrote that it "has sometimes been said that the success of the *Origin* proved 'that the subject was in the air,' or 'that men's minds were prepared for it.'"[3]

Although Darwin had his key ideas at least as early as 1838, he deliberated for twenty years before publishing his theory of evolution. The pressure to complete the work was building, however. Darwin was painfully aware that Wallace was ready to publish his own near-identical version of natural selection.[4]

Developing the subject was forcing Darwin to choose between two worldviews. Was life created or did it evolve from chance alone? Was

Moses' account of creation in Genesis true or false? His wife found evolution disturbing, and he was very aware of the implications.

Many of his close, lifelong friends were unaware of the conclusions he had already arrived at, as he used many of his "creation friends" as a sounding board. Darwin became absolutely committed to the theory, but Emma never felt comfortable with the theory. In a letter to Darwin in 1839, Emma wrote that the "state of mind that I wish to preserve with respect to you, is to feel that while you are acting conscientiously and sincerely wishing and trying to learn the truth, you cannot be wrong, but there are some reasons that force themselves upon me, and prevent myself from being always able to give myself this comfort."[5]

Emma reasoned that if Darwin continued to pursue and accept the theory, they might not be together in eternity: "Everything that concerns you concerns me and I should be most unhappy if I thought we did not belong to each other forever."[6]

When the publication was finally released in 1859, all of the original 1,250 copies were sold on the first day of release. Commenting on the sales, Darwin wrote that it "was from the first highly successful."[8] Darwin sent a complementary copy of the book to Karl Marx. In a letter to the German philosopher Friedrich Engels, co-author of the *Communist Manifesto*, Marx, commenting on *The Origin of Species*, stated that although "it is developed in the crude English style, this is the book which contains the basis in natural history for our view."[9]

Although Darwin was a prolific writer, completing twenty-five books and contributing to ten others, Darwin admitted that the discipline of writing was a challenge for him because there "seems to be sort of a fatality in my mind leading me to put first my statement and proposition in a wrong or awkward form."[10]

Even with all the writing, though, Darwin never became a skilled writer. One of Darwin's greatest champions, Thomas Huxley, had to struggle through Darwin's "awkward form." Huxley concludes: "Exposition was not Darwin's forté—and his English is sometimes wonderful. But there is a marvelous dumb sagacity about him and he gets to the truth by ways as dark as those of the Heathen Chinese."[11]

Rising to the challenge of writing, though, unquestionably became Darwin's most rewarding exercise. The book was translated in Darwin's lifetime into Danish, Dutch, French, German, Hungarian, Italian,

Polish, Russian, Serbian, Spanish, and Swedish, and it has appeared in an additional eighteen languages since. Darwin described the book as just "one long argument from the beginning to the end."[12]

Today though, *The Origin of Species* is only rarely included in academic science curriculum for good reasons. Huxley writes that not only is the *Origin* difficult to read, it "is one of the hardest books to understand thoroughly that I know of" and "the *Origin of Species* is one of the hardest of books to master." [13, 14]

Darwin was the first one in Western civilization to comprehensively, cohesively, and convincingly explain how the theory of evolution might work. Darwin proposed a completely naturalistic and simplistic explanation mechanism for the development of new species and by inference, the origin of life.

Darwin reasoned that given variation in all species, in times of limited food resources only the ones with favorable variations wound have a greater chance to survive and reproduce, and drive out the less favored ones.

Since surviving favorable variations are then inherited and accumulate along with other variations over time, the net effect is species change—evolution. Over longer periods of time, new species would emerge "by means of natural selection." Darwin had drawn a major victory—a mechanism for evolution: random variation sifted by natural selection.

Six Editions

Over a period of thirteen years, *The Origin of Species* evolved through five different editions, each with substantial changes after the first edition. Darwin began the second edition on December 8, 1859, just twelve days after the release of the first edition. Even the title was changed. By the sixth edition, the first word in the title, "On," was deleted, rendering a more emphatic title.

The sixth edition was published on February 19, 1872. With each successive edition, the length of the book increased, from 3,878 in the first edition to 5,088 sentences in the last edition, as calculated by the director of the University of Pennsylvania Press, Morse Peckham. One of the reasons the book is difficult to read is the length of each sentence.

The average sentence contains more than thirty-two words. It was in the sixth edition that Darwin used the term "evolution."

Table I
Summary of Text Changes between Editions[60]

Year	Edition	Copies	Sentences Eliminated	Sentences Re-Written	Sentences Added	Total Sentences	% Change
1859	1st	1,250				3,878	
1860	2nd	3,000	9	483	30	3,899	7
1861	3rd	2,000	33	617	266	4,132	14
1866	4th	1,500	36	1,073	435	4,531	21
1869	5th	2,000	178	1,770	227	4,580	29
1872	6th	3,000	63	1,699	517	5,088	29

Sixteen years after the publication of *The Origin of Species*, over 16,000 copies had been sold in England alone. In Darwin's lifetime, the British printings alone sold more than 27,000 copies. From 1859 until the time of his death in 1882, Some 25,000 English copies had been published in Britain.

Darwin used the successive editions to address a wide range of criticisms and to align with progressive ideologies. A key phrase associated with *The Origin of Species*, "survival of the fittest," did not appear until the fifth edition. In the first edition, the title of chapter four was "Natural Selection." By the fifth edition, the title was changed to "Natural Selection; or the Survival of the Fittest."

In the text of chapter nine in the fifth edition, Darwin defines "natural selection" as a process of "preservation" and the "rejection" of variations, and by the fifth edition, natural selection was redefined as "preservation" and the "destruction" of variations.[1, 17]

Herbert Spencer was the one who originally coined the term "survival of the fittest," in his book *Principles of Biology* in 1864. Spencer was a prominent and progressive English philosopher. As editor of the journal *The Economist*, Spencer published his first book in 1851, *Social Statics,* which included the radical prediction that humanity was soon to completely adapt to the requirements of living in society, resulting in the destruction of the state. The concept of the survival of the fittest was derived from economics, not the scientific method.

Today, the sixth edition is considered Darwin's final word. Like an artist, though, a creative work is never complete. In 1869, Darwin expressed in a letter to Joseph Hooker that if "I lived twenty more years and was able to work, how I should have to modify the *Origin*, and how much the views on all points will have to be modified!"[18]

When writing *The Origin of Species*, Darwin characteristically exhibited a measure of caution, using phrases such as "we must be cautious in attempting," "we may look with some confidence," and "we must not overrate the accuracy of." This is because Darwin presented *The Origin of Species* as a theory, yet to be proved.

Sometimes Darwin made small emendations that shifted the language to more metaphoric phrases; for example, he revised "since the first creature ... was created" to "since the first organic beings appeared on the stage." Writing about where life first came from, Darwin originally wrote, "into which life was first breathed," which surprisingly was changed to "into which life was first breathed by the Creator." In light of human intellectual limits, Darwin was not intimidated to reflect on a divine role in nature. In a letter to the American biologist Asa Gray, Darwin wrote:

blind Watch-maker

> I am inclined to look at everything as resulting from designed laws, with the details, whether good or bad, left to the working out of what we may call chance. Not that this notion at all satisfies me. I feel most deeply that the whole subject is too profound for the human intellect. A dog might as well speculate on the mind of Newton. Let each man hope and believe what he can.[19]

The Chapters

The cleverest and most devastating critique of natural selection in Darwin's lifetime was *On the Genesis of Species,* by George Jackson Mivart, published in January 1871. This critique motivated Darwin like none other. From April 1871 to the end of the year, Darwin made extensive revisions to *The Origin of Species.*

While *The Origin of Species* started with fourteen chapters, to respond to Mivart, Darwin expanded the number of chapters in the sixth edition with the addition of "Objections to the Theory of Natural Selection," along with the existing "Difficulties of the Theory." It was in this last edition Darwin used the term "evolution" for the first time.

The first five chapters introduce the subject of selection mechanisms. Darwin begins with domestic breeding to develop an analogy between breeding selection by man and natural selection.

intelligence

Table II
Chapter Comparisons Between First and Sixth Edition

Chapter	First Edition	Sixth Edition
	Quotations from William Whewell and Sir Francis Bacon	Quotations from William Whewell and Sir Francis Bacon
		Additions and Correction
		An Historical Sketch
	Introduction	*Introduction*
1	*Variation Under Domestication*	*Variation Under Domestication*
2	*Variation Under Nature*	*Variation Under Nature*
3	*Struggle for Existence*	*Struggle for Existence*
4	*Natural Selection*	*Natural Selection; or the Survival of the Fittest*
5	*Laws of Variation*	*Laws of Variation*
6	*Difficulties on Theory*	*Difficulties on Theory*
7	*Instinct*	*Miscellaneous Objections to the Theory of Natural Selection*
8	*Hybridism*	*Instinct*
9	*On the Imperfection of the Geological Record*	*Hybridism*
10	*On the Geological Succession of Organic Beings*	*On the Imperfection of the Geological Record*
11	*Geographical Distribution*	*On the Geological Succession of Organic Beings*
12	*Geographical Distribution – continued*	*Geographical Distribution*
13	*Mutual Affinities of Organic Beings: Morphology: Embryology: Rudimentary Organs*	*Geographical Distribution – continued*
14	*Recapitulations and Conclusions*	*Mutual Affinities of Organic Beings: Morphology: Embryology: Rudimentary Organs*
15		*Recapitulations and Conclusions*

To introduce the reader to the subject, Darwin highlights the history of the evolutionary debate that was well under way by quoting first from a contemporary, William Whewell (1794–1866), and the renowned Sir Francis Bacon (1561–1626) in the preamble. Whewell, a well-known Anglican priest, philosopher, and historian of science, and well trained in inductive reasoning, essentially abandoned inductive reasoning for the emerging contemporary English practice of empiricism, in response to popular criticism of the day. Whewell became an adherent follower of Immanuel Kant who, like Aristotle, promoted deductive reasoning. Darwin takes a Whewell quotation from Bridgewater Treatise (1833) that argues against "Divine" intervention in the "material world."[20]

By contrast, and balance, Darwin quotes from the Englishman Sir Francis Bacon. Bacon popularized the use of inductive reasoning, known as the Baconian method, which is now known as the scientific method, in performing scientific inquiry a century earlier. Bacon envisioned that man should continue to "endeavor [for] an endless progress" using inductive reasoning and the scientific method, and at the same time continue to search "the book of God's works" for answers.[21]

The dynamics of the evolutionary debate today are essentially the same as they were 150 years ago, which long preceded Darwin. Following these quotations by Whewell and Bacon, in the preamble of the first edition of *Origin of Species,* Darwin gives a brief historical account of credits, including the "Author of the *Vestiges of Creation"* (the author was later revealed as Charles De Secondat Montesquieu), Joseph Dalton Hooker, Thomas Robert Malthus, Alfred Russel Wallace, and Charles Lyell.

By the sixth edition, Darwin expands the preamble by including *Additions and Corrections* and *An Historical Sketch,* expanding the preamble remarks and table of contents from ten pages to twenty-one pages. In the *Historical Sketch,* Darwin discusses the contributions of forty-four authors writing on the subject of evolution.

In chapter one, Darwin starts the "one long argument" with discussing the causes of variation, focusing on domestic variation. The first paragraph opens with the statement "variability may be partly connected with excess of food" over many generations, which reflects the influence of *On Population,* written by Thomas Malthus.[22]

Darwin continues that domestic variation serves as a model for variation in nature. Divided into twelve sections, chapter one concludes with a statement by Darwin that variation is caused by the conditions of life, which are "governed by many unknown laws, of which correlated growth is probably the most important."[23, 24]

In chapter two, Darwin envisions that the vast expanse of variations within nature is the raw materials available for natural selection. In distinguishing between varieties and species, Darwin states that the difference will eventually be established with the "discovery of intermediate linking forms."[25]

In chapter three, "The Struggle for Existence," Darwin explains how "the doctrine of Malthus" is the actual driving force of evolution in both the animal and vegetable kingdoms: "It is the doctrine of Malthus applied with manifold force to the whole animal and vegetable kingdoms."[26]

Darwin tempers the harsh realities of "The Struggle for Existence," which directly contradicts any concept of harmony in nature by stating that "death is generally prompt," and those that survive are "happy and multiply."[27]

Chapter four outlines Darwin's theory of natural selection. Darwin starts the chapter by drawing an analogy between domestic breeding by the hand of man and natural selection alone: "Can the principle of selection, which we have seen is so potent in the hands of man, apply under nature? I think we shall see that it can act most efficiently."[28]

In the selection process, there must be something to select. The central question is where do these selections come from? Darwin takes on this question in chapter five, "Laws of Variation." The origin of variation is the fuel and the crux of evolution. But Darwin argues that the origin of variation is not from "mere chance."[29] For Darwin to ascribe chance to the origin of variation only "acknowledge(s) plainly our ignorance."[30] In the summary of chapter five, Darwin explains that our "ignorance of the laws of variation is profound. Not in one case out of a hundred can we pretend to assign any reason why this or that part has varied."[31]

Expecting objections to the theory of natural selection, Darwin acknowledges in chapters six and seven that while objections may not be "fatal to the theory," even he is "in some degree staggered" by the "crowd of difficulties."[32] Even looking at how the eye could develop with natural selection is a problem. Darwin concedes, "the belief that an organ so perfect as the eye could have been formed by natural selection is enough to stagger any one."[33]

While Darwin envisions natural selection acting slowly over time, he yields to objections on the issue by stating, "some have been developed in a different and abrupt manner."[34]

Opening the world of sociobiology, Darwin ventures to explain the role of instincts and behavior in chapter eight. However, the chapter raises more questions than answers. Darwin could not define even the term "instinct": "I will not attempt any definition of instinct."[35]

Even more interestingly, Darwin addresses the definition of species in chapter nine. Since the book is entitled *Origin of Species*, the definition of "species" is central to understanding the concepts of natural selection in the book. Today, while sterility is the litmus for a species, Darwin takes exception to that definition: "There is no more reason to think that species have been specially endowed with various degrees of sterility to prevent their crossing and blending in nature."[36] By the turn of the century, Darwin's concept of blending was finally abandoned following acceptance of Mendelian genetics.

Turning to geology, Darwin examines in chapters ten and eleven the fossil record for evidence of the "transitional links" that are also referred to as "missing link" or "intermediate forms." Darwin states, "we should always look for forms intermediate between each species and a common but unknown progenitor,"[37] because at the time Darwin could not verify any known intermediates.

Evidence to support the theory of natural selection, Darwin argues, should be found in the fossil record. The difficulty Darwin encountered was that the links that should join the species had yet to be discovered. Darwin confides that the "several difficulties here discussed, namely, that, though we find in our geological formations many links between the species which now exist and which formerly existed, we do not find infinitely numerous fine transitional forms closely joining them all together."[38]

Not only were the joining links between species missing, Darwin acknowledged that the known geological formations discoveries only support the "sudden manner" of species appearing in the fossil record.

Since the evidence contradicts the theory of evolution through successive, slight changes, Darwin recognized that the lack of available evidence is "all undoubtedly the most serious nature."[39] Based on what was known at the time, Darwin argues that one "will rightly reject the whole theory."[40]

Of all the objections to the theory of natural selection, joining the links between the species, "and their not being blended together by innumerable transitional links, is a very obvious difficulty."[41] In looking for answers, Darwin reasons that it is because only "a small portion of the surface of the Earth has been geologically explored, and no part with sufficient care, as the important discoveries made every year in Europe prove."[42] Darwin envisions that further explorations would eventually discover the missing links and validate the theory.

Darwin revisits issues with the fossil evidence in geological formations in chapters twelve and thirteen. Ironically, while geology originally captivated Darwin's interest in evolution as a young naturalist aboard the HMS *Beagle,* he later recognized that the "geological record is extremely imperfect; that only a small portion of the globe has been geologically explored with care; that only certain classes of organic beings have been largely preserved in a fossil state; that the number of both specimens and of species preserved in our museums is absolutely as nothing compared with the number of generations which must have passed away even during a single formation."[43]

Moving beyond the problems in the fossil record, Darwin, in chapter fourteen, focuses on integrating morphology, embryology, and rudimentary organs. Of these, Darwin argues that the "leading facts in embryology … are second to none in importance."[44] In concept, the developmental process in the embryo was thought to replay evolution: the successive, slight changes in the embryo correspond to the changes in evolution.

This concept was popularized by the German zoologist, Ernest Haeckel, in the phrase "ontology recapitulates phylogeny." Embryology was thought to define the evolutionary line of descent. Darwin wrote, "If they pass through closely similar embryonic stages, we may feel assured that they all are descended from one parent-form."[45] For Darwin, evidence in the embryo is thought to be of ever-greater importance than evidence found in mature adult forms of life: "On this view we can understand how it is that, in the eyes of most naturalists, the structure of the embryo is even more important for classification than that of the adult."[46]

In summarizing the chapter, Darwin explains that he has "attempted to show that the arrangement of all organic beings throughout all time in groups under groups—that the nature of the relationships by which all living and extinct organisms are united by complex, radiating, and circuitous lines of affinities into a few grand classes."[47] This forward-sounding statement avoids one technical issue: evidence was still sketchy.

In the final chapter, "Recapitulations and Conclusions," Darwin conceptualizes the implications of the development of new species, concluding in the last sentence of *The Origin of Species* that life, "having been originally breathed by the Creator ... [is] still being evolved."[48]

While the last word in all of the editions of *The Origin of Species* is "evolved," in the first edition, the last sentence in the book is the only time Darwin uses any derivative of the term evolution. By the sixth edition, Darwin uses evolution eight times.

How Darwin envisions the origin of man in *The Origin of Species* in the scheme of evolution is only found in the following statement in the last chapter: "Much light will be thrown on the origin of man and his history."[49] Not until just a year before the publication of the sixth edition of *The Origin of Species* did Darwin specifically address the origin of man in *The Descent of Man and Selection in Relation to Sex*. Darwin explains that man evolved from a lower form: "man bears in his bodily structure clear traces of his descent from some lower form."[50]

Early in the fateful month of November 1859, after becoming anxious about the book's acceptance, Darwin began sending copies of the book for review and feedback. Copies were sent to Louis Agassiz, Hugh Falconer, Asa Gray, J. S. Henslow, and A. De Candolle on the eleventh, John Lubbock on the twelfth, L. Jenynson and Russel Wallace on the thirteenth, W. D. Fox on the sixteenth, and W. B. Carpenter on the eighteenth. Darwin's note to Carpenter read:

> My Dear Carpenter,—I beg pardon for troubling you again. If, after reading my book, you are able to come to a conclusion in any degree definite, will you think

me very unreasonable in asking you to let me hear from you. I do not ask for a long discussion, but merely for a brief idea of your general impression. From your widely extended knowledge, habit of investigating the truth, and abilities, I should value your opinion in the very highest rank. Though I, of course, believe in the truth of my own doctrine, I suspect that no belief is vivid until shared by others. As yet I know only one believer, but I look at him as of the greatest authority, viz. Hooker. When I think of the many cases of men who have studied one subject for years, and have persuaded themselves of the truth of the foolishest doctrines, I feel sometimes a little frightened, whether I may not be one of these monomaniacs.[51]

On the day before releasing the publication, Huxley wrote a fateful warning letter arguing against Darwin's view of natural selection. Darwin had written in Latin *"natura non facit saltum,"* meaning that nature takes no leaps; natural selection can only act slowly. Huxley laments, "You have loaded yourself with an unnecessary difficulty in adopting *natura non facit saltum* so unnecessarily."[52] Darwin and Huxley's anxieties were well founded; unfortunately, the warning came too late.

Point of View

The Origin of Species is popularly thought of as one of the finest applications of the scientific method. Darwin, however, termed his approach a "scientific point of view," not the scientific method: "Under a scientific point of view, and as leading to further investigation, but little advantage is gained by believing that new forms are suddenly developed in an inexplicable manner from old and widely different forms, over the old belief in the creation of species from the dust of the Earth."[53] A scientific point of view is not the scientific method.

Darwin clearly states that the scientific point of view was selected to exclude the concept of "suddenly developed." Establishing a point of view as an *a priori* effectively limited Darwin's range of possible interpretations. Subjectivity triumphed over objectivity. Darwin had literally predetermined the final answer: "Whoever is led to believe that species are mutable will do good service by conscientiously expressing his conviction; for thus only can the load of prejudice by which this subject is overwhelmed be removed."[54]

Having a predetermined final answer was central to Darwin's approach in writing *The Origin of Species*. Darwin, like his grandfather, was a relentless theorizer, but the "observation must be for or against some view." In a letter, Darwin wrote, "About thirty years ago there was much talk that geologists ought only to observe and not theorize; and I remember someone saying that a man might as well go into a gravel pit and count the pebbles and describe the colors. How odd it is that any one should not see that all observations must be for or against some view if it is to be of any service."[55] In Darwin's view, the process of collecting of evidence must be guided by "some view." Darwin's approach was subjective, not objective: deductive not inductive.

How Darwin used a point of view to draw conclusions can be seen in his discussion on the variation of species. Darwin's science is just an argument of exclusion: "On the view that each species has been independently created, with all its parts as we now see them, I can see no explanation. But on the view that groups of species are descended from some other species, and have been modified through natural selection, I think we can obtain some light."[56]

Darwin even stated this clearly in the introduction, when he said that the belief that "each species has been independently created—is erroneous." Darwin's point of view effectively closed the door to objectivity. The reason Darwin rejects "independently created" in this discussion on variation of species is because he had already excluded creation as a possible theory.

This approach was paralyzing. In concluding the chapter fourteen, Darwin clearly declares his subjective point of view: "The several classes of facts which have been considered ... innumerable species ... are all descended ... even if it were unsupported by other facts or arguments."[57] For Darwin, the evidence (facts) must be interpreted in subjection to the theory.

Or as Charles Darwin's brother, Erasmus, put it in a letter to Charles on November 23,1859, one day before the publication of *The Origin of Species*: "In fact the *a priori* reasoning is so entirely satisfactory to me that if the facts won't fit, why so much the worse for the facts, in my feeling."[58]

Use of a point of view allowed Darwin to abandon the scientific method, which is based only on inductive reasoning, to develop a theory through deductive reasoning. While Darwin provides exhaustive information in *The Origin of Species*, the book does not contain a single reproducible measurement or any comparative measurements. The only measure Darwin used was a "remarkable degree," which is not a scientific standard unit of measure. The purpose of the scientific method is to avoid subjectivity and bias, but Darwin clearly embraced subjectivity.

The Origin of Species is simply a logical, long argument and not a scientific analysis supported by objective data. Darwin even confirms this by stating, "this whole volume is one long argument" and not a work of science.[59] An argument without any scientific measurements is simply not a scientific analysis. The scientific method requires that the hypothesis must be developed based on objective data alone. From the hypothesis, a theory can be developed which can be tested. Darwin never tested his theory.

How Darwin's point of view works is illustrated in Darwin's following explanation: "There may truly be said to be a constant struggle going on between, on the one hand, the tendency to reversion to a less perfect state, as well as an innate tendency to new variations, and, on the other hand, the power of steady selection to keep the breed true."[60] The explanation is simply a process of subjective reasoning and is not based on any verifiable or reproducible measurement. Expounding on his theory as a point of view allowed Darwin to rationalize the use deductive reasoning.

The scientific method uses inductive and not deductive reasoning. Copernicus was considered one of the first to use objective evidence along with inductive reasoning to propose his hypothesis in the sixteenth century.

Copernicus' use of the scientific method eventually opened the door to the truth and turned upside down the view that the Earth is the center of the universe, which was held by ruling institutions of the day. Expanding on this approach, later in the seventeenth century Francis Bacon became known as the founder of the scientific method by establishing the use of objective data only and espousing that any conclusions made about natural laws must be drawn only from inductive reasoning.

Inductive is different from deductive reasoning in that inductive reasoning is a process of reasoning that starts with evidence from which a hypothesis can be developed to a theory that can be tested for the purpose of eventually establishing a natural law. This is a process that moves from the specific (evidence) to the general (natural law).

Deductive reasoning progresses in the opposite direction. Deductive reasoning moves from the general to the specific, a pattern that Darwin used over his lifetime. In his autobiography, Darwin recalls that from "my early youth I have had the strongest desire to understand or explain whatever I observed, that is, to group all facts under some general laws."[61] In other words, the interpretation of evidence is subject to the hypothesis.

The origins of the Scientific Revolution, founded through use of the scientific method approach, culminated with the publication of the *Philosophiae Naturalis Principia Mathematica,* by Isaac Newton, in 1687. Darwin claims to have initially started with the scientific method. Darwin wrote in his autobiography, "My first notebook was opened in July 1837. I worked on true Baconian principles [scientific method], and without any theory collected facts on a wholesale scale." But Darwin continues in the same sentence and explains what he means: "more especially with respect to domestication productions, by printed enquiries, by conversation with skilful breeders and gardeners, and by extensive reading."[62] The question is what scientific measurements were taken through conversation? The answer is none.

In his autobiography, in the section entitled "Charles Darwin and His Grandfather"—written by his granddaughter, Nora Barlow—Darwin's bent to the deductive approach is clearly declared: "the facts are useless without the frame of the theory to receive them.... For Darwin came to believe that the value of fact-finding lies solely in relation to theory."[63] Emma Darwin often used one of Darwin's sayings, "It is a fatal fault to reason whilst observing, though so necessary beforehand and so useful afterwards."[64]

Darwin knew the problems inherent in following deductive reasoning. In a letter to J. D. Hooker in 1844, Darwin wrote, "I must be allowed to put my own interpretation on what you say of 'not being a good arranger of extended views'—which is that you do not indulge in loose speculations so easily started by every smatterer and wandering collector. I look at a strong tendency to generalise as an entire evil."[65]

Even Francis, Darwin's son, wrote, "He said that no one could be a good observer unless he was an active theorizer. This brings me back to what I said about instinct for arresting exceptions: it was as though he were charged with theorizing power ready to flow into any channel on the slightest disturbance, however small, could avoid releasing a stream of theory, and thus the fact became magnified into importance."[66] In other words, Darwin was captivated by creating theories.

Reading an essay by an economist is how Darwin came across the theory of natural selection. Darwin wrote, in "October 1838 ... I happened to read for amusement Malthus *On Population*.... Here, then, I had at last got a theory by which to work."[67] The question is how did Darwin employ the scientific method to develop the theory of natural section? Darwin's clear answer is that the scientific method had no role in the development of his theory of natural selection.

In reality, Darwin, like his father, "speculated" on nearly every subject encountered. He reasoned that observations should only be in the context of a hypothesis: "I can have no doubt that speculative men, with a curb on, make far the best observers."[68]

Credit should actually be given to Darwin for stating openly he used a point of view rather than the scientific method. Confirming Darwin's position that the ends justify the means, Francis Darwin, Darwin's son, wrote, "For Darwin came to believe the value of fact finding lies solely in relation to the theory."[69]

Darwin even advised colleagues that the collection of data should be guided by the theory. In a letter to J. Scott in 1865, Darwin wrote, "I would suggest to you the advantage ... let the theory guide your observations."[70]

Deductive reasoning was foundational to Darwin's approach. In his autobiography, Darwin comments on writing the *Coral Reef* paper: "No other work of mine was begun is so deductive a spirit as this; for the whole theory was thought out on the west coast of S. America before I had seen a true coral reef."[71]

Even Niles Eldredge of the American Museum of Natural History concedes that though Darwin initially used inductive reasoning, he could not leave deductive reasoning: "At first Darwin was just collecting data, letting nature come to him, and then he formed hypotheses that could be tested."[72] *? which? how?*

Focusing on the same problem, in 1995, Daniel C. Dennett of Tufts University echoes the same conclusion: "The basic deductive argument is short and sweet, but Darwin himself described *The Origin of Species* as 'one long argument' … He bolsters up his logical demonstration with thought experiments—'imaginary instances' that show how these conditions might actually account for the effects he claimed to be explaining."[73]

Even evolutionary biologist Stephen Gould, in his book *Evolution and the Triumph*, said that Darwin used a method of "inferring history from its results."[74] For Darwin, to explore and investigate without a theory is akin to walking aimlessly in a desert without a map. In a letter to Wallace in 1857, Darwin wrote, "I am a firm believer that without speculation there is no good and original observation."[75]

With the deductive reasoning approach, Darwin used analogies in developing his theory. Darwin stated, "all the organic beings which have ever lived on this Earth may be descended from … one primordial form. But this inference is chiefly grounded on analogy, and it is immaterial whether or not it be accepted."[76] The use of analogies is clearly not compatible with the scientific method approach.

Darwin was not alone in abandoning the scientific method. Thomas Huxley, Darwin's "bulldog," echoes the same sentiments. Huxley declares that "Those who refuse to go beyond fact rarely get as far as fact; and anyone who has studied the history of science knows that almost every great step therein has been made by the 'anticipation of nature,' that is, by the invention of hypothesis which, though verifiable, often had very little foundation to start with; and not infrequently, in spite of a long career of usefulness, turned out to be wholly erroneous in the long run."[77] Clearly, in Darwin's inner circle, the evidence was allowed to take second place in anticipation that someday the evidence will be discovered to support the theory.

Darwin was interested in what other scientific philosophers of the day had to say about *The Origin of Species*. Darwin had sent a copy of

The Origin of Species to John Herschel and John Stuart Mill. Hershel supported inductive reasoning, and Mill supported deductive reasoning, or empiricism. In a letter to Charles Lyell concerning Herschel, the "great philosopher," Darwin wrote, "I should excessively like to hear whether I produce any effect on such a mind."[78] Apparently, while Herschel never replied to Darwin, Darwin later wrote to Lyell "I have heard, by a round-about channel, that Herschel says my book 'is the law of higgledy-piggeldy.'"[79]

On the other hand, liberal thinker Stuart Mill responded by writing that he thought *The Origin of Species* was "in the most exact accordance with the strict principles of logic."[80] The problem is logic; logic is not proof. Science historian David L Hull, in 1983, commented that on "closer examination, however, Mill's endorsement can be seen to be not nearly reassuring. Darwin had properly used the Method of Hypothesis, but this method belonged to the logic of discovery, not proof. In spite of twenty years of labor, Darwin had failed to provide proof for his theory of evolution."[81]

When questioned about his method of investigation, Darwin, in 1859, wrote a letter to Asa Gray stating that his work could not be called inductive: "What you hint at generally is very, very true: that my work is grievously hypothetical, and large parts are by no means worthy of being called induction, my commonest error being probably induction from too few facts."[82]

The Origin of Species, as Darwin had suggested all along, is only a theory. Nearly every chapter contains at least one statement that challenges his own theory. Even in the *Introduction* Darwin writes, "I am well aware that scarcely a single point is discussed in this volume on which facts cannot be adduced, often apparently leading to conclusions directly opposite to those at which I have arrived."[83] In *Chapter 8* entitled *Instinct* Darwin writes, "I do not pretend that the facts given in this chapter strengthen in any great degree my theory."[84]

In the final chapter of *The Origin of Species*, Darwin actually only lends hedging confidence to the theory of evolution, noting that the "whole volume is one long argument.... We ought to be extremely cautious in saying that any organ or instinct, or any whole structure, could not have arrived at its present state by many graduated steps."[85]

The personal advantage to Darwin using his point-of-view perspective was troubling. Deciphering between the competing perspectives of chance and design was unsettled his mind. In 1870, Darwin confided in a letter to J. D. Hooker: "My theology is a simple muddle; I cannot look at the universe as a result of blind chance, yet I can see no evidence of beneficial design, or indeed of design of any kind."[86]

Darwin was torn between chance and design. In the end, Darwin calls himself a theist by rejecting blind chance and accepting the existence of an intelligent God:

> Another source of conviction in the existence of God, connected with the reason and not with the feelings, impresses me as having much more weight. This follows from the extreme difficulty or rather impossibility of conceiving this immense and wonderful universe, including man with his capacity of looking far backwards and far into futurity, as the result of blind chance or necessity. When thus reflecting I feel compelled to look to a First Cause having an intelligent mind in some degree analogous to that of man; and I deserve to be called a Theist.[87]

The question surfaces—is Darwin's theory the result of scientific investigation? Clearly, the answer is no. Since the theory stood at odds with the consortium of known evidence, Darwin was eventually forced to drift away from inductive reasoning. Echoing the negative effect science was having on the theory of evolution, Thomas Huxley pined that the "great tragedy of science [is] the slaying of a beautiful hypothesis by an ugly fact."[88] Julian Huxley, the grandson of Thomas Huxley, conceded in 1939 that Darwin had indeed abandoned the scientific method by "combining inductive and deductive reasoning in a single argument."[89]

After 150 years, Darwin's point of view has taken a toll on Darwin's theory. Since the study of biology has been inextricably linked to evolution, the scientific status of biology is now in question. Darwin's "bulldog" of the twentieth century, Ernst Mayr, eventually drew the conclusion that "biology, even though it has all the other legitimate properties of a science, still is not a science like the physical sciences."[90]

Darwin knew all along, however, that the evidence did not support the theory. In a letter to one of his closest friends in 1869, J. D. Hooker, Darwin confided, "If I lived twenty more years and was able to work, how I should have modified the *Origin*, and how much the views on all points will have to be modified!"[91]

While *The Origin of Species* is described as a "scientific work," use of the scientific method continues to mount challenging evidence against Darwin's "beautiful hypothesis."

Chapter Six
Species

On The Origin of Species by Means of Natural Selection, or the
Preservation of Favoured Races in the Struggle for Life, 1ˢᵗ Edition
—by Charles Darwin

The Origin of Species stands as one of the most well-known books in the world and is arguably the pivotal work in establishing the field of evolutionary biology. Looking at the title gives a perspective into Darwin's theory.

Exploring the Title

The twenty-one-word title summarizes the key elements Darwin uses in the development of "my theory." The most common word from the title that is used throughout the book is the word "species," appearing a total of 1,926 times, followed by "natural" and "selection," appearing 764 and 563 times, respectively. In the sixth edition, Darwin eliminated the word "on."

The most commonly recognized phrase besides "origin of species" is "natural selection," which appears 408 times in the sixth edition. Ironically, the phrase "preservation of favoured races" never appears in the text of the entire book.

While the term "species" is the most common term from the title used in the book, defining "species" became one of Darwin's great challenges. From the start, Darwin recognized that among naturalists of the day, the term "species" had no consistent definition: "No one definition has satisfied all naturalists; yet every naturalist knows vaguely what he means when he speaks of a species. Generally the term includes the unknown element of a distinct act of creation."[1]

Table III
Use of Key Words and Phrases

Words and Phrases	Number of Times Used
Species	1926
Natural	764
Selection	563
Natural Selection	408
Life	387
Origin	183
Races	76
Preservation	36
Struggle for Life	26
Origin of Species	18
Favoured Races	2*
Preservation of Favoured Races	2*

* only used as a title

Species

The naturalists of the nineteenth century had continued largely in the tradition of the founder of modern biology, the Swedish naturalist Carolus Linnaeus. Linnaeus considered species uniquely created but classifiable according to recognizable similarities. Darwin, however, viewed this as a limitation of subjectivity: "Hence, in determining whether a form should be ranked as a species or a variety, the opinion of naturalists having sound judgment and wide experience seems the only guide to follow."[2]

Defining "species" is not only a recognized problem by Darwin. Naturalist Henry Alleyne Nicholson explains: "No term is more difficult to define than 'species,' and on no point are zoologists more divided than as to what should be understood by this word."[3] Even the distinction between varieties and species has no known measure. Darwin wrote, "it is not pretended that we have any sure criterion by which species and varieties can be discriminated."[4]

Unlike Newton's measurable laws of gravity, defining a species is not measurable. Classification into species is determined by the subjective interpretation of the naturalist, without a corresponding unit of measure. According to Darwin, there "is no possible test but individual opinion to determine which of them shall be considered as species and which as varieties."[5]

This subjective approach to measuring differences between species was of no small concern for Darwin: "It is all-important to remember that naturalists have no golden rule by which to distinguish species and varieties."[6] Darwin recognized that science cannot exist without a verifiable measurement.

The reality of this problem became acutely apparent while analyzing the specimens from the Galápagos Islands. Darwin concluded that the differentiation between a species was dubious at best, even on the Galápagos Islands. Darwin acknowledges in looking back many "years ago, when comparing, and seeing others compare, the birds from the closely neighbouring islands of the Galápagos Archipelago, one with another, and with those from the American mainland, I was much struck how entirely vague and arbitrary is the distinction between species and varieties."[7]

Darwin pined that there was no known method to scientifically measure and define what distinguishes the difference between one species and another: "Nevertheless, no certain criterion can possibly be given by which variable forms, local forms, sub species, and representative species can be recognised."[8] To resolve this problem, Darwin attempted to determine the definition that distinguishes the difference between species.

Numbers to Struggle

Species is a numbers game. Darwin envisioned that as a species began to evolve, new varieties would be formed. In the pursuit of a definition of species, Darwin reasoned that as a variety begins to exceed the number of the parent species, the newly emerging populous variety should be defined as a distinct species: "If a variety were to flourish so as to exceed in numbers the parent species, it would then rank as the species."[9] This difference between species would be dependent on the population size of the varieties.

With a certain population size, however, the question is, is there a measure of difference between species or varieties? In taking one approach, Darwin reasons that the answer may be found in the varieties, the intermediate links, between species. Darwin writes: "Finally varieties cannot be distinguished from species—except, first, by the discovery of intermediate linking forms; and, secondly, by a certain indefinite amount of difference between them.... but the amount of difference considered necessary to give to any two forms the rank of species cannot be defined."[10] In taking this approach, Darwin acknowledges that not only can the varieties not be distinguished from species, but also that there is no definable distinction between species.

Without a method to define the difference between species, Darwin was resigned "to look at the term species as one arbitrarily."[11] Likewise, in extending the discussion to varieties, Darwin writes that the "term variety, again, in comparison with mere individual differences, is also applied arbitrarily, for convenience sake."[11]

In attempting to work around the problem of distinguishing varieties from species, Darwin suggests an alternative term, "incipient species": "Nevertheless according to my view, varieties are species in the process of formation, or are, as I have called them, incipient species."[12]

The question is how can introducing the term "incipient species" create a distinction between varieties and species? Actually, Darwin asks the same question: "Again, it may be asked, how is it that varieties, which I have called incipient species...?"[13] Darwin continues to ask the same question in the same paragraph, blurring the definition, further stating: "Again ... How do those groups of species, which constitute what are called distinct genera and which differ from each other more than do the species of the same genus, arise?"[14] With more questions than answers, Darwin continues the argument by changing the subject to the struggle for life, suggesting that what really defines a species is the result of the struggle for life—survival: "Again ... All these results, as we shall more fully see in the next chapter, follow from the struggle for life."[15] Today, how "struggle for life" constitutes a definition of species remains elusive.

Later in *The Origin of Species*, grasping for a definition, Darwin flips and blurs the argument further by suggesting that the different varieties of the same species is greater than between other genera. Darwin explains: "The amount of variation in the individuals of the same species is so great that it is no exaggeration to state that the varieties of the same species differ more from each other in the characters derived from these important organs, than do the species belonging to other distinct genera."[16] For Darwin, each attempt to define species ended with more questions than answers.

Darwin's goal was to discover the natural laws that operate evolution in the same way Newton discovered the physical laws governing motion. Without a definition of species and without a measurement of species, Darwin was forced to work with the "arbitrary" species. Without a measurable variable for species, Darwin was forced to abandon the scientific method.

Given that the term species is not definable, Darwin dismisses the importance of the term species, the key term in *The Origin of Species*, stating that it "is immaterial for us whether a multitude of doubtful forms be called species or sub-species or varieties."[17]

Sterility

The question of how to define "species" continues even today. The problem is now referred to as the "species problem." Sterility has traditionally been used by naturalists to define a species. Darwin counters by noting: "The view commonly entertained by naturalists is that species, when intercrossed, have been specially endowed with sterility, in order to prevent their confusion. This view certainly seems at first highly probable, for species living together could hardly have been kept distinct had they been capable of freely crossing."[18]

Breaking from the traditional sterility definition of species, Darwin envisions sterility could not be a fixed characteristic of a species. Darwin argues that sterility could be overcome, writing that we "must, therefore, either give up the belief of the universal sterility of species when crossed; or we must look at this sterility in animals, not as an indelible characteristic, but as one capable of being removed by domestication."[19]

At the same time, Darwin concedes that the interbreeding between species almost always results in biological problems, including sterility.

Darwin recognized that "interbreeding continued during several generations ... almost always [leading] to decreased size, weakness, or sterility."[20] Despite Darwin's efforts to remove sterility as a characteristic for defining "species," the evidence that sterility defines a species continues. Today, sterility is a cardinal feature that defines a species.

Order of Nature

Natural laws intrigued Darwin. Noticing that the laws of physical elements allow for classification in groups, as in the periodic table, Darwin reflects on using the same approach: "We know, for instance, that minerals and the elemental substances can be thus arranged. In this case there is of course no relation to genealogical succession, and no cause can at present be assigned for their falling into groups. But with organic beings the case is different, and the view above given accords with their natural arrangement in group under group; and no other explanation has ever been attempted."[21]

While naturalists used this approach, known as the natural system of classification, ironically Darwin was skeptical: "Naturalists, as we have seen, try to arrange the species, genera, and families in each class, on what is called the Natural System. But what is meant by this system? Some authors look at it merely as a scheme for arranging together those living.... It seems to me that nothing is thus added to our knowledge."[22] Darwin wanted to discover the mechanism that actually operated evolution of species; going beyond just arrange the species like chemicals on the periodic table.

Darwin thought that the natural system was problematic: "As we have no written pedigrees, we are forced to trace community of descent by resemblances of any kind."[23] Therefore, Darwin suggests that embryology will eventually be seen to hold the keys to species classification: "There can be no doubt that embryonic [specimens], excluding larval characters, are of the highest value for classification, not only with animals but with plants."[24] Darwin proposes that evidence from embryology would provide the most important data in classifying species and giving the order of descent, which is the evolution order of each species.

While embryology may be valuable, Darwin never argued again that embryology alone should be used to define or classify a species. In a discussion of intermediate gradations, Darwin finally concedes that defining the term "species" is just a "vain search." Darwin concedes: "we shall be compelled to acknowledge that the only distinction between species and well-marked varieties is that the latter are known or believed to be connected at the present day by intermediate gradations, whereas species were formerly thus connected.... This may not be a cheering prospect; but we shall at least be freed from the vain search for the undiscovered and undiscoverable essence of the term species."[25]

At no point in *The Origin of Species* does Darwin define the term "species." Interestingly enough, however, this did not deter Darwin from arguing his point of view that species cannot be a distinct creation only because it is not scientific explanation: "On the ordinary view of the independent creation of each being, we can only say that so it is; that it has pleased the Creator to construct all the animals and plants in each great class on a uniform plan; but this is not a scientific explanation."[26]

Without a definition of species, how Darwin considers his approach scientific is certainly open to question. Darwin rigidly held to the theory of descent (evolution) to a fatal fault in objectivity, since the evidence was immaterial to theory. Darwin wrote, "the innumerable species, genera and families, with which this world is peopled, are all descended … from common parents, and have all been modified in the course of descent, that I should without hesitation adopt this view, even if it were unsupported by other facts or arguments."[27] Darwin clearly chose to abandon any facts that did not support the concept that all species have evolved from common parents. For Darwin, theory played the paramount role over evidence. Darwin's theory is clearly based on a philosophy and not the scientific method.

Even in the discussion of tracing the origin of species, the bottom line is that Darwin never defined the term species. Darwin concedes: "We have seen that there is no infallible criterion by which to distinguish species and well-marked varieties."[28]

The plague of defining "species" continued into the twentieth century. In 1937, leading evolutionary biologist Theodosius Dobzhansky bemoaned the vague status of the term "species": "Of late, the futility of attempts to find a universally valid criterion for distinguishing species has come to be fairly generally, if reluctantly, recognized."[29]

According to evolutionary geneticist Jody Hey, in an article in *Trends in Ecology and Evolution* in 2001, attempts to define Darwin's key term, "species," have been a failure: "The species problem is the long-standing failure of biologists to agree on how we should identify species and how we should define the word 'species.'"[30]

Darwin's key word, "species," ironically continues to defy a precise definition. The problem is now known as "the species problem."

Chapter Seven
Natural Selection

The old argument of design in nature, as given by Paley, which formerly seemed to me so conclusive, fails, now that the law of natural selection has been discovered.

—Charles Darwin[1]

Evolutionary biologist and American paleontologist Niles Eldredge, in the book *Darwin, Discovering the Tree of Life* (2005), credits Darwin with formulating the essence of evolution: "When [Darwin] formulated the principle of natural selection, he had discovered the central process of evolution."[2]

Coining the Term

Darwin coined the term "natural selection" from the phrase "natural process of selection" that had actually been developed twenty years earlier by English zoologist and chemist Edward Blyth. While Darwin and Russell Wallace were the first to use the phrase "natural selection," they were not the first to entertain the same basic tenets. In the first chapter of *The Origin of Species*, Darwin gives credit to Blyth: "Mr. Blyth, whose opinion, from his large and varied stores of knowledge, I should value more than that of almost any one."[3]

The popularity of evolution is now axiomatically linked to the development of natural selection. Today, the terms Darwinism and natural selection are essentially synonymous. By successfully introducing the concept of natural selection, Darwin, like no one else, opened a new chapter for the acceptance of evolution.

In the same way that Newton's discovery of gravity revolutionized the physical sciences, Darwin wanted natural selection to revolutionize the life sciences. Newtonian physics paved the way for space exploration; natural selection opened exploration into the origins of life. Like physics, Darwin envisioned evolution as the result of causes and effects; natural laws caused life to evolve from a sequence of biological steps.

Natural selection is now considered more than just a theory. As evolutionary biologist, Stephen J. Gould from Harvard University wrote, "The essence of Darwinism lies in a single phrase: natural selection is the creative force of evolutionary change."[4]

The American Museum of Natural History, in the New York presentation of the *Darwin* exhibit organized by Niles Eldredge, declares that evolution occurs by natural selection: "A century and a half ago, Charles Darwin offered the world a single, simple scientific explanation for the diversity of life on Earth: evolution by natural selection."[5]

The acceptance of evolution as a "scientific explanation" of life extends beyond the hallowed halls of academia and into the bastion of governmental support. The Smithsonian Institute's exhibit on "how evolution takes place" reflects the pinnacle stature of natural selection: "Darwin did recognize natural selection as the most important mechanism by which evolution takes place."[6]

In April 2006, even the leading conservative paper, *The Wall Street Journal,* published an editorial, by Kevin Shapiro of Harvard University, on the merits of natural selection in the wake of the Dover trial in Pennsylvania, declaring: "There is no longer any serious dispute about the evidence for natural selection."[7]

"Evolution takes place" encapsulates the new doctrine of Western civilization. Natural selection operates Darwin's theory of evolution— "by means of natural selection." Life through evolution is thought to be the key to unlock the greatest mystery of all—how did we get here?

Lurking Shadows

In the shadow of this solidarity façade lingers one hard-core reality question—is natural selection valid? Is natural selection based on a natural law or even a series of natural laws? The answer is surprising. Did Darwin arrive at natural selection through any physical measurements, like Newton, when he discovered the laws of gravity? The answer is no.

Even Darwin willingly acknowledges that natural selection is "grounded" on a collection of beliefs and arguments rather than scientific measurements. Darwin wrote:

> The theory of natural selection is grounded on the belief
> that each new variety and ultimately each new species,
> is produced and maintained by having some advantage
> over those with which it comes into competition; and
> the consequent extinction of less-favoured forms almost
> inevitably follows.[8]

By the sixth edition of *The Origin of Species*, Darwin eventually entangles the theory of natural selection into a maze of inconsistencies and tacit contradictions. Even Darwin concluded, "natural selection … is by far the most serious special difficulty which my theory has encountered."[9]

In recognizing an array of unresolved issues with natural selection, Darwin gave clear instructions in *The Origin Species* regarding the pivotal evidence for accepting or rejecting the theory of natural selection: "If numerous species, belonging to the same genera or families, have really started into life at once, the fact would be fatal to the theory of evolution through natural selection."[10]

Darwin gives patent instructions, natural selection should be rejected if the evidence is found that supports the sudden explosion of species. In chapter nine, "Fossils," we will examine the Cambrian explosion, but in this chapter, we will examine how Darwin answers the following four basic questions on natural selection in *The Origin of Species*:

1. What is natural selection?
2. Does natural selection have a purpose?
3. How does natural selection work?
4. What do "rudimentary" structures tell us about natural selection?

Defining Natural Selection

Darwin was captivated with the concept of natural selection, and at the same time perplexed by the scope of its actions: "Can we believe that natural selection could produce, on the one hand, an organ of trifling importance, such as the tail of a giraffe, which serves as a fly-flapper, and, on the other hand, an organ so wonderful as the eye?"[11]

In theory, natural selection acts through a series of natural laws. In the sixth edition of *The Origin of Species,* Darwin wrote, "So again it is difficult to avoid personifying the word Nature; but I mean by nature, only the aggregate action and product of many natural laws, and by laws the sequence of events as ascertained by us."[12]

In the first edition, Darwin presented natural selection acting simply through the natural law of preservation: "This principle of preservation, I have called, for the sake of brevity, Natural Selection."[13]

Immediately after publication of the first edition, though, natural selection became the focal point of criticism. As more problems emerged, the theory became more complicated. By the sixth edition, Darwin had eliminated the phrase "for the sake of brevity." Darwin's theory of natural selection is not as simple, as we will soon discover.

Along with preservation is the law of struggle. In the complete title of *The Origin of Species,* Darwin links natural selection with *Struggle for Life*: *The Origin of Species by means of Natural Selection, or the Preservation of Favoured Races in the Struggle for Life.*

Darwin often rephrases the same or similar concept in different ways, often complicating the theory. For example, in the text, while defining natural selection, Darwin paraphrases the title of *The Origin of Species* to include the terms "variation," "destruction," and "survival of the fittest": "This preservation of favourable individual differences and variations, and the destruction of those which are injurious, I have called Natural Selection, or the Survival of the Fittest."[14]

In essence, though, Darwin envisions natural selection to preserve what in the end survives. Thomas Hunt Morgan, the eminent geneticist, pioneer of fruit-fly research, and winner of the 1933 Nobel Prize in Physiology or Medicine, did not hesitate to comment that natural selection is simply no more than a tautology—circular reasoning. Early in the twentieth century, Morgan noted, "it may be little more than a truism to state that the individuals that are best adapted to survive have a better chance of surviving than those not so well adapted to survive."[15]

Historically, natural selection defined by "survival" has not faired well with even more contemporary scientists. The leading twentieth century developmental biologist, C. H. Waddington from Edinburgh University, came to the same conclusion as Morgan: "There, you do come to what is, in effect, a vacuous statement: Natural selection is that some things leave more offspring than others; and you ask, which leave more offspring than others; and it is those that leave more offspring; and there is nothing more to it than that."[16]

Morgan and Waddington are not alone. In 1982, British evolutionist Francis Hitching wrote: "Darwinism, as Darwin wrote it, could be simply but nonsensically stated: survivors survive. Which is certainly a tautology; and tells us nothing about how species originate, as even Darwin's supporters admit."[17]

Ironically, Darwin agrees. In defining natural selection, Darwin sometimes does not even use the term "preservation," using instead the phrase "struggle for existence," preserving only those that are not destroyed: "This, indeed, might have been expected; for as natural selection acts through one form having some advantage over other forms in the struggle for existence, it will chiefly act on those which already have some advantage."[18]

Darwin's differing views of natural selection stem from one fundamental issue—what are the natural laws that operate natural selection? At no point in *The Origin of Species* does Darwin even claim that the theory of natural selection is a "scientific" theory. Natural selection was certainly Darwin's "most serious special difficulty."

For Darwin, natural selection became a belief, not a science. This approach is pervasive in *The Origin of Species*: ", as illustrated "I believe that the hive-bee has acquired, through natural selection, her inimitable architectural powers."[19]

In presenting evidence for natural selection, at times the evidence is based only on "imaginary illustrations": "In order to make it clear how, as I believe, natural selection acts, I must beg permission to give one or two imaginary illustrations."[20]

Darwin also developed belief in natural selection through analogy: "But how, it may be asked, can any analogous principle apply in nature? I believe it can and does apply most efficiently."[21] Darwin envisioned that life was simply a logical sequence of laws.

Even though, a simple and consistent definition of natural selection eluded Darwin. The popular emerging societal confidence in the concept of evolution superseded the unforeseen looming technical difficulties. In the wake of the Enlightenment Age, as the mysteries of nature were vanishing, the miraculous was just mechanical; the hope was that solutions to nature's most difficult details were soon to be discovered. Faith held that man would soon discover the mechanics of nature, and in the end, life would prove to be simply an accumulation of the laws of nature.

Darwin was in a race against a line of competing naturalists, including Russell Wallace, to complete a working theory for evolution. Like no one else before or since, Darwin successfully garnered widespread acceptance of evolution by presenting natural selection as the mechanism for evolution, using a litany of "imaginary illustrations," analogies, and logical explanations.

The long family legacy of the "Darwin" social status, similar to twentieth century America's "Kennedy" status, sealed Darwin's success. Underlying the motivation was a relentless passion for theorizing, just like his father. Darwin wrote, "my father's mind was not scientific ... yet he formed a theory for almost everything which occurred."[22]

In the tradition of a true "Darwin," wanting to leave a mark on history, Charles Darwin was driven to find the key that turns on the power of nature. For Darwin, the key was natural selection: "Natural Selection, we shall hereafter see, is a power incessantly ready for action."[23]

Power is foundational to the actions of natural selection. The nature of this power is so vast that Darwin equates the actions of natural selection to that of a deity: "It has been said that I speak of natural selection as an active power or Deity; but who objects to an author speaking of the attraction of gravity as ruling the movements of the planets?"[24]

Not only is natural selection all powerful—omnipotent—but Darwin envisions natural selection to be ever-present—omnipresent: "Further we must suppose that there is a power, represented by natural selection or the survival of the fittest, always intently watching each slight alteration." [25]

To Darwin, this power and presence of natural selection is unlimited: "there is no limit to this power." Using a logical line of argument, Darwin writes:

> If man can by patience select variations useful to him,
> why, under changing and complex conditions of life,

should not variations useful to nature's living products often arise, and be preserved or selected? What limit can be put to this power, acting during long ages and rigidly scrutinising the whole constitution, structure, and habits of each creature, favouring the good and rejecting the bad? I can see no limit to this power, in slowly and beautifully adapting each form to the most complex relations of life.[26]

Eventually, in characteristic style, Darwin hedges and retracts his statement that he "can see no limit to this power," and writes, "natural selection will be powerless."[27]

While the basic tenets of natural selection initially appear simple, the simplicity quickly crescendos into complexity as Darwin develops, then hedges, his arguments with inconsistencies and blatant contradictions. The second question is—does natural selection have a purpose?

Purposes of Natural Selection

Eldredge, writing for the American Museum of Natural History, states natural selection is "the result of random mutations." The University of California at Berkley state-sponsored Web site curriculum designed to instruct educators on evolution echoes the same theory that evolution is a random process: "mutations are random—whether a particular mutation happens or not is unrelated to how useful that mutation would be."[28]

Contrary to today's popular opinion though, Darwin actually envisions natural selection to be a nonrandom and purpose-driven process. In *The Origin of Species,* Darwin develops seven key purposes of natural selection:

1. Natural selection acts for the overall good: "Natural selection will never produce in a being any structure more injurious than beneficial to that being, for natural selection acts solely by and for the good of each."[29]
2. Natural selection acts for developing organizational economy: "I suspect, also, that some of the cases of compensation which have been advanced, and likewise some other facts, may be

merged under a more general principle, namely, that natural selection is continually trying to economise in every part of the organisation."[30]

3. Natural selection acts for the development of efficiency: "For the best definition which has ever been given of a high standard of organisation, is the degree to which the parts have been specialised or differentiated; and natural selection tends towards this end, inasmuch as the parts are thus enabled to perform their functions more efficiently."[31]

4. Natural selection acts for the development of perfection: "Although the belief that an organ so perfect as the eye could have been formed by natural selection, is enough to stagger any one; yet in the case of any organ, if we know of a long series of gradations in complexity, each good for its possessor, then under changing conditions of life, there is no logical impossibility in the acquirement of any conceivable degree of perfection through natural selection."[32]

5. Natural selection acts for protection: "So, conversely, modifications in the adult may affect the structure of the larva; but in all cases natural selection will ensure that they shall not be injurious: for if they were so, the species would become extinct."[33]

6. Natural selection acts for profitability: "Natural selection will ... adapt the structure of each individual for the benefit of the whole community; if the community profits by the selected change."[34]

7. Natural selection acts for domination in the struggle for life: "He who believes in the struggle for existence and in the principle of natural selection, will acknowledge that every organic being is constantly endeavouring to increase in numbers; and that if any one being varies ever so little, either in habits or structure, and thus gains an advantage over some other inhabitant of the same country, it will seize on the place of that inhabitant, however different that may be from its own place."[35]

Starting with the first purpose, with natural selection acting to "seize on the place," Darwin creates a conflicting dynamic between domination and the development of greater diversity. The theory of natural selection, then, is not founded on a single law, but a number of even divergent natural laws.

By integrating divergent purposes, Darwin creates a dynamic tension between the purposes of natural selection. An example of this reasoning is seen in the following quotation, as Darwin envisions how "domination" exists with "divergence of character," which causes rise to new species, and at the same time envisions how the "extinction of less improved" causes the elimination of species:

> We have seen that it is the common, the widely diffused, and widely ranging species, belonging to the larger genera within each class, which vary most; and these tend to transmit to their modified offspring that superiority which now makes them dominant in their own countries. Natural selection, as has just been remarked, leads to divergence of character and to much extinction of the less improved and intermediate forms of life.[36]

Divergence and extinction seem to be inconsistent purposes, like other aspects of natural selection: mutual good and domination, life and death. Darwin wrote, "natural selection acts by life and death."[37]

Darwin clearly defines purposes for natural selection. But, these purpose-driven actions of natural selection highlight one fact: natural selection is not simple. With these conflicting purposes, the third question is, how does natural selection work?

Natural selection acts on existing species. Darwin never proposed that natural selection accounts for the origin of life: "Finally, I believe that many lowly organised forms now exist throughout the world, from various causes. In some cases variations or individual differences of a favourable nature may never have arisen for natural selection to act on and accumulate."[38]

Darwin envisions natural selection to be "produced by laws acting around us," after having "been originally breathed by the Creator." Darwin identifies these laws as 1) growth, 2) inheritance, 3) variability,

4) ratio of increase, 5) struggle for life, 6) natural selection, 7) divergence of character, and 8) extinction.

> It is interesting to contemplate a tangled bank, clothed with many plants of many kinds, with birds singing on the bushes, with various insects flitting about, and with worms crawling through the damp Earth, and to reflect that these elaborately constructed forms, so different from each other, and dependent upon each other in so complex a manner, have all been produced by laws acting around us. These laws, taken in the largest sense, being Growth with reproduction; Inheritance which is almost implied by reproduction; Variability from the indirect and direct action of the conditions of life, and from use and disuse; a Ratio of Increase so high as to lead to a Struggle for Life, and as a consequence to Natural Selection, entailing Divergence of Character and the Extinction of less improved forms. Thus, from the war of nature, from famine and death, the most exalted object which we are capable of conceiving, namely, the production of the higher animals, directly follows. There is grandeur in this view of life, with its several powers, having been originally breathed by the Creator into a few forms or into one; and that, whilst this planet has gone circling on according to the fixed law of gravity, from so simple a beginning endless forms most beautiful and most wonderful have been, and are being, evolved.[39]

During the twentieth century, the presentation of Darwin's eight purposes of natural selection has been reformatted. The most popular format has been encapsulated in the VISTA format presented by Niles Eldredge's *Darwin* exhibit.

VISTA—The *Darwin* Exhibit

During the twentieth century, a range of neo-Darwinian versions of natural selection emerged, modifying and simplifying Darwin's original laws of natural selection, and completely eliminating any reference to "the Creator." In the 2005 *Darwin* exhibit at the American Museum of Natural History, the curator, Niles Eldredge, envisions natural selection operating by just five "simple" mechanisms: "In fact, it is so simple that it can be broken down into five basic steps, abbreviated here as V.I.S.T.A.: 1) Variation, 2) Inheritance, 3) Selection, 4) Time, and 5) Adaptation."[40]

> VISTA: Variation + Inheritance + Selection + Time + Adaptation = New Species

To explore how natural selection works, we will expand on this VISTA format, focusing on Darwin's statements. This brings us to the third question: How does natural selection work? Eldredge's first topic is variations—the central driving force in evolution.

Similar to shopping at the mall and selecting from a wide variety of options, natural selection takes variations from a wide range of options in nature. Natural selection can only select, and not produce, the variations. Darwin wrote, "natural selection can do nothing until favourable individual differences or variations occur."[41]

From the available variations, "natural selection will pick out with unerring skill each improvement."[42] Darwin envisions that "natural selection ... only takes advantage of variations." [43] Darwin writes: "Natural selection acts solely through the preservation of variations in some way advantageous, which consequently endure."[44] "Only those variations which are in some way profitable will be preserved or naturally selected."[45]

Darwin envisions natural selection acting as an omnipresent power "insensibly working, whenever and wherever opportunity offers," not limited by space or time, acting "daily and hourly scrutinizing, throughout the world, the slightest variations."[46]

Central to the process of selection is the selection criteria. Darwin envisions the purpose of natural selection working through a series of at least five "complex contingencies at work." The criteria for selection include: 1) beneficial nature of the variations, 2) freedom of the new variations intercrossing, 3) the environmental conditions, 4) the influence of the other new "immigrants," and 5) nature of the existing species.

> Whether such variations or individual differences as may arise will be accumulated through natural selection in a greater or less degree, thus causing a greater or less amount of permanent modification, will depend on many complex contingencies—on the variations being of a beneficial nature, on the freedom of intercrossing, on the slowly changing physical conditions of the country, on the immigration of new colonists, and on the nature of the other inhabitants with which the varying species come into competition.[47]

The purpose of natural selection through a complex selection process is to increase the complexity of the species, from mold to man. Central to the process is the origin of these new variations. Without new variations, evolution would not exist. Evolution is dependent on the production of new variations in nature. New variations are the cornerstone of evolution.

Variations

New variations are the fuel natural selection uses to formulate the future of evolution. Since natural selection can only select from available variations, according to Darwin, nature must have a way to cause these new variations to arise.

New variations must originate in nature, but the question is, how does nature cause new variations? The *Darwin* exhibit at the American Museum of Natural History indicates that new variations are "often the result of random mutations, or 'copying errors.'"[48]

To infer that Darwin theorized that "random mutations" and "copying errors" account for the origin of variations is a fabrication. A synonym for random is chance, but Darwin had clearly excluded that possibility: "I was so convinced that not even a stripe of colour appears from what is commonly called chance."[51]

In ascribing a role for chance in the origin of variation, Eldredge contradicts Darwin's theory. Darwin had written, "Mere chance ...

would never account for so habitual and large a degree of difference as that between the species of the same genus."[52] Darwin continues, "I have hitherto sometimes spoken as if the variations ... were due to chance. This, of course, is a wholly incorrect expression, but it serves to acknowledge plainly our ignorance of the cause of each particular variation."

Actually, Darwin never used the terms "random" or "copying errors" in *The Origin of Species*. While Darwin uses the term "mutation" three times in *The Origin of Species,* "mutation" is used only in reference to the fossil record, and not to "random mutations" from genetic "copying errors."

At first, Darwin uses the term "mutations" in reference to the lack of fossil evidence for evolution: "Why does not every collection of fossil remains afford plain evidence of the gradation and mutation of the forms of life?"[49] The two other uses of "mutation" are in the same sentence in which Darwin ironically again questions the value of the fossil record as evidence for evolution: "The geological record is so perfect that it would have afforded us plain evidence of the mutation of species, if they had undergone mutation."[50]

In addressing the origin of variations, Darwin combines two causes: the "nature of the organism and that of the surrounding conditions."[54] The "nature of the organism" includes the pre-existent range of variations within a species—inherent variations. Darwin envisions that the range of inherent variations within a species even exceeds the variations observed between species: "The amount of variation in the individuals of the same species is so great that it is no exaggeration to state that the varieties of the same species differ more from each other in the characters derived from these important organs, than do the species belonging to other distinct genera."[55] According to Darwin, these inherent variations are the second cause (source) of new variations.

For Darwin, this vast range of inherent variations accounts for the origin of variations. Darwin continues: "That species have a capacity for change will be admitted by all evolutionists; but there is no need, as it seems to me, to invoke any internal force beyond the tendency to ordinary variability, which ... give rise by graduated steps to natural races or species."[56]

A third cause of variations is from the "surrounding conditions"—the environment. Darwin envisions that as the environment changes, variations will increase. Darwin writes, "We have good reason to believe, as shown in the first chapter, that changes in the conditions of life give a tendency to increased variability; and in the foregoing cases the conditions changed, and this would manifestly be favourable to natural selection, by affording a better chance of the occurrence of profitable variations."[57]

While Darwin envisions new variations arising, these are not random or chance variations. These new variations are spontaneous in the "right direction"—designed variations: "For all spontaneous variations in the right direction will thus be preserved; as will those individuals which inherit in the highest degree the effects of the increased and beneficial use of any part."[58] Designed variations are the fourth cause (source) of new variations.

In discussing the eye, Darwin suggests that the origin of new variations can be considered analogous to man's inventions. Darwin writes:

> In living bodies, variation will cause the slight alteration, generation will multiply them almost infinitely, and natural selection will pick out with unerring skill each improvement. Let this process go on for millions of years; and during each year on millions of individuals of many kinds; and may we not believe that a living optical instrument might thus be formed as superior to one of glass, as the works of the Creator are to those of man?[59]

In essence, Darwin envisions variations to produce improvements "as the works of the Creator." The Creator originates new variations in the same way that man makes the lenses for an optical instrument. As the telescope is a product of man, Darwin envisions that the eye "must be the work of the Creator." For Darwin, the designed new variations are analogous to the acts of a Creator.

The question is, was Darwin content with these possible scenarios for the origin of variations? The answer is no. As the famous Dutch botanist Hugo de Vries explained in 1905: "Natural selection may explain the survival of the fittest, but it cannot explain the arrival of the fittest."[60]

In 1982, the late English paleontologist Colin Patterson placed the origin of variations into perspective: "No one has ever produced a species by the mechanisms of natural selection. No one has ever got near it, and most of the current argument in neo-Darwinism is about this question: how a species originates. And it is there that natural selection seems to be fading out, and chance mechanisms of one sort or another are being invoked."[61]

Actually, Darwin agrees: "But we are far too ignorant to speculate on the relative importance of the several known and unknown causes of variation."[62] Darwin continues: "Our ignorance of the laws of variation is profound. Not in one case out of a hundred can we pretend to assign any reason why this or that part has varied."[63]

The reason Darwin gives for not deciding on how variations arise is that they cannot be "proved by clear evidence."[64] Darwin concludes: "Variability is governed by many unknown laws, of which correlated growth is probably the most important."[65]

In looking for an answer to the origin of variation, Darwin introduces a contradiction inferring that natural selection causes variability: "In one sense, the conditions of life may be said not only to cause variability, either directly or indirectly, but likewise to include natural selection."[66] This statement contradicts Darwin's statement: "Some have even imagined that natural selection induces variability, whereas it implies only the preservation of such variations."[67]

Certainly, the origin of variations was a problem for Darwin. Eventually Darwin even contradicts the unlimited nature of inherent variations within a species, stating, "We have seen that species at any one period are not indefinitely variable, and are not linked together by a multitude of intermediate gradations."[68]

Darwin knew that natural selection would be out of business without a continuous supply of new variations. The development of new variations is foundational to the theory of evolution by means of natural selection.

Actually, Darwin never identified the laws associated with the origin of variations. Italian geneticist Giuseppe Sermonti concurs with

Darwin's origin of new variation problem: "Natural selection could perhaps be invoked as a mechanism accounting for the survival of the species. But the claim that natural selection is creative of life, of life's essence and types and orders, can only leave one dumbstruck."[69]

In light of the fact that there is no scientific evidence, Sermonti even questions the logic of natural selection's role in the origin of new variation: "Natural selection only eliminates, and its adoption as a mechanism of origin is like explaining the 'appearance' by 'disappearance.'"[70] In the end, Darwin did not discover one natural law for the origin of variations—the cornerstone of evolutionary biology.

Even though Darwin never identified the origins of variations, Darwin envisioned "descent with modification through variation and natural selection" as the pathway to evolution.[71] To support this theory, Darwin suggests that eventually the "great leading facts of paleontology" (fossil record) will provide the evidence.[71]

In essence, evolution is a process of variation selection. By selecting available variations, Darwin envisions the resulting modifications leading to greater diversity within the species. This diversity eventually leads to the development of new species—the building blocks of biological evolution:

> An extraordinary amount of modification implies an unusually large and long-continued amount of variability, which has continually been accumulated by natural selection for the benefit of the species. But as the variability of the extraordinarily developed part or organ has been so great and long-continued within a period not excessively remote, we might, as a general rule, still expect to find more variability in such parts than in other parts of the organisation which have remained for a much longer period nearly constant.[72]

Darwin envisions that new variations and modifications will provide the species a better "chance of success in the battle of life."[73] Darwin calls this effect the "the law of correlation": "It is also necessary to bear in mind that, owing to the law of correlation, when one part varies and the variations are accumulated through natural selection, other modifications, often of the most unexpected nature, will ensue."[74]

While the phrase "law of correlation" is only used twice in *The Origin of Species,* the concept is used throughout the book: "I mean by this expression that the whole organisation is so tied together, during its growth and development, that when slight variations in any one part occur and are accumulated through natural selection, other parts become modified."[75] The process of selecting variations, by means of natural selection, leads to modification.

Evidence Darwin cites for representing the "laws of correlation" is the flower: "I know of no case better adapted to show the importance of the laws of correlation and variation, independently of utility, and therefore of natural selection, than that of the difference between the outer and inner flowers in some Compositous and Umbelliferous plants."[76]

The question is what evidence was actually measured? The answer is none. Clearly, this example of "evidence" is not based on any specific measurement, like Newton's measurements of gravity. Darwin had completely abandoned the scientific method.

Darwin extends this concept and further speculates that natural selection can function to produce "similar organs," even in other distant species: "As two men have sometimes independently hit on the same invention ... natural selection, working for the good of each being ... has produced similar organs, as far as function is concerned, in distinct organic beings, which owe none of their structure in common to inheritance from a common progenitor."[77]

Again, Darwin's argument for new variations and modifications is based on an analogy, "as two men ... [so] natural selection." Curiously, analogies are foundational evidences for natural selection in *The Origin of Species,* and in Darwin's approach to science. For Darwin, acceptance of the facts is "immaterial," for the whole theory is "grounded on analogy":

> Therefore, on the principle of natural selection with divergence of character, it does not seem incredible that, from some such low and intermediate form, both animals and plants may have been developed; and, if we admit this, we must likewise admit that all the

organic beings which have ever lived on this Earth may be descended from some one primordial form. But this inference is chiefly grounded on analogy, and it is immaterial whether or not it be accepted.[78]

Hypothetical speculations like these are in part the reason why scientists of the twentieth century have largely abandoned true "Darwinism" for varying forms of neo-Darwinism in an effort to salvage the theory of evolution.

Speculating on the timing of the variations and modifications in the life cycle, Darwin entangles modifications with a new set of problems. First, Darwin envisions modifications developing from any phase of the life cycle: "Natural selection, on the principle of qualities being inherited at corresponding ages, can modify the egg, seed, or young as easily as the adult."[79]

For modifications that develop early in the life cycle, Darwin envisions the modifications to affect the parent, again in a two-way effect: "Natural selection will modify the structure of the young in relation to the parent and of the parent in relation to the young."[80] According to Darwin, the parents can modify the young and the young can modify the parents.

While evidence for two-way inheritance has never been demonstrated, Darwin further elaborates, stating that "modifications in the adult may affect the structure of the larva; but in all cases natural selection will ensure that they shall not be injurious: for if they were so, the species would become extinct."[81] If this were true, genetic counseling would become obsolete.

Complicating the hypothesis further, Darwin suggests that a modification in one species must render an advantage to another species: "What natural selection cannot do, is to modify the structure of one species, without giving it any advantage, for the good of another species; and though statements to this effect may be found in works of natural history, I cannot find one case which will bear investigation."[82] While this speculation is compatible with the purpose of natural selection to promote "mutual good," Darwin continues that there was no supporting evidence: "I cannot find one case which will bear investigation."[60]

Even Darwin's logic in the natural selection "argument" is debatable. At times, natural selection "gives" an advantage to another species, and at other times it "gains" an advantage: "He who believes in the struggle for existence and in the principle of natural selection, will acknowledge that every organic being is constantly endeavouring to increase in numbers; and that if any one being varies ever so little, either in habits or structure, and thus gains an advantage over some other inhabitant of the same country, it will seize on the place of that inhabitant, however different that may be from its own place."[83]

How Darwin reconciles natural selection as a natural law that gives an advantage to others and at the same time "gains an advantage over" follows a pervasive two-way logic pattern that is characteristic of Darwin.

Sidestepping natural selection, Darwin ironically minimizes the importance of the exclusive role of natural selection in evolution: "Furthermore, I am convinced that natural selection has been the most important, but not the exclusive, means of modification."[84] Darwin could have entitled *The Origin of Species* as "The Origin of Species Mostly by Means of Natural Selection."

Modification through acquiring new variations, as part of Darwin's theory of natural selection, is clearly not as "simple" as Eldredge presents in the *Darwin* exhibit. Darwin even retracts further from the role of natural selection, envisioning that "many morphological changes may be attributed to the laws of growth and the inter-action of parts, independently of natural selection."[85]

The evidence Darwin presents for variation is only an argument. Not a single concept on how variations lead to modifications through natural selection is supported by any measurable and reproducible evidence. Even Darwin acknowledges that the "specialization of organs" by natural selection, while a popular concept, is difficult to prove:

> So that by this fundamental test of victory in the battle
> for life, as well as by the standard of the specialisation of
> organs, modern forms ought, on the theory of natural
> selection, to stand higher than ancient forms. Is this the
> case? A large majority of paleontologists would answer

in the affirmative; and it seems that this answer must be admitted as true, though difficult of proof.[86]

Not one of Darwin's modifications through natural selection has been reproducible. But, Darwin never claimed evolution to be founded by the scientific method—it was just an argument. Darwin confides, "this whole volume is one long argument."[87]

The idea that Darwin's theory was developed by using the scientific method is a modern-day myth. Nor is natural selection as "simple" a theory as suggested by Eldredge. The origin of new variations was certainly not one of Darwin's strong suits.

Inheritance

The next letter in Eldredge's *Darwin* exhibit is the letter "I." Eldredge, writing for the *Darwin* exhibit, implies that Darwin envisioned inheritance acting through the transfer of deoxyribonucleic acid (DNA) from parents to offspring, stating: "When organisms reproduce, they pass on their DNA—the set of instructions encoded in living cells for building bodies—to their offspring. And since many traits are encoded in DNA, offspring often inherit the variations of their parents. Tall people, for example, tend to have tall children."[88]

One point that Eldredge evades is that Darwin had no concept of DNA or of the modern understanding of genetics. The principles of Mendelian genetics used today were not widely accepted by scientists until later in the early twentieth century.

During his lifetime, Darwin adhered to the widely popular theory of inheritance known as "blending inheritance." Blending inheritance views inheritance as a fusion of both paternal and maternal elements in the offspring in an inseparable mixture that results in features intermediate between the two. Ironically, blending is impossible to reconcile with natural selection, and is clearly at odds with Mendelian genetics.

To circumvent the obvious difficulty of blending inheritance, Darwin applied an economic behavioral approach to inheritance. Darwin developed the "principle of inheritance" from a paper entitled *An Essay on the Principle of Population,* written in 1798, by the British economist Thomas Malthus.

Malthus argues that populations tend to increase faster than the available resources unless the population is controlled by means of a struggle. In applying "principle of inheritance" to natural selection, Darwin wrote:

> This is the doctrine of Malthus, applied to the whole animal and vegetable kingdoms. As many more individuals of each species are born than can possibly survive; and as, consequently, there is a frequently recurring struggle for existence, it follows that any being, if it vary however slightly in any manner profitable to itself, under the complex and sometimes varying conditions of life, will have a better chance of surviving, and thus be NATURALLY SELECTED. From the strong principle of inheritance, any selected variety will tend to propagate its new and modified form.[89]

In this struggle for existence, Darwin envisions that only those modifications "profitable to itself" are selected by natural selection: "Natural selection acts only by the preservation and accumulation of small inherited modifications."[90]

As in the case of a seed from a bush traveling across the ocean on to a distant island, Darwin envisions that the plant seed through "natural selection would tend to add to the stature of the plant, to whatever order it belonged, and thus first convert it into a bush and then into a tree."[91]

This principle of inheritance and preservation of new variations is central to Darwin's theory of natural selection, known as the "the survival of the fittest":

> If variations useful to any organic being ever do occur, assuredly individuals thus characterised will have the best chance of being preserved in the struggle for life; and from the strong principle of inheritance, these will tend to produce offspring similarly characterised. This principle of preservation, or the survival of the fittest, I have called natural selection.[92]

Beyond inheritance through the "survival of the fittest," Darwin envisions a complex relationship between inheritance and natural selection, and at times envisions inheritance as a part of natural selection: "the strong principle of inheritance ... [t]his principle of preservation ... I have called natural selection."[92] Inheritance is a part of the natural selection process.

At other times, though, Darwin envisions inheritance as a process distinct from natural selection, "through inheritance and the complex action of natural selection."[93] Despite clouding the roles of inheritance and natural selection, Darwin remains constant on envisioning natural selection as continuously "accumulating variations": "I can see no difficulty in natural selection preserving and continually accumulating variations."[94]

While "accumulating variations" is an active process, Darwin limits the role of natural selection to acting only on the available variations; natural selection "only takes advantage of variations as [they] arise":

> On our theory the continued existence of lowly organisms offers no difficulty; for natural selection, or the survival of the fittest, does not necessarily include progressive development—it only takes advantage of such variations as [they] arise and are beneficial to each creature under its complex relations of life.[95]

In making the theory even more complicated, in the classic style Darwin contradicts the "continually accumulating variation" purpose of natural selection by envisioning natural selection actually stopping the further accumulation of variations. Darwin wrote, "natural selection has succeeded in giving a fixed character to the organ, in however extraordinary a manner it may have been developed."[96] Darwin continues arguing that the actions of natural selection are "fixed" and paradoxically changeable: "That the struggle between natural selection on the one hand, and the tendency to reversion and variability on the other hand, will in the course of time cease, and that the most abnormally developed organs may be made constant, I see no reason to doubt."[97]

As the unrelenting theorizer, Darwin envisions natural se... not only to accumulate and simultaneously block the accumu... variations, but to reverse evolution—"reversions to long-lost characte... [98] Darwin's theory of natural selection is a *de facto* contradiction.

These multiple actions of natural selection bring into question— what natural law can act to simultaneously accumulate new variations and block the accumulation of new variations, and yet at other times delete new variations and bring about the "reversions (of) long-lost characters"? The answer is none. The popular impression that evolution is a simple process is far from Darwin's theory of natural selection.

Like variation, the theory of inheritance was not derived using the scientific method. The theory of inheritance is an inconsistent at best. In the end, Darwin clearly acknowledges the obvious: "laws governing inheritance are for the most part unknown."[99]

Selection

"S" is the third letter in Eldredge's acronym, which stands for selection. Selection is the act of making a choice. Darwin envisioned the selection process to be analogous to domestic selective breeding: "Can the principle of selection, which we have seen is so potent in the hands of man, apply under nature? I think we shall see that it can act most efficiently."[100] For selection, like variation, Darwin employs analogies. Darwin does not use any observed measurements as evidence for defining selection.

Darwin envisions the "survival of the fittest" process to be compatible with domestic breeding.[101] For Darwin, selective breeding and "survival of the fittest" are compatible processes, in the same way that natural selection acts by life and death: "Natural selection acts by life and death, by the survival of the fittest, and by the destruction of the less well-fitted individuals."[102]

Envisioning selective breeding is a purposeful and directed process. Darwin encompasses "the whole community" into this complex selection process. He limits selection to only act for the benefit the whole community: "Natural selection will … adapt the structure of each individual for the benefit of the whole community, if the community profits by the selected change."[103]

ion era, the most important evolutionary
on" was "Lamarckism." Lamarckian evolu-
ie French naturalist, Jean-Baptiste Lamarck,
cquiring new characteristics while interacting
and selectively passing these on to the next

w characteristics are acquired through the process
of "us. ." Darwin's grandfather, Erasmus Darwin, was a
Lamarckian ... itionist. Charles Darwin, however, in pursuit of a
"scientific theory" of evolution, initially opposed Lamarckian evolution,
only granting the theory marginal support. In a letter written to J. D.
Hooker in 1844, Darwin wrote, "Heaven forefend me from Lamarck
nonsense of a 'tendency to progression.' ... But the conclusions I am
led to are not widely different from his, though the means of change
are wholly so."[104] "With respect to books on this subject," Darwin
continues, "I do not know any systematic ones, except Lamarck's, which
is veritable rubbish."[105]

When Lyell suggested that Darwin's theory resembled Lamarck's,
Darwin, in 1859, wrote a rebuffing letter to Lyell, stating, "You often
allude to Lamarck's work; I do not know what you think about it, but
it appeared to me extremely poor; I got not a fact or idea from it."[106]

Darwin should have marked his words. The development of the
long giraffe neck is the most famous example of Lamarckism. Lamarck
envisioned that the long neck of the giraffe developed by successive
generations feeding on higher and higher branches of trees. The long
neck gave the giraffe a selective and survival advantage. Initially Darwin
did not view Lamarck's view of "use and disuse" to be a natural law. In
the "Historical Sketch" in *The Origin of Species*, Darwin writes about the
"erroneous" opinions of Lamarck as well as his own grandfather: "It is
curious how largely my grandfather, Dr. Erasmus Darwin, anticipated
the views and erroneous grounds of opinion of Lamarck in his *Zoönomia*
... published in 1794."[107]

Darwin was the first to successfully challenge Lamarck by rephrasing
and popularizing the theory of "natural selection" over "use and disuse."
But despite all of Darwin's intentions to identify other natural laws that
operate selection, the appeal to incorporate Lamarck's concepts proved

to be far too great, and Darwin eventually turned back towards the opinions of Lamarck.

While Darwin included concepts of "use and disuse" eleven times, starting in the first edition, use of Lamarck's concept increases to ninety-seven times in the sixth edition. Additionally, Darwin refers to Lamarck himself three times in the first edition, and seven times in the sixth edition. Thomas Morgan, in his book *Evolution and Adaptation*, wrote, "Despite the contempt with which Darwin referred to Lamarck's theory, he himself, as we have seen, often made use of the principle of inheritance of acquired characteristics, and even employed the same illustrations cited by Lamarck."[108] By the sixth edition, Darwin used Lamarck's same version of how the long neck of the giraffe evolved—by reaching for the tree leaves.

In his book *From the Greeks to Darwin*, Professor H. F. Osborn wrote of Darwin's drift back toward Lamarck: "Starting with some leaning towards the theories of modification of Buffon and Lamarck, he reached an almost exclusive belief in his own theory, and then gradually inclined to adopt Buffon's and then Lamarck's theories as well."[109]

Darwin, in a letter to Francis Galton in 1875, wrote, "If this implies that many parts are not modified by use and disuse during the life of the individual, I differ widely from you, as every year I come to attribute more and more to this agency."[110]

At the very least, "use and disuse" through the selection process gave Darwin a greater measure of reasons why and how new variations could develop and be accumulated through the selection process. Eventually, the concepts of "use and disuse" became foundational to natural selection, too. In the sixth edition, Darwin wrote:

> I have now recapitulated the facts and considerations which have thoroughly convinced me that species have been modified, during a long course of descent. This has been effected chiefly through the natural selection of numerous successive, slight, favourable variations; aided in an important manner by the inherited effects of the use and disuse of parts; ... and by variations which seem to us in our ignorance to arise spontaneously.[111]

Unfortunately, the incorporation "use and disuse" into the theory launched Darwin into another entire series of consistencies and contradictions. First, Darwin envisions natural selection to be strengthened by "use and disuse."[112] Then Darwin flips the argument: "use of parts, and perhaps of their disuse, will be strengthened by natural selection."[113]

As an example of how "use and disuse" works, Darwin argues that a hoofed animal could be "converted" into a giraffe: "With the inherited effects of the increased use of parts, it seems to me almost certain that an ordinary hoofed quadruped might be converted into a giraffe."[114]

In yet a third approach, Darwin envisions equivocal roles for natural selection and "use and disuse," concluding: "How much to attribute in each particular case to the effects of use, and how much to natural selection, it seems impossible to decide."[115]

Incorporating "use and disuse" created glaring contradictions for Darwin. These problems emerged for Darwin because, as he had originally acknowledged, the arguments and analogies for Lamarckian evolution are not based on any scientific evidence. The question is, is natural selection a consistent theory? The answer is no.

A consistent theory for the "process of selection" remained completely evasive to Darwin. Eventually Darwin comes to terms with "use and disuse," stating, "I will not pretend to decide" how the process of selection works.… Through what agency the glands over a certain space became more highly specialised than the others, I will not pretend to decide, whether in part through compensation of growth, the effects of use, or of natural selection."[116]

Darwin did apply concepts of "use and disuse" to his own life. In his autobiography, Darwin, in lamenting his loss of pleasure from reading and music, wrote: "A man with a mind more highly organized or better constituted than mine, would not I suppose have thus suffered; and if I had to live my life again, I would have made a rule to read some poetry and listen to some music at least once a week; for perhaps the parts of my brain now atrophied could thus have been kept active through use."[117]

Of all the theories Darwin introduces to support evolution, it is only the theory of sexual selection that is actually original to Darwin. Darwin developed sexual selection hypothetically, and not with scientific evidence. Later, in a letter to A. R. Wallace in 1868, Darwin wrote how the importance of sexual selection was maintained: by feelings, not by scientific evidence. "You will be pleased to hear that I am undergoing severe distress about protection and sexual selection; this morning I oscillated with joy towards you; this evening I have swung back to the old position, out of which I fear I shall never get."[118]

Darwin envisions sexual selection to be the struggle "for the possession of the other sex."[119] Like the theories of "use and disuse" though, Darwin entangles sexual selection into a series of inconsistencies and contradictions. First, Darwin argues that sexual selection is mainly the process for changing behavior, color, and the structure between the males and females: "Thus it is, as I believe, that when the males and females of any animal have the same general habits of life, but differ in structure, colour, or ornament, such differences have been mainly caused by sexual selection."[120] Continuing this same thought process, Darwin envisions sexual selection to account for the competition between males for the females, with the acquired features by the males "transmitted to their male offspring alone."[121]

Blurring the distinction between sexual and natural selection, Darwin explains that the "green colour … had been acquired through natural selection … the colour is probably in chief part due to sexual selection."[122] In explaining competition between the males for the females, Darwin undermines the definition of sexual selection by stating that natural selection primarily controls the competition between males.[123] Perhaps recognizing no difference between sexual and natural selection, Darwin hypothesizes that the roles may be equivocal, "by natural and sexual selection."[124]

Certainly one of Darwin's most controversial sexual comments concerns the superior intellectual of the males over the females: "The chief distinction in the intellectual powers of the two sexes is shown by man's attaining to a higher eminence in whatever he takes up, than can a woman—whether requiring deep thought, reason, or imagination, or merely the use of the senses and hands."[125]

Sexual selection, Darwin's only original contribution to the theory of evolution, is inconsistent and contradictory; the differential role between natural and sexual selection becomes completely vague. Contrary to Eldredge's *Darwin* exhibit, Darwin's selection hypothesis is complex.

Time

The fourth letter in Eldredge's acrostic is the letter "T." The emerging widespread acceptance of a long geological timetable of the Earth during the early nineteenth century fostered Darwin's initial interest in evolution. During long periods of time Darwin envisions natural selection acting in concert as a very slow process: "But I do believe that natural selection will generally act very slowly, only at long intervals of time, and only on a few of the inhabitants of the same region."[126]

British geologist Charles Lyell, Darwin's professor at Oxford University, fueled the fire of a long timetable for the Earth. Lyell challenged the current concepts of the Earth's age by arguing against evidence of the flood and for small, gradual changes over the lifetime of the Earth. For Lyell, the same processes that shape the Earth in the past are shaping the Earth now: geological processes have not changed throughout Earth's history. This is known as the uniformitarianism theory.

A long geological timetable is essential for natural selection. Darwin envisions that "natural selection acts solely by accumulating slight, successive, favourable variations; it can produce no great or sudden modifications."[127] Natural selection acts only to either accept or reject variations over long periods of time.

Darwin envisions natural selection making no leaps—*Natura non facit saltum*; the process of evolution through natural selection occurs only over long periods of time. Darwin writes, "it must on this theory be strictly true."[128] Since natural selection "can never take a great and sudden leap," Darwin envisions natural selection acting only by "slight successive variations." Natural selection "must advance by the short and sure, through slow steps."[129]

From these long time periods, with the accumulation of "slight successive variations," Darwin was convinced that evidence would be found for the "inconceivably great" number of fossils that represent all the transitional links "such as have lived upon the Earth."[130] So persuaded in the existence of the "inconceivably great" number of transitional links yet to be found, Darwin argues that if evidence of "numerous species, belonging to the same genera or families, have really started into life at once, the fact would be fatal to the theory of evolution through natural selection." [131]

Fossil evidence for these transitional links, though, was even an issue in Darwin's time. Darwin concedes: "Although geological research has undoubtedly revealed the former existence of many links, bringing numerous forms of life much closer together, it does not yield the infinitely many fine gradations between past and present species required on the theory, and this is the most obvious of the many objections which may be urged against it."[132]

A keen interest of Darwin was the evolution of the eye even though no transitional eyes were known: "He who will go thus far, ought not to hesitate to go one step further, if he finds on finishing this volume that large bodies of facts, otherwise inexplicable, can be explained by the theory of modification through natural selection; he ought to admit that a structure even as perfect as an eagle's eye might thus be formed, although in this case he does not know the transitional states."[133]

In acknowledging the lack of "inconceivably great" links and "infinitely numerous fine transitional forms," Darwin concedes that this problem is "undoubtedly of the most serious nature."[134]

Challenged by the lack of fossil evidence, which Darwin argued would be "inconceivably great," eventually the British Empire, in 1872, commissioned the HMS *Challenger* for the largest fossil-finding expedition in world history.

Since the early twentieth century, Eldredge along with most of the modern scientific community has largely abandoned Darwin's "old canon in natural history, *Natura non facit saltum*," because life-forms are now known to have appeared suddenly, as observed in the Cambrian explosion. Eldredge concedes, "Evolution has no single schedule. Sometimes, new species or varieties arise in a matter of years or even days."[135]

Clearly, natural selection as defined by Darwin is no longer valid; the fossil record clearly displays evidence of the sudden appearance of species.

Adaptation

The last letter in the VISTA acronym stands for "adaptation." *The American Heritage Dictionary of the English Language* defines adaptation in the context of biology as "an alteration or adjustment in structure or habits, often hereditary, by which a species or individual improves its condition in relationship to its environment." Likewise, Darwin defines adaptation as acquiring "the sum of many inherited changes."[136] Eldredge concurs with Darwin, writing that the "result is a population that is better suited—better adapted—to some aspect of the environment than it was before. Legs once used for walking are modified for use as wings or flippers. Scales used for protection change colors to serve as camouflage."[137]

Darwin issues a caution though, when attributing a structure's modification through adaptation. What may at first appear to be a direct result of adaptation, Darwin concludes may not be adaption:

> The naked skin on the head of a vulture is generally considered as a direct adaptation for wallowing in putridity; and so it may be, or it may possibly be due to the direct action of putrid matter; but we should be very cautious in drawing any such inference, when we see that the skin on the head of the clean-feeding male turkey is likewise naked.[138]

How do these adaptations develop? Darwin envisions adaptations developing through the "successive adaptations" of natural selection, which goes "hand in hand" with extinction, such that "all the transitional varieties will generally have been exterminated by the very process of the formation and perfection of the new form."[139]

Specifically, Darwin envisions natural selection acting through a purpose-driven process in which "natural selection almost inevitably causes much extinction of the less improved forms of life."[140]

What scientific evidence did Darwin use in the development of this hypothesis? The answer is none. Darwin clearly states that the theory is only "grounded on the belief,"[141] not on any measurable evidence.

Darwin views the extinction of species as a natural result of evolution. Since the extinction of species is not popular, Eldredge conveniently skirts Darwin's vision that natural selection and extinction go "hand in hand."

Eldredge contradicts Darwin by envisioning extinction to stem from "rapid environmental" changes and not from natural selection. Arguing for rapid environmental changes contradicts Lyell and Darwin's theory of uniformitarianism. Eldredge writes, "At least five times in the past 500 million years, rapid environmental change drove much of life on Earth to extinction. But such mass extinctions offered surviving species an opportunity, as regions that were once inhabited became vacant and environments changed."[142]

Darwin never envisioned massive destruction and rapid extinction as the result of rapid global environmental changes. Destruction and extinction, like preservation, is just natural selection in action, but acting over long periods of time: "natural selection will preserve and thus separate all the superior individuals, allowing them freely to intercross, and will destroy all the inferior individuals."[143]

In the book accompanying the *Darwin* exhibit, *Darwin, Discovering the Tree of Life*, Eldredge ironically writes that Darwin would find the rate of extinction alarming—Darwin would be "heartbroken." However, Darwin would certainly not be heartbroken, since extinction is the result of the "principle of inheritance" at work. Darwin wrote, "This is the doctrine of Malthus ... [that] many more individuals of each species are born than can possibly survive; and ... consequently, there is a frequently recurring struggle for existence."[144]

Eldredge never mentions that Darwin envisions extinction as the result of natural selection, ready to "destroy any individuals departing form the proper type."[145] Not only are those "departing from the proper type" destroyed by natural selection, Darwin envisions natural selection acting to even exterminate "the less improved parent-form.... Thus extinction and natural selection go hand in hand."[146]

Darwin continues the argument that we "must suppose that there is a power, represented by natural selection or the survival of the fittest, always intently watching each slight alteration ... until a better is produced, and then the old ones to be all destroyed."[147]

With natural selection destroying "the old" structures over long periods of time, Darwin envisions old useless structures becoming "rudimentary." This brings us to the fourth question—what do rudimentary structures tell us about natural selection?

Rudimentary Structures

The perceived existence of rudimentary organs has served to support the philosophical justification for the theory of evolution during the twentieth century. In the *Darwin* exhibit, which uses the term "vestigial features" rather than "rudimentary structures," Eldredge writes, "Humans also have vestigial features, evidence of our own evolutionary history. The appendix, for instance, is believed to be a remnant of a larger, plant-digesting structure found in our ancestors."[148]

Darwin never used the appendix as evidence for evolution in *The Origin of Species*, but he popularized the concept in the *Descent of Man*. Eldredge, though, should have referred to the thirty-sixth edition of *Gray's Anatomy*, which paints a different picture of the appendix: "In view of its rich blood supply and histological differentiation, the vermiform appendix is probably more correctly regarded as a specialised [rather] than ... a degenerate vestigial structure."[149]

Whereas it once symbolized a vestigial organ (involution with evolution), the appendix is now recognized to be a highly specialized, well-differentiated organ with maximal development in man. The appendix is not a remnant—a leftover from evolution. Not having *Gray's Anatomy* as a reference though, Darwin argues: "Rudimentary organs plainly declare their origin" and the evidence for the theory of evolution. [150]

Darwin envisions that these rudimentary organs, which have "the plain stamp of inutility," are an "extremely common" occurrence in nature.[151] Darwin introduces an array of plastic roles for natural selection: "Rudimentary organs plainly declare their origin and meaning in various ways."[152] In chapter fourteen, Darwin argues that natural selection may "reduce any part of the organization" or "conversely ... natural selection may perfectly well succeed in largely developing an organ."[153] Natural selection can act to change a "useless or injurious" organ to perform "another purpose."[154] Even "an organ, useful under certain conditions, might become injurious under others ... in this case natural selection will have aided in reducing the organ, until it was rendered harmless and rudimentary."[155]

Just when natural selection appears to have risen to untold plastic omnipotent powers, Darwin contradicts his argument by stating that natural selection is, at times, only an aid: "Disuse, aided sometimes by natural selection, will often have reduced organs when rendered useless under changed habits or conditions of life; and we can understand on this view the meaning of rudimentary organs."[156]

Limiting the role of natural selection further, Darwin envisions that once an organ becomes useless, even though it was originally formed by natural selection, the variations of the organ "can no longer even be checked by natural selection."[157] Nor can the resulting rudimentary organs be regulated by natural selection: "Rudimentary organs, from being useless, are not regulated by natural selection, and hence are variable."[158]

Darwin envisions that once an organ becomes rudimentary, even if useful variations exist in the organ, it "would not be affected by natural selection."[159] Again, Darwin argues that this is because once an organ becomes rudimentary, "natural selection ... [has] no power to check deviations in [its] structure."[160]

As evidence, Darwin argues that because the "retained" tail on terrestrial animals was never even formed by natural selection, the tail now remains a rudimentary organ:

> We may, also, believe that a part formerly of high importance has frequently been retained (as the tail of an aquatic animal by its terrestrial descendants), though it has become of such small importance that it could not, in its present state, have been acquired by means of natural selection.[161]

Darwin envisions that the same holds true for "teeth which never cut through the gums" and the "wings of an ostrich."[162] Deciphering these complex and plastic roles for natural selection with "rudimentary organs" eventually became a conundrum for Darwin. Darwin finally concluded that it is not even possible to determine what role natural selection plays in the formation of "rudimentary organs."

> In many cases we are far too ignorant to be enabled to assert that a part or organ is so unimportant for the welfare of a species, that modifications in its structure could not have been slowly accumulated by means of natural selection.[163]

Darwin's vision of a limited role for natural selection stands in sharp contrast with Darwin's more popular limitless version of natural selection: "there is no logical impossibility in the acquirement of any conceivable degree of perfection through natural selection.... I can see no limit to this power, in slowly and beautifully adapting each form to the most complex relations of life." [164, 165]

Reflections

In his characteristically famous wavering style, Darwin envisions both an unlimited and a limited version of natural selection. By 1863, just four years after the publication of *The Origin of Species,* in recognizing the problems with natural selection, Darwin began to waver on the importance of natural selection. This confession was written to one of Darwin's earliest American supporters, the Harvard botanist Asa Gray: "personally, of course, I care much about Natural Selection; but that seems to me utterly unimportant compared to the question of *Creation or Modification.*"[166]

Darwin supports the use of caution when applying a role to natural selection; Darwin had lingering doubts, too: "I have felt the difficulty far too keenly to be surprised at others hesitating to extend the principle of natural selection to so startling a length."[167]

In discussing the distinctiveness of ants, Darwin writes that it "will indeed be thought that I have an overweening confidence in the principle of natural selection, when I do not admit that such wonderful and well-established facts at once annihilate the theory."[168] In discussing the evolution of the eye, Darwin confesses:

> To suppose that the eye with all its inimitable contrivances for adjusting the focus to different distances, for admitting different amounts of light, and for the correction of spherical and chromatic aberration, could have been formed by natural selection, seems, I freely confess, absurd in the highest degree.[169]

Darwin paints the definition of natural selection so broadly, the meaning of natural selection becomes an ocean of meanings. While the term "natural selection" is nearly synonymous with evolution, the question is what does natural selection actually mean? Italian geneticist Giuseppe Sermonti, in 2005, suggests that "natural selection (which should be more accurately termed 'differential survival') ... chiefly eliminates the abnormal, the marginal, the out-of-bounds, and keeps natural populations within the norm."[171]

Even Darwin knew that the arguments in *The Origin of Species* would not stand the test of time. Critical of his own work, in a letter to H. Falconer in October 1862, Darwin wrote, "I look at it as absolutely certain that very much in the *Origin* will be proved to be rubbish; but I expect and hope that the framework will stand."[172] While conceptual framework has stood for 150 years, the evidence has seriously eroded Darwin's original foundation, nearly to extinction.

The VISTA acronym designed by Eldredge is a useful modern-day tool for organizing the concepts behind Darwin's vision of evolution through natural selection. But, even the most ardent evolutionists of the twenty-first century, like Eldredge, have largely abandoned adherence to Darwin's arguments and beliefs, for good reasons.

In evolution circles, Darwinism has now been replaced by neo-Darwinism. Evolution by natural selection, as defined by Darwin, has not stood the test of time. *The Origin of Species* is not studied as a scientific textbook because it is not scientific. Eldredge comments, "Darwin staked out the evolutionary territory so broadly and so thoroughly that, in a general sense, he did literally define the entire content of evolutionary biology right down to the present day."[173]

However, for every VISTA concept—variation, inheritance, selection, time, adaptation—Darwin entangles himself in a maze of inconsistencies and contradictions. A sampling of the top fifteen contradictions is provided at the end of this chapter. In light of 150 years of accumulated evidence, it is no small wonder Eldredge was compelled to modify the mechanics of nearly every one of Darwin's arguments for natural selection.

The ultimate question, though, is, is natural selection valid? Is natural selection based on a natural law or even a series of natural laws? Did Darwin ever find the laws behind the theory of evolution? Darwin clearly gave the answer. The answer is no. Darwin knew that he only opened a door to further investigation. Darwin wrote, "A grand and almost untrodden field of inquiry will be opened, on the causes and laws of variation, on correlation, on the effects of use and disuse, on the direct action of external conditions, and so forth."[174]

In time, history will likely conclude that Darwin's greatest legacy is that he successfully motivated further exploration to test his theory. In the following chapters, the results of these explorations over the past 150 years will be examined.

What Eldredge fails to mention is that Darwin finally conceded that natural selection is actually "a false term": "In the literal sense of the word, no doubt, natural selection is a false term; but who ever objected to chemists speaking of the elective affinities of the various elements?"[175]

Giuseppe Sermonti, the chief editor of one of the longest-running biology journals in the world, *Rivista di Biologica*. In his book written in 2005, *Why a Fly is not a Horse*, Sermonti pines, "The claim that natural selection is creative of life, of life's essence and types of orders, can only leave us dumbstruck. Natural selection only eliminates, and it's adoption as a mechanism of origin is like explaining 'appearance' by 'disappearance.'"[176]

Sermonti continues: "Natural Selection, which indeed occurs in nature…, mainly has the effect of maintaining equilibrium and stability."[177]

In a presentation entitled "Evolutionary Theories" at the World Summit on Evolution held at the Galapagos Islands in June 2005, William Provine, from Cornell University, concluded that natural selection is not a mechanism:

> Natural selection does not shape an adaptation or cause a gene to spread over a population or really do anything at all. It is instead the result of specific causes: hereditary changes, developmental causes, ecological causes, and demography. Natural Selection is the result of these causes, not a cause that is by itself. It is not a mechanism.[178]

The natural law that operates natural selection, the key to evolution, remains as an elusive enigma, even after 150 years of investigation.

Top Fifteen Contradictions

The fact that Darwin wove a wide range of contradictions into *The Origin of Species* is well-known. Contradictions surround nearly every topic of natural selection. In a letter to Wallace in 1868, even Darwin acknowledges his contradictions: "Nevertheless, I myself to a certain extent contradict my own remark."[179]

1. Power. Natural selection has unlimited power.
    "I can see no limit to this power, in slowly and beautifully adapting each form to the most complex relations of life."[26]

    Contradicted by:

    "If the numbers be wholly kept down by the causes just indicated, as will often have been the case, natural selection will be powerless in certain beneficial directions; but this is no valid objection to its efficiency at other times and in other ways; for we are far from having any reason to suppose that many species ever undergo modification and improvement at the same time in the same area."[27]

2. Perfection. Natural selection has no limits in producing perfection.
    "There is no logical impossibility in the acquirement of any conceivable degree of perfection through natural selection."[32]

    Contradicted by:

    "Natural selection will not produce absolute perfection, nor do we always meet, as far as we can judge, with this high standard under nature."[180]

3. Preservation. Natural selection is omnipresent watching everything.
    "Further we must suppose that there is a power, represented by natural selection or the survival of the fittest, always intently watching each slight alteration."[25]

    Contradicted by:

"Natural selection should not have preserved or rejected each little deviation."[181]

4. Mutual Good. Natural selection acts for the mutual good of even other species.
    "What natural selection cannot do, is to modify the structure of one species, without giving it any advantage, for the good of another species."[82]

Contradicted by:

"Natural selection can and does often produce structures for the direct injury of other animals"[182]

5. Economy. Natural selection acts only for profitability.
    "Natural selection acts solely through the preservation of variations in some way advantageous, which consequently endure."[183]

Contradicted by:

"Variations neither useful nor injurious would not be affected by natural selection."[159]

6. Chance. "Mere chance" has no part in the origin of variations.
    "Mere chance, as we may call it, might cause one variety to differ in some character from its parents, and the offspring of this variety again to differ from its parent in the very same character and in a greater degree; but this alone would never account for so habitual and large a degree of difference as that between the species of the same genus."[52] And, "I was so convinced that not even a stripe of colour appears from what is commonly called chance."[51]

Contradicted by:

"As many more individuals of each species are born than can possibly survive; and as, consequently, there is a frequently recurring struggle for existence, it follows that any being, if it vary however slightly in any manner profitable to itself, under the complex and sometimes varying conditions of life, will have a better chance of surviving, and thus be naturally selected. From the strong principle of inheritance, any selected variety will tend to propagate its new and modified form."[184]

7. Variations. Arise infinitely.
    "In living bodies, variation will cause the slight alteration, generation will multiply them almost infinitely, and natural selection will pick out with unerring skill each improvement."[42]

Contradicted by:

"We have seen that species at any one period are not indefinitely variable, and are not linked together by a multitude of intermediate gradations."[68]

8. Variations. Accumulate continuously.
    "I can see no difficulty in natural selection preserving and continually accumulating variations."[94] And, "An extraordinary amount of modification implies an unusually large and long-continued amount of variability, which has continually been accumulated by natural selection for the benefit of the species."[72]

Contradicted by:

"Natural selection has succeeded in giving a fixed character to the organ, in however extraordinary a manner it may have been developed."[96] And, "From the fact of the above characters being unimportant for the welfare of the species, any slight variations which occurred in them would not have been accumulated and augmented through natural selection."[185]

9. Variations. Natural selection causes variability.
    "In one sense the conditions of life may be said not only to cause variability, either directly or indirectly, but likewise to include natural selection."[66]

Contradicted by:

"Some have even imagined that natural selection induces variability, whereas it implies only the preservation of such variations."[67]

10. Variations. Natural selection overcomes variability.
"… natural selection having more or less completely, according to the lapse of time, overmastered the tendency to reversion and to further variability."[186]

Contradicted by:

"For variation is a long-continued and slow process, and natural selection will in such cases not as yet have had time to overcome the tendency to further variability."[187]

11. Efficiency. Natural selection acts efficiency.
"For the best definition which has ever been given of a high standard of organisation, is the degree to which the parts have been specialised or differentiated; and natural selection tends towards this end, inasmuch as the parts are thus enabled to perform their functions more efficiently."[31]

Contradicted by:

"Natural selection either has not or cannot come into full play, and thus the organisation is left in a fluctuating condition."[188]

12. Links. By natural selection, the links between species are "inconceivably great."
"By the theory of natural selection, all living species have been connected with the parent-species of each genus, by differences not greater than we see between the natural and domestic varieties of the same species at the present day … So that the number of intermediate and transitional links, between all living and extinct species, must have been inconceivably great. But assuredly, if this theory be true, such have lived upon the Earth."[130]

Contradicted by:

"We have seen that species at any one period are not indefinitely variable, and are not linked together by a multitude of intermediate gradations."[68]

13. Extinction. Natural selection causes extermination.

"Lastly, looking not to any one time, but at all time, if my theory be true, numberless intermediate varieties, linking closely together all the species of the same group, must assuredly have existed; but the very process of natural selection constantly tends, as has been so often remarked, to exterminate the parent-forms and the intermediate links."[189]

Contradicted by:

"On our theory, the continued existence of lowly organisms offers no difficulty; for natural selection, or the survival of the fittest, does not necessarily include progressive development—it only takes advantage of such variations as [they] arise and are beneficial to each creature under its complex relations of life."[43]

14. Natural Selection. Gives clear understanding.

"On the theory of natural selection, we can clearly understand the full meaning of that old canon in natural history."[128]

Contradicted by:

"Natural selection, and likewise because this is by far the most serious special difficulty which my theory has encountered."[190]

15. Paleontology. Evidence agrees with theory.

"Passing from these difficulties, the other great leading facts in paleontology agree admirably with the theory of descent with modification through variation and natural selection."[71]

Contradicted by:

"Existence of many links ... does not yield the infinitely many fine gradations between past and present species required on the theory."[132]

Chapter Eight
The Challenge

By the theory of natural selection all living species have been connected....
So that the number of intermediate and transitional links, between
all living and extinct species, must have been inconceivably great. But
assuredly, if this theory be true, such have lived upon the Earth.
—Charles Darwin[1]

On the campus of Oxford University on Saturday, June 30, 1860, just nine months after the publication of *The Origin of Species,* one of the greatest events in the history of science was about to begin. The occasion was the British Association for the Advancement of Science annual meeting.

The Legendary Exchange

Darwin was not in attendance at the annual meeting, as he was "taking a cure" at Dr. Lane's Hydropathic Clinic. Thomas Henry Huxley was set to leave after Thursday, but was challenged by evolutionist Robert Chambers, author of *The Vestiges of Creation* (1844), to stay until Saturday for an impending showdown on the implications of *The Origin of Species.* By the end of Thursday, Huxley was exhausted after being grilled on using the similarity of ape and human brains as evidence of evolution.

American professor John William Draper, from New York, was scheduled to present his paper, "On the Final Causes of the Sexuality of Plants, With Particular Reference to Mr. Darwin's Work on Origin of Species," on Saturday. The christening event was in a crowded room filled with an estimated seven hundred people trying to find a seat. Even the windows on the west side, which lighted the room, were packed with ladies anxiously waiting to wave and flutter their handkerchiefs. What was ready to emerge was the legendary encounter between paleontologist Thomas Huxley and ornithologist Samuel Wilberforce, bishop of Oxford.

The events of the legendary event were encapsulated nearly thirty years later in the October 1898 issue of *Macmillan's Magazine*, in an article entitled "A Grandmother's Tales," written by Isabella Sidgwick.[2] Sidgwick recalls Draper setting the stage by asking: "Are we a fortuitous concourse of atoms?"[2]

With the challenge, Wilberforce arose from the crowd, and according to Sedgwick, declared, "There was nothing in the idea of evolution; rock-pigeons were what rock-pigeons had always been."[2]

Drawing on the momentum of the crowd and seeking to score points, Wilberforce, using the same arguments he later published as an anonymous review of *The Origin of Species* for July's *The Quarterly Review,* baited Huxley following a thirty-minute speech by inquiring whether it was "through his grandfather or his grandmother that he [Huxley] claimed descent from a monkey?"[2]

While Wilberforce sat down to a thunderous handkerchief-waving applause with students cheering, Lady Brewster fainted, and according to legend, Huxley whispered to Sir Benjamin Brodie the now famous reply: "The Lord hath delivered him into mine hands."[3]

Waiting until invited to speak, Huxley rose to reply, taking on the mark as "Darwin's Bulldog." Unfortunately, not a single verbatim account of the day's event exists. Just letters and news reports published in such journals as *The Guardian, The Athenaeum,* and *Jackson's Oxford Journal.*

What argument won at the end of Saturday is still a matter of perception. Not everyone was pleased with Huxley's remarks. Nine weeks after the event, in September 1860, Huxley wrote to his colleague Dyster about the event: "It was great fun.... I unhesitatingly affirmed my preference for the ape. Whereupon there was inextinguishable laughter among the people—and they listened to the rest of my argument with the greatest of attention."[3]

Joseph Hooker, Darwin's longtime friend and mentor, wrote to Darwin: "Well, Sam Oxon got up and spouted for half an hour with inimitable spirit, ugliness and emptiness and unfairness ... Huxley answered admirably and turned the tables, but he could not throw his voice over so large an assembly nor command the audience ... he did not allude to Sam's weak points nor put the matter in a form or way that carried the audience."[3]

As a Fellow of the Royal Society, Wilberforce was influential and knowledgeable in his own right. Earlier in 1860, Wilberforce had previously reviewed *The Origin of Species* for the *London Quarterly Review.* While beginning the review by writing about Darwin's scientific attainments, and his insight and carefulness as an observer, Wilberforce ultimately described the 1859 *Origin of Species* as "the most illogical book ever written." In a series of articles in *The Athenaeum,* it was reported, "The Bishop of Oxford stated that the Darwinian theory, when tried by the principles of inductive science, broke down. The facts brought forward, did not warrant the theory."[4]

In the *Jackson's Oxford Journal,* the bishop later wrote, "Mr. Darwin's conclusions were a hypothesis, raised most unphilosophically to the dignity of a causal theory. He was glad to know that the greatest names in science were opposed to this theory."[5]

A measure of anxiety even existed at the Wilberforce estate. It was reported that "when the Bishop of Worcester told his wife what had happened [at Oxford that day], she is said ... to have replied, 'Descended from the apes! My dear, let us hope that it is not true, but if it is, let us pray that it will not become generally known.'" [6]

How pivotal that session of the British Association for the Advancement of Science was, in terms of shifting the weight of popular and scientific opinion to an evolutionary viewpoint, is as unclear as what was actually said. What was clear was a popular movement was developing to explore and suppport Darwin's theory.

Darwin was never a participant in public debates. In a letter to Huxley in July 1860, he wrote, "I would as soon have died as tried to answer the Bishop in such an assembly ... this row is [the] best thing for [the] subject."[7]

The Origin of Species deepened divisions in the scientific community. In 1860, the Harvard University zoologist Louis Agassiz wrote that Darwin's theory was a "scientific mistake, untrue in its facts, unscientific in its methods and mischievous in its tendency."[8]

Months earlier, while *The Origin of Species* gained nearly immediate widespread notoriety, even the authoritative *Athenaeum* was quick to pick out the unstated implications of "men from monkeys." The

Saturday Review realized that the treatise was already "into the drawing-room and the public street." By December 9, John Murray, Darwin's publisher, was organizing a second run of 3,000 copies.

The remarkable spread of evolutionary thought was, in large part, rooted in the Victorian era's fascination with understanding the laws of nature. Whatever the reason, the origin of life was no trivial issue. In 1873, Huxley wrote, "We are in the midst of a gigantic movement, greater than that which preceded and produced the Reformation."[9]

Lending key, unyielding support, the popular *Westminster Review*, a philosophical, radical publication, largely underwrote Darwin. The term "Darwinism" was first put in print by Huxley, in his favourable review of *The Origin of Species* in the April 1860 issue of the *Westminster Review*. The race was on.

Missing Links

Central to Darwin's theory was finding the evidence, and Darwin had given clear instructions. If his theory was true, there were an "inconceivably great" number of missing links representing evolutionary transitions from one species to the next. "If this theory be true, such have lived upon the Earth."[1]

These links must develop over long periods of time. Darwin envisioned that "natural selection generally acts with extreme slowness … by accumulating slight, successive, favorable variations, it can produce no great or sudden modifications; it can act only by short and slow steps."[10]

So convinced in the existence of an "inconceivably great" number of transitional links yet to be found, Darwin argues that if evidence is discovered demonstrating that numerous species ever "started into life at once, the fact would be fatal to the theory of evolution through natural selection."[11]

At the time, Darwin clearly acknowledged the fact that evidence for "slight, successive favorable variations" had not been found and this was emerging as a glaring problem: "The distinctiveness of specific forms and their not being blended together in innumerable transitional links is a very obvious difficulty."[12] Even comprehending how all of the possible gradations, variations, and modifications could be arranged was beyond

Darwin: "It is, no doubt, extremely difficult even to conjecture by what gradations many structures have been perfected."[13]

Not only were possible gradations and transitions not available, the sudden appearance of new species, the Cambrian explosion, was widely known as a glaring problem even during the nineteenth century. Darwin concedes, "it is indisputable that before the lowest Cambrian stratum was deposited ... several of the main divisions of the animal kingdom suddenly appear in the lowest known fossiliferous rocks."[14] This "sudden appearance," Darwin argues in *The Origin of Species,* was "a valid argument against the views here entertained."[15]

Not hiding the problem of the "sudden appearance" and the lack of "transitional links," Darwin grants: "He who rejects this view of the imperfection of the geological record, will rightly reject the whole theory."[16]

In addressing the difficulty with the absence of transitional forms, Darwin suggests two reasons why transitional links may be missing. The first is that exploration of the Earth was limited, and "only a small portion of the surface of the Earth has been geologically explored."[17] The second is that the preservation of the fossil record had been incomplete, "a history of the world imperfectly kept."[18]

Darwin argues, "I believe the answer mainly lies in the record being incomparably less perfect than is generally supposed. The crust of the Earth is a vast museum; but the natural collections have been imperfectly made, and only at long intervals of time."[19] While Darwin continues to argue for finding evidence in the fossil record, stating, "we surely ought to find at the present time many transitional forms," it was with a measure of skepticism. Darwin explains, "Hence we ought not to expect at the present time to meet with numerous transitional varieties in each region, though they must have existed there, and may be embedded there in a fossil condition." [20, 21]

The question is did Darwin have the fossil record evidence to support the theory? The answer is no.

The Best Shot

Even though Darwin did not have the fossil record evidence at the time, in *The Origin of Species,* Darwin suggests that four species may be representative of the transitional links: *Halithermium, Zeuglodon, Hipparion,* and *Archaeopteryx.*

Halithermium and *Zeuglodon* are large sea mammals. In accordance with the earlier Greek philosophers, land animals originated from the sea. Darwin suggests that the *Halithermium* may be an intermediate link because "the extinct *Halitherium* ... makes some approach to ordinary hoofed quadrupeds."[22]

At the time, other large whales, the *Zeuglodon* and *Squalodon,* were transitional links to carnivorous mammals, including Cetacea (dolphins). Darwin explains, "*Zeuglodon* and *Squalodon,* which have been placed by some naturalists in an order by themselves, are considered by Professor Huxley to be undoubtedly Cetaceans, and to constitute connecting links with the aquatic Carnivora."[23]

The concept of land animals arising out of water mirrors the Greek philosopher's concept that life arose from sea deities. Even the origin of the European mermaid can be traced back to the Siren mythology of the legendary aquatic creature appearing with the head and torso of a human female and the tail of a fish. Greek philosopher Anaximander taught that "life evolved from moisture," and that "man developed from fish."

Darwin's most legendary transitional links are the *Hipparion* horse and the bird *Archaeopteryx.* The horse and *Archaeopteryx* eventually became popular examples of the missing links of evolution used in biology textbooks, even throughout the late twentieth century.

Darwin envisioned the extinct three-toed horse, *Hipparion,* as the intermediate to the five-toed horse. Yet, mutant three-toed horses still are born today. Certain characteristics of the *Archaeopteryx* are reptilian, a possible intermediate between modern birds and reptiles. The *Archaeopteryx* is thought to be the best intermediate Darwin identified.

The "inconceivably great" number of intermediate and transitional links, though, was clearly missing. Logically, Darwin reasoned that with further exploration the "inconceivably great" number would be discovered, since at the time "only a small portion of the surface of the Earth [had] been geologically explored."[24]

Darwin was attracting worldwide attention. Karl Marx, then living in England, wanted to align with Darwin's rising popularity. He sought to dedicate the English translation of volume two of *Das Kapital* to Darwin, but Darwin courteously refused. In the refusal letter to Marx, dated October 13, 1880, Darwin paints a classic picture, characterizing his perspective on religion, freedom of thought, and the emerging role of science to argue "against Christianity and theism":

> I am much obliged for your kind letter and the enclosure. The publication in any form of your remarks on my writing really requires no consent on my part, and it would be ridiculous [of] me to give consent to what requires none. I [should] prefer the part or volume not to be dedicated to me (though I thank you for the intended honour) as this implies to a certain extent my approval of the general publication, about which I know nothing. Moreover, though I am a strong advocate for free thought on all subjects … it appears to me (whether rightly or wrongly) that direct arguments against Christianity and theism produce hardly any effect on the public; and freedom of thought is best promoted by the gradual illumination of men's minds, which follow from the advance of science. It has, therefore, always been my object to avoid writing on religion, and I have confined myself to science. I may, however, have been unduly biased by the pain which it would give some members of my family, if I aided in any way direct attacks on religion.[25]

HMS *Challenger*

Though America was on the verge of a Civil War, the emerging popularity of *The Origin of Species* was generating growing interest, even across the Atlantic. The challenge Darwin had given for further exploration rang throughout the halls of universities and elite social circles in America while reverberating through the chambers of Parliament.

At the University of Edinburgh, the newly elected Regius Chair of Natural History, Professor Sir Charles Wyville Thomson, and William Benjamin Carpenter, University of London professor and Secretary of the Royal Society, led the campaign to contract with the Royal Navy to use the HMS *Challenger*. The main purpose of the voyage was to find the missing transitional links that Darwin declared would be "inconceivably great."

The Royal Society, then as now, occupied an enormously powerful position in the scientific world because its members were both bureaucrats and scientists. The timing was perfect—the Society was headed by the now eminent biologist Thomas Henry Huxley.

For more than ten years, the Victorian establishment had been rocked by Darwin's theory. The theory was being debated in meeting halls, parlors, and pubs, and was gaining widespread acceptance from scientists to chimney sweeps. Yet, the burden of proof was still to be discovered.

This scientific expedition was to become historic as the first cooperative venture between the military and the academia and the first international exploration ever convened. Carpenter and Thomson reasoned that given a properly equipped global expedition to investigate the last great geographical unknown on Earth—the oceans of the world—the mysteries of the origin of life might be uncovered according to Darwin's theory. The scope and size of the expedition was unprecedented.

To become a platform for study, the warship HMS *Challenger*, powered by a steam corvette capable of producing over 1,200 units of horsepower, had to be extensively refitted. Fifteen of the seventeen cannons were removed. The interior was modified to include laboratories, extra cabins, and a special dredging platform for the scientific party, and storage for their strange-seeming collection of equipment and supplies.

Room was made to store 12 1/2 miles of piano wire for dredging, and 144 miles of Italian hemp rope for sounding. She was loaded with specimen jars, alcohol for preservation of samples, microscope and chemical apparatus, trawls and dredges, thermometers, water sampling bottles, sounding leads, and devices to collect sediment from the seabed.

By December 7, 1872, the HMS *Challenger* was ready to set sail from Portsmouth with 269 persons aboard: 23 naval officers under the command of Captain George Strong Nares; a team of 6 scientists led by Professor Wyville Thomson; and a crew of 240, known as the "bluejackets."

HMS *CHALLENGER* UNDER SAIL, 1874

Besides England, this international expedition included crew members from Switzerland, Scotland, Canada, and Germany. Charles Thomson, the chief scientist of the civilian scientific staff, was English; Jean Jacques Wild was an artist from Switzerland; John Young Buchanan was a Scottish chemist; zoologists—Henry, Nottidge, and Moseley were Englishmen; John Murray a Scottish-Canadian; and Rudolf von Willimoes-Suhn was a German.

No details were forgotten, including a bandmaster to muster entertainment. By virtue of a letter from one of the deckhands, Joseph Matkin, known as a "bluejacket," the tale of the early life of the band aboard the *Challenger* is painted: "They practice every day in the fore peak of the vessel and the noise is something fearful and causes the Watch below to swear a good deal. The Bandmaster expects to fetch tolerable music in about 6 months."[26] The voyage was launched packed with excitement and high expectations—history was in the making.

Once under way, the expedition developed a standard set of methods for collecting data at each of the observation stations. Since life was thought to have originated in the sea, one of the expectations was to validate Darwin's theory by discovering transitional links in the ocean.

When the long-awaited time had come, excitement and anxiety filled the air. Who was going to see the first new transitional link? Henry Moseley wrote:

> At first, when the dredge came up, every man and boy who could possibly slip away, crowded 'round it, to see what had been fished up ... Gradually, as the novelty of the thing wore off, the crowd became smaller and smaller, until at last only the scientific staff, and perhaps one or two other officers besides the one on duty, awaited the arrival of the net on the dredging bridge.[27]

As the onboard natural scientist and zoologist, Moseley had fitted the *Challenger* with a state-of-the-art zoology laboratory and brought enthusiasm to the voyage. Moseley's lab was just below the upper deck. When a trawl was brought on board the deck, the contents were quickly and carefully washed and examined by the team of scientists.

The work was hard; and the rigors of systematically detailing the collected specimens began to take a toll on the crew and "the novelty of the thing wore off." Eventually, 61 of the 269 crew members deserted at various port calls.

Circumnavigating the globe to trawl the depths of all of the oceans but the Arctic, the *Challenger* logged a total of 110,224 kilometers (68,890 miles). Eventually, after a thousand days at sea, the ship returned to Spithead on Queen Victoria's fifty-first birthday, May 21, 1876.

In total, the expedition collected specimens from 360 "stations" along the route. After the data had been recorded and the samples had been contained, the specimens were sent to Edinburgh University in Scotland for further systematic analysis. In his introduction to the scientific reports, Charles Thomson summarized the collection, which included nearly 5,000 bottles, jars, glass tubes, and tin cases.

Upon *Challenger's* return, the specimens and scientific findings were examined by over 100 scientists. Henry Moseley led the investigation and received many honors for his work, including the Royal Medal of the Royal Society, to which he was also elected a Fellow in 1879.

The pressure of the mission, though, became overwhelming. Appearing to have no immediate answer to the lingering and pervasive question, "where are all the transitional links?" and exhausted by the rigors of investigation, Moseley died in 1891 at the age of forty-seven. The report from the expedition was not released until four years later, in 1895.

Preparation of the report, which was overseen by John Murray, had taken nearly twenty years. The results were finally published in 1895 in a report entitled *Report Of the Scientific Results of the Exploring Voyage of* HMS *Challenger during the years 1873–76.* The report, occupying fifty volumes, was 29,552 pages long, with each page measuring about thirteen by ten inches.

John Murray announced in 1895 that the report was "the greatest advance in the knowledge of our planet since the celebrated discoveries of the fifteenth and sixteenth centuries." In fact, not only was information gathered on the nature of the biosphere, but also the format of the report set the standard for scientific papers presentations during the twentieth century.

The transitional links, though, were still missing. Expectations for finding the transitional links that Darwin said would be "inconceivably great" were largely dashed. Darwin's explanation that "only a small portion of the surface of the Earth has been geologically explored" was evaporating.

Chapter Nine
Fossils

Why then is not every geological formation and every stratum full of such intermediate links? Geology assuredly does not reveal any such finely graduated organic chain; and this, perhaps, is the most obvious and serious objection which can be urged against my theory.

—Charles Darwin[1]

Evidence from the HMS *Challenger* mission raised a big red flag on what Darwin knew all along: "The distinctiveness of specific forms and their not being blended together in innumerable transitional links is a very obvious difficulty."[2]

These "intermediate links" are often now more popularly referred to as the "missing links." Not only were the missing links a problem, Darwin found it difficult even to envision the transitions involved: "It is no doubt difficult even to conjecture by what gradations many structures have been perfected ... in many cases it is most difficult even to conjecture by what transitions organs have arrived at the present state."[3]

Darwin was even "staggered" how an "organ as perfect as the eye could have been formed by natural selection."[4] Continuing, Darwin confides, "I have felt the difficulty far too keenly to be surprised at others hesitating to expand the principle of natural selection to so startling a length."[4]

Despite these doubts, Darwin released *The Origin of Species* for publication. Since then, the challenge has been to find these missing links—the cornerstone of evidence for evolution. In the first edition of *Origin*, Darwin proposed three possible missing links—two large sea mammals, the Dugong *Halithermium* and the *Zeuglodon,* and a three-toed horse, the *Hipparion.*

The fossil race was on; there was no returning. Since 1859 the search for fossil evidence has been so vast, it is estimated that more than

99.9 percent of all paleontology work has been carried out following the publication of *The Origin of Species*.[5] In 1866, Darwin raised the stakes further in the fourth edition of *The Origin of Species* by suggesting that a series of newly discovered fossils represented the eagerly anticipated missing link between the reptile and the bird.

Archaeopteryx

In 1861, just two years after the release of *The Origin of Species,* in the midst of heated controversy, the discovery of a series of fossils seeming to confirm Darwin's theory was announced. The first fossil, a single feather, discovered a year earlier by paleontologist Hermann von Meyer in a German limestone quarry in the town of Solnhofen, was thought to be a feather of a new species, the *Archaeopteryx lithography*.[6] *Archaeopteryx* means "ancient wing."

That same year, the first skeleton, now known as the "London Specimen," was unearthed in 1861 near Langenaltheim, Germany. The specimen was later handed over to Karl Häberlein, a local physician, in return for medical services. Häberlein sold the specimen to the British Museum of Natural History in London, where it remains today.

While missing most of its head and neck, the London Specimen was examined by Richard Owen and given the name *Archaeopteryx macrura* in 1863. The fossil had wings and feathers like a bird, but unlike any modern bird, the *Archaeopteryx* had a long lizard-like tail and claws on its wings. In the framework of Darwin's theory, the specimen emerged as the expected missing link between the reptile and the bird.

In 1877, an even more complete specimen was found. This second find, known as the "Berlin Specimen," is now recognized as the most complete specimen of *Archaeopteryx* and is on display at the Humboldt Museum in Berlin. The discovery of teeth seemed to support the intermediate link status of the *Archaeopteryx*.

The Berlin Specimen is now thought to be one of the most beautiful fossils ever unearthed. With the optimal preservation conditions of Solnhofen limestone, even the finest detail structures of the feather have been preserved. It is the analysis of these fine feather details that has been central to studying the intermediate status of the specimen.

The *Archaeopteryx* quickly became recognized as the most famous missing fossil link ever discovered. The evidence appeared to support

Darwin's theory. The *Archaeopteryx* was announced as the "impeach-able" evidence for evolution. While the "impeachable" status was challenged, the *Archaeopteryx,* without question, has played a leading role in paleontology. In the words of evolutionary biologist Alan Feduccia of the University of North Carolina at Chapel Hill, the Berlin Specimen "may well be the most important natural history specimen in existence … Beyond doubt it is the most widely known and illustrated fossil."[7]

As a missing link, the *Archaeopteryx* filled in the gap between reptiles and birds. Darwin's ardent defender, Thomas Henry Huxley, widely publicized the *Archaeopteryx* as an important missing link, with a measure of cautious optimism: "We have not knowledge of the animals which linked reptiles and birds together historically and genetically"; fossils "only help us to form a reasonable concept of what those intermediate forms may have been."[8]

By the sixth edition of *The Origin of Species*, Darwin concluded that the missing link was at least "partially bridged over" and that the rudimentary elements of the theory were becoming seemingly validated, writing: "Even the wide interval between birds and reptiles has been shown by [Huxley] to be partially bridged over in the most unexpected manner, by the ostrich and extinct *Archeopteryx*."[9]

In the book *Historical Geology*, Carl Dunbar wrote in 1961 that it would be difficult to find a more perfect link or "cogent proof" of the reptilian ancestry of the birds. To paleontologist Pat Shipman, the *Archaeopteryx* is "more than the world's most beautiful fossil … [it is] an icon—a holy relic of the past that has become a powerful symbol of the evolutionary process itself. It is the First Bird."[10]

Since 1866, the missing link status of the *Archaeopteryx* has centered on the characteristics of the feather. According to Darwin, natural selection acts only by "successive, slight modification." The question is how did the feather change by "successive, slight modification"? While there has been a litany of theories on how the scale of the reptile evolved in the feather of the bird, the most popular theories center on whether the reptile was jumping from trees or running for prey. In either case, it is thought that the resulting air friction on the scale initiated the development of the feather.

If the *Archaeopteryx* is the transitional missing link between a reptile scale and a bird feather, then the feather must be a transitional feather—

part scale and part feather: a scale-feather. Or must it? In attempting to answer this question, earlier enthusiasm started to waver. Under scrutiny, the evidence continued to mount against the transitional status of the scale-feather.

By the early 1970s, paleontologists began to seriously question the "transitional link" status of the *Archaeopteryx* scale-feather. In the words of Barbara Stahl in *Vertebrate History: Problems in Evolution* (1974): "How [birds] arose initially, presumably from reptile scales, defies analysis."[11]

Alan Feduccia and colleagues, writing in the journal *Science* in 1979, in the paper entitled "Feathers of the Archaeopteryx: Asymmetric Vanes Indicate Aerodynamic Function," likewise have concluded that the feather was "essentially like those of modern birds" and not a transitional form of the feather.[12]

In examining the evidence to support *Archaeopteryx* as the missing link between the reptile and the bird or even a prototype bird, sometimes called "pro avis," John Ostrom in 1979 published a paper in the *American Scientist,* concluding that not only is the *Archaeopteryx* not a missing link, but that "No fossil evidence exists of any pro-avis. It is purely hypothetical."[13]

Coming to the same conclusion, in a 1979 article in the journal *Science,* Alan Feduccia writes, "I conclude that *Archaeopteryx* was … considerably advanced aerodynamically, and probably capable of flapping, powered flight to at least some degree. *Archaeopteryx* … was, in the modern sense, a bird."[14] The scale-feather is unquestionably a complete feather.

Harvard professor and neo-Darwinist Ernest Mayr, in 1982, even began to weigh in with caution, backpedaling by calling the *Archaeopteryx* discovery "the almost perfect link between reptiles and birds."[15]

By the early 1980s, the early enthusiasm over the missing link status of the *Archaeopteryx* was nearly gone. University of Kansas paleontologist, Larry Martin acknowledged in 1985 that the "*Archaeopteryx* is not ancestral of any group of modern birds."[16] Instead, it is "the earliest known member of a totally extinct group of birds."[16]

In 1984, Alan Feduccia again concluded that not only was the feather like a modern bird, the *Archaeopteryx* could not be the missing link between the reptile and bird.[17] At the International *Archaeopteryx* Conference in 1985, Peter Dodson even concluded that the *Archaeopteryx* was a bird capable of flight and published his conclusion in the *Journal of Vertebrate Paleontology:*

> At the end of the three days of presentations, [Alan] Charig [chief curator of fossil amphibians, reptiles, and birds at the British Museum—BH/BT] orchestrated a concerted effort to summarize the ideas for which consensus exists. The general credo runs as follows: Archaeopteryx was a bird that could fly.[18]

The feathers on the *Archaeopteryx* are designed for flight like a modern bird. There is no evidence to suggest the feather ever evolved from a scale. Leading molecular biologist Michael Denton concluded in 1985 that the *Archaeopteryx* feather is not just a transitional scale, but is a feather fully designed for flight.[19]

Robert L. Carroll, professor of biology at McGill University, in 1997 concluded, "The geometry of the flight feathers of *Archaeopteryx* is identical with that of modern flying birds, whereas non-flying birds have symmetrical feathers. The way in which the feathers are arranged on the wing also falls within the range of modern birds."[20]

To date, no known transitional forms between the scale and the feather have been discovered. Efforts to maintain the missing link status of the *Archaeopteryx* have been nearly completely exhausted, even in popular science magazines. Henry Gee, the chief science writer for *Nature,* wrote in 1999 that the missing link status of the *Archaeopteryx* is only an illusion; a "once upon a time" story.[21]

So what is the truth about *Archaeopteryx?* Colin Patterson, senior paleontologist at the British Museum of Natural History, may have summed it up best in a letter to Luther Sunderland on April 10, 1979, writing, "such stories are not a part of science."[22]

Alan Feduccia, in 2007, drove another nail in the coffin of the alleged scale-to-feather evolution with the publication of his paper "A new Chinese specimen indicates that 'protofeathers' in the Early Cretaceous theropod dinosaur Sinosauropteryx are degraded collagen fibres."

Sinosauropteryx, meaning "Chinese lizard-wing," was thought to be the first and most primitive dinosaur found with the fossilized impressions of feathers. But, as the title alludes, the "protofeathers" of the dinosaur *Sinosauropteryx* are nothing more than collagen fibers. Feduccia concludes that the "proposal that these fibres are protofeathers is dismissed."[23]

In analyzing the fossil remains of a dinosaur specimen, dromaeosaur *Sinornithosaurus,* in 2003, paleontologist Theagarten Lingham-Soliar, of the University of KwaZulu-Natal in South Africa, came to the same conclusion.[24] Further, in examining fossilized collagen from sharks, dolphins, snakes, and turtles, none resemble any prototype form of a feather. Lingham-Soliar concluded that "the overall findings of the study are that the thesis of dinosaur 'protofeathers' requires more substantial support than exists at present."[24]

The issue of birds originating from reptiles or dinosaurs has been among the most contentious issues in paleobiology. Why is it that reptile scales assume a "protofeather" appearance? In 2005, Feduccia concluded that these "protofeathers" are actually scales—or "'meshwork' of the skin"—in degrees of decomposition.[25]

The *Archaeopteryx* as the missing link between reptile and bird has had a long, contentious history. An ongoing debate as to whether the specimen is an intermediate of a bird has been continuing nonstop for over a century; some scientists have even publicly suggested that use of the *Archaeopteryx* as a missing link bordered on blatant fraud.[26]

In 2000, Elaine Kennedy, of the *Geoscience Research Institute* in California, concluded in the end that the *Archaeopteryx* does not represent a missing link: "Despite all the conflicting data with respect to the linkage between dinosaurs/reptiles and birds, it seems clear that although *Archaeopteryx* is the best candidate, it is not the link."[27]

While the search for new scale-to-feather-evolution theories continues, for the time, the *Archaeopteryx* is silently sliding into the halls of history as another lesson learned. The *Archaeopteryx* has failed Darwin's challenge to provide the evidence for "successive, slight modifications" as required by the theory of natural selection. Darwin gives the caution: "If it could be demonstrated that any complex organ exists which could not possibly have been formed by numerous, successive, slight modifications, my theory would absolutely break down."[28]

Geological Columns

While at Edinburgh University, Darwin studied under the geology professor Adam Sedgwick. After Darwin left Edinburgh, Sedgwick and Darwin continued to correspond while Darwin was at Christ's College and later while aboard the *Beagle*—and even after the publication of *The Origin of Species.*

Sedgwick is now recognized as one of the founders of modern geology and was one of the first to recognize the Devonian and later the Cambrian period. At the time, knowing that Sedgwick rejected any concept of biological evolution, Darwin did not discuss his theories with Sedgwick. It was only after receiving a copy of *The Origin of Species* from Darwin that their collegial relationship turned rocky. Within weeks of the book's release, Sedgwick wrote a letter to Darwin, commenting: "I laughed … till my sides were almost sore."[178]

What did Sedgwick find so funny? Perhaps the issue was Darwin's speculations without the fossil evidence. In *The Origin of Species*, Darwin wrote, "I view all beings not as special creations, but as the lineal descendants of some few beings which lived before the first bed of the Cambrian system was deposited."[29] Darwin defended the theory, however, by first reasoning "only a small portion of the surface of the Earth has been geologically explored."[30]

The known geological evidence was limited and "written in a changing dialect; of this history we possess the last volume alone, relating only to two or three countries."[31] This was true. At the time, very little of Russia, Australia, Africa, much less Asia had ever been explored by trained geologists or paleontologists. For this reason, the absence of fossil evidence was not considered a problem for Darwin's theory in 1859—an argument that blunted Darwin's critics, for the time.

The second argument Darwin uses for the lack of missing links was based on the argument that fossils are typically not well preserved, Darwin argues: "The crust of the Earth is a vast museum; but the natural collections have been imperfectly made, and only at long intervals of time."[32] Therefore, Darwin concludes: "Geology assuredly does not reveal any such finely graduated organic chain; and this, perhaps, is the most obvious and serious objection which can be urged against my theory. The explanation lies, as I believe, in the extreme imperfection of the geological record."[1]

In addressing these issues in *The Origin of Species,* Darwin devoted all of chapter ten, "On the Imperfection of the Geological Record," and part of chapter eleven on the geological problems. Darwin concedes: "He who rejects this view of the imperfection of the geological record will rightly reject the whole theory."[33]

To reconcile the evidence with the theory, Darwin suggests that the problem rests with the geological evidence, not with the theory: "The noble science of geology loses glory from the extreme imperfection of the record. The crust of the Earth, with its embedded remains, must not be looked at as a well-filled museum, but as a poor collection made at hazard and at rare intervals."[34]

Rejection of geological evidence highlights how Darwin purposely let the theory take precedence over the evidence. The theory, not the evidence, drove the development of natural selection exemplifies Darwin's classical deductive reasoning approach.

This lack of evidence incited even greater interest in geology and paleontology, since without fossil evidence, Darwin said the theory must be rejected.[35]

The Burgess Shale of British Columbia

From the "deepest valleys or the formation of [a] long line of inland cliffs," Darwin envisions overcoming criticism by finding the evidence to support his theory. Using a parallel between himself and Lyell, Darwin argues that the geological evidence would eventually validate his "doctrine" and invalidate the evidence of creation and the flood:

> I am well aware that this doctrine of natural selection, exemplified in the above imaginary instances, is open to the same objections which were first urged against Sir Charles Lyell's noble views on 'the modern changes of the Earth, as illustrative of geology'; but we now seldom hear the agencies which we see still at work, spoken of as trifling and insignificant, when used in explaining the excavation of the deepest valleys or the formation of long lines of inland cliffs. Natural selection acts only

by the preservation and accumulation of small inherited modifications, each profitable to the preserved being; and as modern geology has almost banished such views as the excavation of a great valley by a single diluvial wave, so will natural selection banish the belief of the continued creation of new organic beings, or of any great and sudden modification in their structure.[36]

For natural selection to "banish the belief," Darwin envisioned that the fossil evidence would be discovered in the Earth's geological columns. These columns were argued to contain evidence of the "intermediate" forms—the "missing links," and Darwin gave instructions: "We should always look for forms intermediate between each species and a common but unknown progenitor."[37]

Darwin theorized that life's nascent period contained only a limited number of the simplest types of organisms, such as fungi and bacteria. According to the theory, during long periods, with "successive, slight modifications," the organisms were expected to increase in complexity, eventually producing species that are more complex. Logically then, lower layers in the geological column should only contain a limited number of simple organisms.

Simple organisms were expected to dominate the lower layers. Over time, as species evolved to become more complex, the overlaying geological layers were expected to contain more organisms that are more complex. The higher geological layers should demonstrate progressively more complex organisms with an ever-expanding number of invertebrate organisms, then vertebrates.

The lower layer of the geological column that Darwin called "the deepest valleys" is also known as the Cambrian layer or stratum. This stratum was thought to represent only the earliest life-forms on the Earth. The term "Cambrian" was named by Darwin's geology professor, Adam Sedgwick at Edinburgh University. "Cambrian" is named after Cambria, which is the classical name for Wales.

In reality, Darwin knew that the Cambrian stratum challenged, even contradicted, his theory of evolution. First, Darwin concedes that no missing links had been discovered.[38] Second, the appearance in the fossil record appears in a "sudden manner," not gradually, as predicted by the theory.[38] Darwin concludes that these problems "are all undoubtedly of the most serious nature."[38]

After the publication of *The Origin of Species,* nearly fifty years elapsed before the next major Cambrian site was discovered. In 1909, paleontologist Charles Doolittle Walcott made the discovery in the Burgess Shale formation in British Columbia. But ironically, just like Darwin had previously seen in Wales, the Burgess Shale fossil discovery contained not only simple and well-known organisms, but also an explosive number of complex and advanced organisms, evidence that contradicts Darwin's theory.

Many of the organisms in the Burgess Shale had never been seen before. The *New Dictionary of Cultural Literacy,* 2002, concludes that the "Burgess Shale fossils provide valuable information about the evolution of early life on Earth."[39] The Burgess Shale does indeed "provide valuable information." However, the evidence again failed to support Darwin's theory of natural selection acting by "successive, slight modifications." In fact, the evidence clearly contradicts Darwin's theory, as highly complex organisms were discovered in the Cambrian strata.

The Cambrian stratum in the Burgess Shale contains some of the most exotic forms of life. One species was appropriately named *Hallucinogenia,* because the appearance was so crazy—unlike anything seen before. The *Hallucinogenia* propelled itself across the seafloor by means of seven tentacles of sharply pointed, stilt-like legs.

Since the evidence did not support Darwin's theory, the Burgess Shale was not widely publicized. The full extent of the "Cambrian" phenomena was not widely publicized until the late 1970s, when fossils from Burgess Shale were reanalyzed by paleontologists Harry Whittington and Simon Conway Morris.

Not only had life appeared suddenly, but also no links were found. In 1979, Whittington and Morris concluded: "the most intriguing problem presented by the Burgess Shale fauna is the 10 or more invertebrate genera that so far have defied all efforts to link them with known phyla. They appear to be the only known representatives of phyla whose existence had not been suspected."[40]

Recognizing that the Burgess Shale failed to find the fossil evidence as expected from Darwin's theory, molecular biologist Michael Denton wrote in 1985, "Altogether the representatives of ten completely new invertebrate phyla were eventually recovered in the Burgess Shale, yet none of them turned out to be links between known phyla."[41]

Evidence from the Burgess Shale was Darwin's worst nightmare; the fossil record did not contain the expected missing links, and life apparently appeared suddenly. Why had Darwin acknowledged the *Origin of Species* dilemma and yet continue to ignore the evidence? Darwin had written, "It is indisputable that before the lowest Cambrian stratum was deposited ... several of the main divisions of the animal kingdom suddenly appear in the lowest known fossiliferous rocks."[42]

Avoiding the fossil evidence demonstrates Darwin's abandonment of the scientific method by selectively excluding evidence. Why had Darwin acknowledged the dilemma in the *Origin of Species* and yet continued to ignore the evidence?"[43]

So profound was the diversity and exotic nature of the Burgess Shale species, the Burgess Shale is now known to represent the Cambrian explosion—not the Cambrian evolution. The evidence stands as a contradiction to Darwin's theory. Writing in *Scientific American* in 1994, paleontologist Stephen Jay Gould wrote that the "Cambrian explosion was the most remarkable and puzzling event in the history of life."[44] The explosion is only puzzling in the light of evolution.

The Ediacara Hills of Australia

Since the Cambrian strata discovery at Burgess Shale did not support Darwin's theory, the search continued in the "lowest Cambrian stratum" to find the first signs of life before an explosion of life. Darwin suggested, "If the theory be true, it is indisputable that before the lowest Cambrian stratum was deposited long periods elapsed ... [in which] the world swarmed with living creatures."[45] Darwin envisioned life arising gradually over long periods of time, not suddenly, so that the simplest forms would be discovered in the lowest level of the geological column with an accumulation of the evolving forms, the missing links, higher in the column.

The prevailing thought was that fossils from this "lowest Cambrian stratum," now known as the Precambrian period, would contain the true first signs of life. It was not until nearly forty years later in 1946 that Australian mining geologist Reginald C. Sprigg discovered the highly prized Ediacaran fossils. This was the first notable diverse and well-preserved assemblage of Precambrian fossils ever discovered.

Fossils from Ediacara Hills in Southern Australia were considered Precambrian because the discovery at the site yielded a vast array of soft-bodied organisms without skeletons. British paleontologist Simon Conway wrote in 1998 that the "Ediacaran fossils look as if they were effectively soft-bodied."[46] Most amazingly, the fossil preservation was exquisite. Conway continues that these "remarkable fossils reveal not only their outlines but sometimes even internal organs such as the intestines or muscles."[46] The find demonstrated that the intricacies of ancient fossils could be exquisitely preserved.

Unexpectedly, however, Precambrian era fossils were mixed with Cambrian era fossils. As a result, the expected simple-to-complex fossil sequence in the geological column was not demonstrated. Perplexing to Darwin's theory further was the fact that that even the "earliest" soft-bodied specimens were complex organisms, not simple as originally postulated.

The Ediacara fossils were ironically even more bizarre than the Burgess Shale fossils. In 1961, paleontologist Martin Fritz Glaessner, commenting on the newly found strange fossil named *Tribrachidium heraldicum* in *Scientific American,* wrote, "Nothing like it has ever been seen among the millions of species of animals."[47]

Not only were the organisms discovered in Ediacara Hills complex specimens, these exotic forms appeared in the fossil record suddenly, and in unexpected numbers. Clearly, the fossil evidence demonstrated a sudden appearance of vast numbers of complex life-forms. Again, the evidence uncovered another contradiction to the theory of natural selection that life could not have started into "life at once." Darwin had conceded that if "numerous species, belonging to the same genera or families, have really started into life at once, the fact would be fatal to the theory of evolution through natural selection."[48]

The sudden abundance of complex life was a surprise discovery, and the preservation of the soft-bodied specimens was equally surprising to twentieth century paleontologists. At the time, it was thought that the delicate Precambrian soft tissues could never have remained in existence after being fossilized for hundreds of millions of years. Finding soft tissue in Precambrian fossils was a completely unexpected finding.

Geologist William Schopf, in 1994, wrote that the "long held notion that Precambrian organisms must have been too small or too delicate to have been preserved in geological materials ... [is] now recognized as incorrect."[49]

Most puzzling was that not one of the bizarre forms discovered were thought to represent any of the elusive missing links, as predicted by Darwin. For this obvious reason, the Ediacaran fossils garnered little attention initially. The organisms were written off as either gas-escaping structures or inorganic concretions since no similar soft-bodied organisms had ever been discovered anywhere else in the world.

That was until 1957, when Roger Mason, an English schoolboy in the Charnwood Forest, unearthed soft-bodied fossils. It was a wake-up call. The discovery meant that the Ediacaran fossils were not a fluke and could not be written off. The Precambrian status of the frond-shaped Charnia fossils was later confirmed to be Precambrian by the British Geological Survey.

In 1959, paleontologist Martin Glaessner made the connection between the Charnia and Ediacaran fossil discoveries.[50] Again, these frond-shaped fossils were complex and exotic—not simple. Like the Burgess Shale and Ediacara Hill discoveries, these bizarre organisms appeared in the fossil record suddenly and unexpectedly. Worst of all for Darwin's theory, none of the specimens could be entertained as missing links.

The overall character and meaning of the fossil record at the time was encapsulated in an article presented by leading paleontologist George Gaylord Simpson in 1959 for the Darwin Centenary Symposium held in Chicago. As for "successive, slight modifications" gradually taking place from one species to another, as Darwin predicted, Simpson concludes: "They are not, as a rule, led up to by a sequence of almost imperceptibly changing forerunners such as Darwin believed should be usual in evolution."[51]

Life suddenly appearing in the fossil record with no missing links—"intermediate forms"—continues to be a glaring problem for Darwin's theory of natural selection. Robert Barnes, in 1980, summed up the current situation in an article published in *Paleobiology*: "The fossil record tells us almost nothing about evolutionary origin of phyla and classes. Intermediate forms are non-existent, undiscovered, or not recognized."[52]

Further explorations in the 1980s eventually lead to two other Cambrian fossil discoveries in Sirius Passer, located in northern Greenland, and in the Chengjiang County, located in southern China. The Chengjiang fossils are now recognized as some of the best-preserved fossils. However, vertebrate fossils were unexpectedly found alongside invertebrates. Like all of the other previous Precambrian and Cambrian fossil discoveries, the fossils recovered in Greenland and China demonstrate a bewildering variety of animals appearing suddenly—a now insidious problem for Darwin's theory.

The renowned paleontologist Harry Whittington, whose 1985 examination of fossil evidence first revealed the extent of the Cambrian explosion, cast a long shadow of uncertainty on the cornerstone of Darwin's theory: "I look skeptically upon diagrams that show a branching diversity of animal life through time, and come down at the base to a single kind of animal."[53]

With the fossil record evidence racking up evidence against Darwin's theory, molecular biologist Michael Denton, in 1985, concluded in his *Evolution, a Theory in Crisis* that the "absence of transitional forms is dramatically obvious."[54] Rather than supporting Darwin's theory, the unveiling of the fossil fiasco continued to unravel Darwin's theory. Michael Denton declares:

> Fossils have not only failed to yield the host of transitional form demanded by evolutionary theory, but because nearly all extinct species and groups revealed by paleontology are quite distinct and isolated as they burst into the record, then the number of hypothetical connecting links to join its diverse branches is necessarily greatly increased.[55]

Evolutionary theorist Jeffrey Schwartz, observing that the sudden appearance of organisms is more aligned with Greek mythology than with Darwin's theory of natural selection, wrote in 1999 that species "appear[ed] in the fossil record as Athena did from the head of Zeus—full grown and raring to go."[56]

Echoing the same conclusion, paleontologists James Valentine, Stanley Awramik, Philip Signor, and Peter Sadler, in 1991, concurred that the "single most spectacular phenomena evident in the fossil record is the abrupt appearance and diversification of many living and extinct phyla" near the beginning of the Cambrian period.[57] The Cambrian explosion "was even more abrupt and extensive than previously envisioned."[58]

In *The Panda's Thumb,* evolutionary paleontologist Stephen Gould, recognizing the agony Darwin experienced over the known disconnection between the theory of natural selection and the reality of the fossil record, points out that the "fossil record had caused Darwin more grief than joy. Nothing distressed him more than the Cambrian explosion, the coincident appearance of almost all complex organic designs."[59]

Rising Out of the Water

In Greek mythology, the Sirens were sea deities who lived on Sirenum scopuli, a cluster of three small rocky islands. Sometimes portrayed as mermaids, the Sirens became legendary as aquatic Chiron creatures rising out of the water with the head and torso of a human female and the tail of a fish.

Underscoring the influence of mythology on philosophy, the Greek philosopher Anaximander taught, "life evolved from moisture," and "man developed from fish." While the image of the mermaid on the rock continues as a Greek myth, fossil evidence for the mythological mermaid continued to be illusive. Since early in the twentieth century, the fossilized coelacanth was touted as a textbook example of life arising out of water, representing the missing link between the fish and the reptile.

Evidence supporting this theory was based on the configuration of the pectoral fins. In the fossilized form, the pectoral fins of the coelacanth appeared to be a transitional form between a fish fin and a reptile leg and foot. Therefore, the coelacanth was thought to be the missing link between the fish and the amphibians.

Prior to 1938, the coelacanths were only known as fossils after becoming extinct approximately sixty-five million years ago. But on December 23, 1938, the fishing ship *Nerine* unveiled a decisive moment. After trawling off the mouth of the Chalumna River in South Africa, the fishing crew went ashore, unknowingly leaving one of the worlds most unusual catches stacked on the dock.

During the course of that afternoon, as local curator Marjorie Latimer went down to the dock to wish the crew a Merry Christmas, she noticed a blue fin protruding beneath a pile of rays and sharks on the deck. Pushing the overlaying fish aside revealed, she later wrote, "the most beautiful fish I had ever seen … It was five foot long, a pale mauvy blue with faint flecks of whitish spots; it had an iridescent silver-blue-green sheen all over. It was covered in hard scales, and it had four limb-like fins and a strange puppy dog tail."[60]

Measuring about five feet long and weighing 126 pounds, the live fish, previously only known as a fossil, was hailed a scientific sensation. Since this first discovery in 1938, coelacanths are now known to live at a depth of one thousand feet and deeper, with a territorial range from South Africa to Indonesia.

Subsequent to the initial excitement came the reality question, is the coelacanth actually a missing link? In comparing the fossilized form to the live form, the coelacanth was found to be much different from what was originally expected. On examination, the pectoral fins of the coelacanth are not a transition between a fish fin and foot. The pectoral fins are simply a typical fish fin—not similar to any hand or foot that would be capable of walking on land. Rather than the textbook example of a missing link, the coelacanth is now considered a unique and distinct species of fish.

Surprisingly, it is now known that unlike any fish or a reptile, the female coelacanths deliver live births. Similar to mammals, the coelacanth does not reach sexual maturity until the age of twenty years and can give birth to between five and twenty-five babies after a gestation period of approximately thirteen months. The newborn coelacanths are capable of swimming and surviving on their own immediately after birth. How Darwin's theory of "successive, slight modifications" is demonstrated in the fish-to-reptile sequence defies any explanation.

The coelacanth was initially thought to be the ultimate fish-amphibian link. However, after studying the soft anatomy of the coelacanths in 1974, Barbara Stahl concluded in *Vertebrate History: Problems in Evolution,* "The modern coelacanth shows no evidence of having internal organs preadapted for use in a terrestrial environment."[61]

Studies of the "extinct" coelacanth highlight the limitations of drawing conclusions from the fossil evidence alone. Since most of the biology of any organism is in the soft anatomy, limited evidence is gained by studying only fossilized remains in determining ancestry, as suggested by Darwin. In 1985, molecular biologist Michael Denton highlighted the limitations of examining only the skeletal remains:

> The systematic status and biological affinity of a fossil organism is far more difficult to establish than the case of living form, and can never be established with any degree of certainty. To begin with, ninety-nine percent of the biology of any organism resides in the soft anatomy, which is inaccessible in a fossil.[62]

The coelacanth highlights how enormous gaps can exist between studying a live specimen and fossilized remains. Prior to 1938, it was not even remotely expected that the coelacanth gave live births. Fish, reptiles, and amphibians typically lay eggs. While amphibians lay their eggs in water and their larvae undergo a complex metamorphosis, reptiles develop inside a hard, shell-encased egg and are perfect replicas of the adult on first emerging from the shell. Contrary to Darwin's theory, there are no "successive, slight variations" in the coelacanth.

While textbooks on evolution assert that reptiles evolved from amphibian, they do not explain how the amphibian egg gradually evolved into the amniotic egg of the reptile. In fact, the two eggs are entirely different. Michael Denton explains: "there are hardly two eggs in the whole animal kingdom which differ more fundamentally."[63]

What are the transitional links from fish to amphibian–reptile? Most biology textbooks consider the *Rhipidistian,* which includes the coelacanth, as ancestors to amphibians and reptiles. Weighing in on that theory, vertebrate paleontologist Robert L. Carroll, professor of biology at McGill University, concedes: "We have no intermediate fossils between Rhipidistian fish and early amphibians."[64]

This was not a new conclusion. In 1969, writing in *Biological Reviews of the Cambrian Philosophical Society*, Lewis L. Carroll concludes: "Unfortunately not a single specimen of an appropriate reptilian ancestor is known prior to the appearance of true reptiles. The absence of such ancestral forms leaves many problems of the amphibian—reptile transition unanswered."[65]

Distinguished vertebrate paleontologist Edwin Harris Colbert, curator of Vertebrate Paleontology at the American Museum of Natural History, concluded in 1991: "Despite these similarities, there is no evidence of any Paleozoic amphibians combining the characteristics that would be expected in a single common ancestor."[66]

Studying fossils exclusively can be a tricky business. In *The Natural History of the African Elephant*, 1971, biologist Sylvia K. Sikes concurs with Denton that in studying morphology in fossils, the more important physiological features are overlooked.[67]

At the very least, the coelacanth was expected to exhibit at least some hint of walking behavior, but nothing of the kind has ever been observed. Coelacanths have been observed swimming forward, backward, upside—down, and even standing on their heads, but they have never been observed to walk on land or in the sea. The failed missing link status of the coelacanth has left Darwin's theory of natural selection through "successive, slight modifications" facing foreclosure on the fish to amphibian evolution arena.

The Horse Story

By the mid-twentieth century, the evolution of the horse had taken center stage as the leading example of evolution in biology textbooks, which was supported by Darwin. Using a backdoor approach in support of the three-toed horse as a transitional link to the modern horse, Darwin presents his argument in *The Origin of Species:*

> For instance, he [Mr. Mivart] supposes that the differences between the extinct three-toed Hipparion and the horse arose suddenly. He thinks it difficult to believe that the wing of a bird 'was developed in any other way than by a comparatively sudden modification of a marked and important kind ...' This conclusion, which implies great breaks or discontinuity in the series, appears to me improbable in the highest degree.[68]

The horse legacy as an evolutionary icon can be traced back to 1841, with the discovery of the earliest so-called "horse" fossil in the clay ground surrounding London. Paleontologist Richard Owen, a colleague of Darwin, had unearthed a fossil resembling a fox, but the skull was incomplete. Owen named the fossil *Hyracotherium,* but made no speculation on any connection with the modern-day horse. Owen and Darwin heatedly sparred over the theory of evolution.

In 1874, just two years after the release of the sixth edition of *The Origin of Species,* American paleontologist Othniel Marsh, of Yale University, published a paper in *American Naturalist* describing fossils found while exploring Wyoming and Utah.

In these specimens, which had complete skeletons, Marsh noticed that some of the fossils seemed to be similar to Owen's fox-like specimen, the *Hyracotherium.* Marsh named one of these skeletons *Orohippus,* later known as *Eohippus,* or "the dawn horse."

These American skeletons were of different sizes, with different numbers of toes, and different degrees of skeletal variations in the forearms and legs. Three years before Darwin's death in 1882, Marsh published a drawing in the *American Journal of Science* to show how the modern one-toed horse evolved from a small four-toed ancestor. The one-toed large horse was thought to have a survival advantage, allowing the horse to gallop faster.

Eventually *Hyracotherium,* the first published name given by Owen, became the official name. *Eohippus* is now recognized as a synonym for *Hyracotherium.* After reviewing evidence, Thomas Henry Huxley, visiting Marsh at Yale University, and soon to become Darwin's bulldog of the nineteenth century, concluded that the collection "demonstrated the evolution of the horse beyond question, and for the first time indicated the direct line of descent of an existing animal."[69]

The horse sequence became an influential and powerful illustration of evolution. In 1879, Marsh and Huxley collaborated in delivering a popular public lecture series in New York. During the next 100 Years, variations of the diagram have been reprinted countless times in publications and school textbooks to support Darwin's theory.

While the concept of Marsh's sequence reached celebrity status, difficulties in each sequence quickly emerged with a number of irreconcilable inconsistencies. Since 1879, more than twenty different sequences of horse evolution have been developed to work around the obvious sequence difficulties. The only commonality between the different sequences was the starting point. Except for *Hyracotherium* as the starting point, there has never been a universal consensus on the sequence of horse evolution.

The problem begs the question—do all of the fossil features of the horse found in the fossil record follow the "successive, slight modification" scheme that Darwin proposed should be found in the fossil record? If the answer is yes, then the earliest horse, the *Hyracotherium,* should be found in the lowest geological layers. But this is not the case.

Since the beginning of explorations to find the fossil evidence that Darwin said should be found in "inconceivably great numbers," there is not a single geological site in the world that demonstrates the evolutionary succession of the horse with the "earliest" horse on the bottom and the "modern" horse above. In fact, fossils of modern horse species, *Equus nevadensis* and *Equus occidentalis,* have been discovered in the same layer as the "earliest" horse, the *Hyracotherium.* Evidence from the fossil record demonstrating that the "dawn horse" lived alongside the modern horse contradicts Darwin's theory that natural selection acts through "successive, slight modifications."

Along with the horse never appearing sequentially at any geological site, unresolved issues extend to the evolutionary sequences of bones beyond just the sequences of toes. In attempting to correlate the sequence by the number of vertebra or by the number of ribs, glaring problems emerge. If the horse is sequenced by the number of ribs, from fifteen to nineteen, any evolutionary sequence in the forearm and leg disintegrates. In the same way, if sequencing is done by the number of vertebra, from six to eight, any evolutionary sequence in the toe simply evaporates.

The idea that the "earliest" horse was small and progressively became the "modern" large horse by "successive, slight variations" through natural selection is an obvious problem. Not only have the small horses not become extinct, both small and large horses exist today as they did eons ago. Veterinarians are quick to point out that three-toed horses continue to be born today.

Cutting to the chase, George Gaylord Simpson, in 1953, wasted no time in summing the evolution of the horse, writing the "uniform, continuous transformation of *Hyracotherium* into *Equus*, so dear to the hearts of generations of textbook writers, never happened in nature."[70]

Since that time, scientists have weighed in on the horse evolution as originally suggested by Darwin and popularized by Huxley and Marsh. In 1954, biologist Heribert Nilsson pointed out that the horse evolution sequence "cannot be a continuous transformation series" since the "family tree of the horse is beautiful and continuous only in the textbooks."[71]

In 1954, Normal D. Newell of Columbia University concluded that the fossil evidence for the horse sequence in evolution is filled with discontinuities and gaps. In *The Nature of the Fossil Record,* Newell continues to explain, "Experience shows that the gaps which separate the highest categories may never be bridged in the fossil record. Many of the discontinuities tend to be more and more emphasized with increased collecting."[72]

Since the 1950s, neo-Darwinian paleontologists have been raising more questions than answers. There is no consensus that the *Hyracotherium* was ever the original horse. In 1960, Gerald A. Kerkut notes in *Implications of Evolution,* "In the first place, it is not clear that *Hyracotherium* was the ancestral horse."[73]

In 1979, David Raup, at the Field Museum of Natural History in Chicago, acknowledged that any evolutionary sequence is light-years more complex than originally thought:

> Well, we are now about 120 years after Darwin, and knowledge of the fossil record has been greatly expanded.... Ironically, we have even fewer examples of evolutionary transition than we had in Darwin's time. By this I mean that some of the classic cases of Darwinian change in the fossil record, such as the evolution of the horse in North America, have had to be discarded or modified as a result of more detailed information—what appeared to be a nice, simple progression when relatively few data were available now appears to be much more complex and much less gradualistic. [74]

In 1980, a four-day symposium held at the Field Museum of Natural History in Chicago with 150 evolutionists in attendance convened to discuss the problems with the evolutionary theory. Atop the discussion list was the evolution of the horse. In addressing the meeting, evolutionist Boyce Rensberger highlighted the contradiction between the theory and reality of the horse evolution. In a November 1980 *Houston Chronicle* article, Boyce Rensberger concluded: "Throughout the history of horses, the species are well-marked and static over millions of years."[75]

In the same year, Robert Barnes published an article in *Paleobiology*, stating that the "fossil record tells us almost nothing about evolutionary origin of phyla and classes. Intermediate forms are non-existent, undiscovered, or not recognized."[76]

Using the horse as an example of this little-acknowledged truth in his book *The Great Evolution Mystery* (1984), evolutionist science writer Gordon R. Taylor explains:

> But perhaps the most serious weakness of Darwinism is the failure of paleontologists to find convincing phylogenies or sequences of organisms demonstrating major evolutionary change... The horse is often cited as the only fully worked-out example. But the fact is that the line from Eohippus to Equus is very erratic. It is alleged to show a continual increase in size, but the truth is that some variants were smaller than Eohippus, not larger. Specimens from different sources can be brought together in a convincing-looking sequence, but there is no evidence that they were actually ranged in this order in time.[77]

very Hardcastle

On the topic of the gradual horse sequence, in 1996 Gould used strong words in his book *Full House: The Spread of Excellence From Plato To Darwin* to conclude that the "popularly told example of horse evolution, suggesting a gradual sequence of changes … has long been known to be wrong."[78] Rather than "slight, successive changes" with transitional links as envisioned by Darwin, Gould points out that "fossils of each intermediate species appear fully distinct, persist unchanged, and then become extinct. Transitional forms are unknown."[78] In *The Origin of Species*, Darwin clearly stated that given the discovery of this type of evidence, his theory was invalid.

Bemoaning the continued use of what he termed "misinformation," such as horse evolution, Gould, in 1999, pined: "Once ensconced in textbooks, misinformation becomes cocooned and effectively permanent, because … textbooks copy from previous texts."[79]

Finally, in 1879, after several decades of knowing that the facts do not support horse evolution, textbooks have begun abandoning the horse sequence legacy originally developed by Marsh to supporting Darwin's theory.

wrong date?

Man

One of the most contentious and controversial aspects of Darwin's theory is the origin of man. Dancing around the topic, Darwin sidestepped the issue by providing only one brief statement in *The Origin of Species,* "much light will be thrown on the origin of man and his history."[80]

Then in 1871, twelve years following the first edition of *The Origin of Species,* Darwin published his theory on the origin of man in *The Descent of Man and Selection in Relation to Sex,* clarifying the lingering ambiguity: "Man is constructed on the same general type or model as other mammals."[81]

Even though there was only scant fossil evidence for "successive, slight modifications" at the time, Darwin stated, "Man bears in his body structure clear traces of descent from some lower form."[82] Not only is man's "structure" descended from an animal, but also Darwin envisioned that "there is no fundamental difference between man and higher animals."[83]

Darwin drives the point even further, claiming that man is no different from an animal: "the difference in mind between man and higher animals, great as it is, certainly is one of degree and not of kind."[84] Animal and man are essentially the same, differing only by a matter of "different degrees." [85]

Genesis days "Kinds"

What Darwin really needed was fossil evidence. The search was on to find the fossils to support the animal-to-man theory. Darwin's timing could hardly have been better. In 1829, the first Neanderthal skull had been discovered in Belgium.

Neanderthal

The stage was set with evolution "in the air." In August 1856, just three years before the publication of *The Origin of Species*, the "original" Neanderthal man was discovered in a German limestone quarry in the Neander Valley.

At the time, Darwin did not consider the Neanderthal man as the missing link to man and never mentions the Neanderthal man in *The Origin of Species*. In fact, Darwin only mentions "Neanderthal" once in the *Descent of Man*. Darwin even argues against the "Neanderthal" status as the missing link to man based on comparable skull sizes.[86]

Overlooking the Neanderthal man at the time, the emerging evolutionists envisioned apes as the missing link proxy to man. In the 1863 book *Evidence for Man's Place in Nature*, Thomas Huxley, comparing the skeletons of apes to that of man, suggests "that man might have originated ... by gradual modifications of a man-like ape."[87] Continuing the same line of logic, Huxley concludes: "Man is, in substance and in structure, one with the brutes."[88]

Even in Darwin's circles, the idea that the ape was the missing link to man drew close scrutiny. On closer examination, the gaps between ape and man grew to gargantuan proportions. The fact that the "missing links" from animal to man were missing continued to be a glaring problem even during Darwin's lifetime. The second Neanderthal skull was not found until more than 100 years later, in 1948, in Gibraltar.

Java Man

Things seemed to take a positive turn when Dutch anatomist Eugene Dubois discovered manlike fossil bones on the island of Java, Indonesia, in 1891. The claim was based on an unearthed skullcap, a femur, and a few teeth. No complete skeleton was found. The evidence was hotly disputed, but Dubois continued to contend that the Java man was an intermediate species in between humans and apes.

Today, Java man, although once given the name *Pithecanthropus erectus* by Dubois, is known as *Homo erectus*—a distinct species with no direct link to man. It was not until 1912 that amateur paleontologist Charles Dawson announced that the missing link to man was finally discovered in a gravel pit at Piltdown, England.

Piltdown Man

In the autumn of 1911, Charles Dawson unearthed fragments of a human skull and fragments of a lower jaw with two teeth. In February 1912, Dawson wrote to Arthur Smith Woodward, vertebrate paleontologist and keeper of geology at the British Museum of Natural History, about the find. Woodward reconstructed an entire skull from fragments and reported the missing link discovery to the Geological Society in December 1912.

While the Piltdown man was challenged, subsequent discoveries at the same site seem to confirm Smith Woodward's conclusion that "Dawson's Dawn Man" was indeed the missing link between ape and man, just as Darwin suggested. The Piltdown man was given the name *Eoanthropus dawsoni.* The evidence seemed to support the "fact" that man had evolved from apes. Little did museum gazers expect that the Piltdown was a prescription for a meltdown.

On November 21, 1912, *The Guardian* newspaper announces the discovery: "One of the most important prehistoric finds of our time has been made in Sussex."[89]

In 1913, the fossils were placed on display at the British Museum of Natural History as evidence of the evolution of man. In all the excitement, knowing that the Piltdown specimen fit the prediction so well, nobody checked to determine whether the skull and jaw fragments belonged to the same individual.

Excavations continued, and the remains of a second Piltdown man finally were found in 1915; however, they only consisted of parts of the brain case, a molar tooth, and a lower Pleistocene rhinoceros tooth. From 1915 to 1944, no other evidence was unearthed.

During the summer of 1938 at Barkham Manor, Piltdown, Sir Arthur Keith, leading anthropologist on human evolution, unveiled a memorial to mark the site where Piltdown man was discovered by Dawson. Sir Arthur finished his speech stating:

So long as man is interested in his long history, in the
vicissitudes which our early forerunners passed through,
and the varying fare which overtook them, the name of
Charles Dawson is certain of remembrance. We do well
to link his name to this picturesque corner of Sussex—
the scene of his discovery. I have now the honor of
unveiling this monolith dedicated to his memory.[90]

Using a new dating method, Kenneth Oakley, geologist and paleontologist at the British Museum, tested the dates of the Piltdown man
in 1949. Using a fluorine-ageing test, Oakley concluded that the jaw
and the teeth could not be more than 15,000 years old.

At an assembly of paleontologists in London during the summer
of 1953, smelling a rat, Joseph Weiner, a South African anatomist, and
Professor Le Gros Clark, of Oxford University, requested permission
from the British Museum to carry out further testing.

After gaining official approval, the men demonstrated through an
array of tests that the Piltdown man was actually a composite of an
orangutan jaw and the skull of a man. The array of tests included testing
for the presence of fluoride, iron, nitrogen, collagen, organic carbon,
organic water, radioactivity, and crystal structure.

What had obviously not been known was that the orangutan jaw
had been chemically treated to make it look like a fossil, and its teeth
had been deliberately filed down to make them look human. Weiner
and his colleagues concluded that after forty years, Dawson's Piltdown
man was a forgery, the work of fraudulent paleontologists.

Finally, after decades of display in the museum, a November 1953
officially illustrated British Museum bulletin announced that the
Piltdown man exhibition was the result of an elaborate hoax. The
exhibition was quickly adapted to explain how the scientific world
had been hoaxed by Piltdown man. In 1997, reflecting on the hoax,
paleontologist Roger Lewin lamented:

> Given all the many anatomical incongruities in the
> Piltdown remains, which of course are glaringly obvious
> from the vantage of the present, it is truly astonishing
> that the forgery was so eagerly embraced.[91]

Paleontologists clearly lost sight of the evidence in order to interpret the evidence to fit the theory, an approach incompatible with the scientific method. In 1997, Lewin pointed out that the lesson learned was "how those who believed in the fossil saw in it what they wanted to see."[92]

According to historian of biology Jane Maienschein in 1997, the Piltdown meltdown demonstrates "how easily susceptible researchers can be manipulated into believing that they have actually found just what it is they had been looking for."[93]

In 1982, American evolutionary paleontologist Niles Eldredge, along with Ian Tattersall, pined that fossil discoveries have been failures in supporting "the story of human evolution." Eldredge and Tattersall continued by noting, "One could confidently expect that as more hominid fossils were found, the story of human evolution would become clearer. Whereas if anything, the opposite has occurred."[94]

Henry Gee, chief science writer for *Nature*, regretting the nonscientific approach that emerged in the scientific community to support evolution, concluded: "To take a line of fossils and claim that they represent a lineage is not a scientific hypothesis that can be tested, but an assertion that carries the same validity as a bedtime story—amusing, perhaps even instructive, but not scientific."[95]

At the meeting of the British Association of the Advancement of Science in the early 1980s, Oxford historian John Durant asked for the removal of mythology from the realm of science:

> Could it be that, like 'primitive' myths, the theories of human evolution reinforce the value-systems of their creators by reflecting historically their image of themselves and of the society in which they live? ... As things stand at the present time, we are in urgent need of the de-mythologization of science.[96]

In 1996, Berkley evolutionary biologist F. Clark Howell, bemoaning Darwin's ape-to-man dead end, wrote that there "is no encompassing theory of [human] evolution.... Alas, there never really has been."[97]

In 1997, echoing the disparate state of man's evolution, Arizona State University anthropologist Geoffrey Clark, with more than a century's worth of evidence, was left with only a question: "Scientists have been trying to arrive at a consensus about modern human origins for more than a century. Why haven't they been successful?"[98]

Making a 180-degree about-face from the Piltdown meltdown, the 1980 British Museum of Natural History publication entitled *Man's Place in Evolution,* perhaps patching errors from the past finally conceded, "we assume that none of the fossil species we are considering is the ancestor of another."[99]

Peking Man

While the Piltdown man was on display at the British Museum, a new discovery, the "Peking Man," was found during excavations in China that had started in the early 1920s. The first evidence for the Peking man was two human-similar molars. Later, a lower jaw and several teeth and skull fragments were discovered in November 1928.

The original study on the Peking man fossils was performed by anatomist Davidson Black, who thought the specimens belonged to a new human species and gave them the name *Sinanthropus pekinensis.* Black published the findings in the journal *Nature.*

The discovery garnered international attention and the support of the Rockefeller Foundation for continued exploration. Over the next several years, more than forty fossil specimens, including six nearly complete skullcaps, were uncovered. While being shipped to the United States in 1941 for safety during World War II, the original fossils disappeared. Today, only the casts and descriptions remain.

While originally thought to be a missing link, the Peking man, after critical analysis, like the Java Man, is now classified as *Homo erectus*—a distinct species, with no known link to man.

Drawing conclusions on the origins of man with limited evidence can be a tricky business. In 2001, Henry Gee, senior science writer of the leading British journal *Nature,* concedes that "hominid evolution—[is] as mysterious as ever," and cautioned against the pervasive use of scant evidence in drawing conclusions.[100]

Nebraska Man

The promotion of the Nebraska man was a different story, with a different spin. The Nebraska man, originally described by Henry Osborn in 1922, had been validated based on a single tooth found in Nebraska by rancher and geologist Harold Cook in 1917. The popular press named the new find *Hesperopithecus haroldcookii*.

By means of extrapolation, British anatomist Grafton Elliot Smith assigned the tooth specimen of the *Hesperopithecus* as the third known genus of extinct hominids, along with *Eoanthropus* and *Pithecanthropus*.[101]

From a single tooth, the *Illustrated News of London* published an artist's rendering how the man would look based on the tooth. But even Osborn was appalled, calling the illustration "a figment of the imagination of no scientific value, and undoubtedly inaccurate."[102]

The fieldwork continued at the site, as the tooth became recognized as evidence of the "Ape of the Western World." By 1925, it was known that the tooth belonged neither to man nor ape, but to an extinct pig-like species. In 1927, the journal *Science* retracted their identification of the fossil as that from an ape.[103]

The retraction made front-page news in *The New York Times* in 1928, with the title "Nebraska ape tooth proved a wild pig's," and was reported on page sixteen in *The Times of London,* with the more abstract title, "Hesperopithecus dethroned." [104, 105] Editorial writers for both papers jumped at the chance to extract a lesson from the affair. The *New York Times* pined:

> Professor Henry Fairfield Osborn and his colleagues can snatch consolation from the extinct jaws of the toothsome wild peccary. For science, as this incident shows, demands proof from even its most exalted. Nothing ever went through so many tests as this peccary molar from Nebraska. It survived them all, but then science went digging in the ancient river-bed again … after which the whole business was "on the hog."[106]

Following in the long line of fossil fiascos, the Nebraska man has embarrassingly faded into extinction.

Lucy

The concept of man evolving from some ape or chimpanzee ancestor is central to the evolutionary theory of man. If missing links to man cannot be found, Darwin's theory is posed with a distressing conundrum. Later in the twentieth century, the discovery of "Lucy," *Australopithecus afarensis,* was thought to be the perfect link from ape or chimpanzee to man.

The first *A. afarensis* skeleton was discovered in Ethiopia in November 1974 by Tom Gray, who affectionately named the skeleton "Lucy." That same year, on the other side of the hill, Michael Bush unearthed more than 300 fragments of Australopithecus afarensis. The site of the findings is now known as "site 333," which corresponds to the number of fossil fragments discovered. Thirteen adults were uncovered, apparently all dying instantly, perhaps from a flash flood.

On analyzing the skeletal features, while the *Australopithecus afarensis* likely did walk upright, it did not walk like a man and was best suited for tree-climbing. *Science News,* in 1982, published an article by paleontologist Herbert Wray, who explained, "Lucy's limb proportions indicate that she had not yet developed an efficient upright gait."[107]

With doubts about the upright walking abilities of *Australopithecus afarensis,* the larger question surfaces—is "Lucy" the actual ancestor to man as originally proposed? After conducting a quantitative study of the skeletons, Charles E. Oxnard concluded in 1984 in his book *The Order of Man* that it "is now being recognized widely that the australopithecines [*Australopithecus afarensis*] are not structurally ... similar to humans."[108]

In the journal publication *Natural History,* Stephen Gould, in 1986, took the same stand against the human ancestry of *A. afarensis*: "In short, he [Oxnard] sees Australopithecines [*Australopithecus afarensis*] as uniquely different from apes and humans, not as imperfect people on the way up."[109]

Today, while thousands of fossils have been cataloged as possible missing links between ape or chimpanzee and man, what still evades paleontologists is whether any of these are the actual "missing link" ancestors to man. The gap between animal and man still exists, and the debunked evidence is even wider than in Darwin's day.

Biology textbook portrayals of a Tree of Life diagramming a consistent sequence of monkey to man is conspicuously nonexistent for one good reason—there is no evidence that man as the end product of the evolution ever happened. In the journal *Natural History*, Stephen Gould, in 1987, acknowledged that problems with the fossil evidence for human evolution overwhelm any cohesive theory since "we do not know which branch on the copious bush of apes budded off the twig that led to our lineage … no fossil evidence exists at all."[110]

Anthropologist, science writer, author of twenty books, and co-author of three books with Richard Leaky, Roger Lewin concludes in the book *Bones of Contention* that *Australopithecus* cannot be an ancestor to man.[111]

With the accumulating fossil evidence, the case for the monkey-to-man scenario gets fuzzier by the find. These problems were not unknown or unrecognized even as early as the 1970s. Paleontologist Alec John Kelso wrote is his book, *Physical Anthropology*, in 1974: "Clearly the fossil documentation of the emergence of the Old World monkeys could provide key insights into the general evolutionary picture of the primates, but, in fact, this record simply does not exist."[112]

The problem is not that there is a lack of fossils to analyze. The problem is that the evidence does not point to a monkey-to-man sequence as suggested by Darwin. In the textbook *Primate Origins and Evolution* in 1990, Robert Martin acknowledges that even with the abundance of fossils, there is no evidence of human evolution. Martin explains: "It should be noted at the outset that substantial fossil remains are known for all of the species listed below … but that there is virtually no fossil evidence relating to human evolution."[113]

Ann Gibbons published in the journal *Science* in 1996 how convoluted the evidence for human evolution has become "The story of human evolution has lately become as complicated as a Tolstoy novel."[114]

How man could be the result of evolution is more of a mystery today than at the turn of the twentieth century. Biologist Lyall Watson published a paper in *Science Digest* in 1982 entitled "The Water People," stating: "Modern apes, for instance, sprang out of nowhere. They have no yesterday, no fossil record. And the true origin of modern humans—of upright, naked, toolmaking, big-brained beings—is, if we are to be honest with ourselves, so equally mysterious matter."[115]

Early in the twenty-first century, human evolution was thought to have developed first from *H. habilis* and then through *H. erectus,* but this sequence scenario has even now eroded into extinction. In 2007, Fred Spoor, M. G. Leakey, and colleagues published in *Nature* that the latest fossil evidence from Africa demonstrates that *H. habilis* and *H. erectus* have been distinct and unchanged, indicating no evidence of evolution. [116]

In essence, *H. erectus* could not have evolved from *H. habilis* because they lived together for "half a million years." Rather than being ancestors, *H. habilis* and *H. erectus* are distinct species. There is no evidence that there was ever any interbreeding between *H. habilis* and *H. erectus.* The long-held theory of the human evolution sequence from *H. habilis* to *H. erectus* to *H. sapiens* is now known to be incompatible with the fossil evidence.

Even the *San Francisco Chronicle* noticed the report and ran the story. Science Editor David Pearlman stated, "Scientists have long believed that two species of humanlike creatures were direct forebears of our own Homo sapiens tribe, but the discoverers of the newly described fossils suggest the two species were not directly ancestral at all."[117]

On the last frontier, what is still debated is whether the Neanderthal man is a distinct species or the same as man—*H. sapiens.* When the Neanderthals were first discovered, they were considered a separate species and named *H. neanderthalensis.* Since sustainable reproduction can occur only within the same species, is was assumed that the Neanderthals were incapable of sustained reproduction with modern humans, because by definition they were a different species.

In the 1960s, new studies on the skeletal distinctions of Neanderthals began to challenge their status as a separate species. The Neanderthals were then reclassified as a subspecies along with modern humans, *H. sapiens neanderthalensis.*

Now fossil remains of more than 490 Neanderthal individuals have been recovered, and the accumulation accounts for more fossils and fossil artifacts than of any other fossil group. The evidence demonstrates that the Neanderthals incorporated a range of practices and physical characteristics essentially indistinguishable from the modern human.

Like modern humans, the Neanderthals demonstrated incredible hunting prowess, burial practices, designated specific spatial areas in their dwellings, and used tool kits. The cultural inventory of the Neanderthals exceeds that of the extinct Tasmanian Aboriginals in Australia or the Fuegians Darwin encountered in Tierra del Fuego, South America, during the voyage of the *Beagle*, both of which are agreeably *H. sapiens*.

In November 2006, *Science Daily* published genetic scientific tests comparing the Neanderthals and modern humans. Scientists at the U.S. Department of Energy's Lawrence Berkeley National Laboratory and the Joint Genome Institute who sequenced genomic nuclear DNA (nDNA) from a fossilized Neanderthal femur have concluded, "The Neanderthal and human genomes are at least 99.5% identical."[118]

Genetic testing is an emerging, frontline technique in developing a better understanding of our origins. The question is what is the meaning of this information? In order to obtain a meaning, assumptions must be applied.

The problem is, once the sequence is known, the next question is, how can the distinctiveness of a species be measured in DNA terms? Maryellen Ruvolo of Harvard University highlighted this problem in an article published in *Science* in 1997, stating, "There isn't a yardstick for genetic difference upon which you can define a species."[119]

Currently, it is not known how to apply DNA information to define a distinct species. Since species still cannot be defined by the number of DNA differences, species distinctions are still based on sustainable mating compatibility in family groups. Therefore, gene flow in families trumps gene differences in defining a species. On the issue of gene flow, Edward Rubin, director of both the Joint Genome Institute and the Lawrence Berkeley National Laboratory Genomics Division concluded in 2006 that they were "unable to definitively conclude that interbreeding between the two species of humans did not occur."[120]

In the 2006 article entitled "Archaic Admixture in the Human Genome," in the journal *Current Opinion in Genetics and Development,* the authors maintain that the Neanderthals did interbreed since "Recent work suggests that Neanderthals and an as yet unidentified archaic African population contributed to at least 5% of the modern European and West African gene pools, respectively."[121]

Researchers at the Department of Human Genetics at Howard Hughes Medical Institute also came to the same conclusion that genetic evidence suggests interbreeding between the Neanderthals and modern humans: "This finding supports the possibility of admixture between modern humans and archaic Homo populations (Neanderthals being one possibility)."[122]

The lack of species distinction has even reached the popular media. National Geographic published an article in *National Geographic News* on August 2, 2007, acknowledging that the Neanderthals indeed did interbreed with modern humans based on the distinctive features of the skull between man and the Neanderthals.[123]

Despite the known issues with DNA evidence, the cumulative evidence points toward the Neanderthal and modern humans existing as a single species. In essence, the Neanderthal man is no different from modern man—they are both human. Now, after 150 years, evidence of natural selection acting through "successive, slight variations" in the evolution of man appears more distant by the day.

The *Archaeoraptor* Disaster

Every fossil discovery has a unique story, and the story of the *Archaeoraptor* is no exception. In November 1999, an article in *National Geographic* titled "Feathers for *T. Rex*?" played out to be one of the worst debacles in the now storied history of the new fossil discoveries. The article claimed to provide "a true missing link in the complex chain that connects dinosaurs to birds."[124]

Discovered at Xiasanjiazi in China's northeastern Liaoning Province, the fossil named *Archaeoraptor liaoningensis* appeared to have the body of a bird with the teeth and tail of a small, terrestrial dinosaur.

The "discovery" seemed to fit the missing link criteria by filling in the gap of the popular reptile/dinosaur-to-bird scheme. The *Archaeoraptor* was displayed to have a long, bony tail like that of dinosaurs along with the specialized shoulders and chest of birds.

The *Associated Press* was the first to notice the story, and soon the major news networks were reporting the discovery of the new missing link that looked like a "fierce turkey-sized animal with sharp claws and teeth."[125]

The celebration was on. Philip Currie of the Royal Tyrrell Museum in Alberta, Canada, weighed in, proclaiming the *Archaeoraptor* to be the first dinosaur capable of flying. The story had barely broken before questions about the fossil started flying, leaving the *National Geographic* suddenly embroiled in one of the hottest scientific controversies in decades.

The questioning was started by Storrs Olson, the eminent curator of birds at the prestigious Smithsonian Institution National Museum of Natural History. In a letter to the National Geographic Society, Olson stated that the story reached "an all-time low for engaging in sensationalistic, unsubstantiated, tabloid journalism."[126]

Olson was on target, and the *National Geographic* found itself in the embarrassing position of having to retract the entire article because, as it turned out, the *Archaeoraptor* fossil was a fake—a neatly contrived composite of a bird and a dinosaur tail.

In reflecting on the incident, Olson laid blame for the fossil fiasco clearly on "zealous scientists" that have abandoned the scientific method to become "proselytizers of the faith" promoting "scientific hoaxes," and "the paleontological equivalent of cold fusion."[127]

Several months later in the March 2000 issue of *National Geographic*, the magazine published a letter to the editor from Xu Xing, one of the scientists who had first examined and discussed the fossil discovery. The letter stated, "After observing a new, feathered dromaeosaur specimen … [t]hough I do not want to believe it, *Archaeoraptor* appears to be composed of a dromaeosaur tail and a bird body."[128]

Seven months later in October 2000, *National Geographic* published a five-page article by veteran investigative reporter Lewis Simons describing how the hoax evolved. In the article "*Archaeoraptor Fossil Trail*," Simons pined on the painful discovery: "An investigative reporter does some digging to unearth the truth behind a case of fossil fraud."[129]

Simons explained how farmers in China had developed a profitable hobby of selling the fossils they "discovered." They doctored the fossils to follow basic market economics to increase the value of their "discoveries," but not through any scientific method or philosophy.

The *Archaeoraptor* illustrates the problem when the theory dominates a scientific investigation. Darwin touted this same approach in a letter to J. Scott in 1863: "I would suggest to you the advantage … let the theory guide your observations."[130] Darwinism has continued as recommended by Darwin—the theory guides the interpretation of the facts.

Even in an era with unsurpassed technological advances, fraud in science continues to invade deep into the ranks of esteemed institutions. Storrs Olson, of the Smithsonian Institution National Museum of Natural History, in 2000 lamented that there "probably has never been a fossil with a sadder history than this one."[131] Proof of the hoax was not long in coming. Later in March 2001, *Nature* published the results of the fossil investigation. Using high-resolution X-ray computed tomography (CT), the investigators concurred that the fossil was a forgery built in three layers.[132]

Rowe concluded that *Archaeoraptor* represents two or more species and that it was assembled from at least two, and possibly five, separate specimens. If there is any light at the end of the tunnel, Rowe gave a positive spin in the *Nature* article on the *Archaeoraptor* forgery, saying that technology may prevent future forensic fraud.[133]

Fossil Fiasco   *Key Point*

What good is a scientific theory without evidence? No wonder Darwin said that *The Origin of Species* was just "one long argument." Even in 1859, the evidence did not support the theory. But Darwin was convinced that the "argument" would soon be supported with the newly emerging evidence.

What is surprising is that Darwin did not change his theory even though he knew that the fossil evidence did not support the theory. Natural selection was envisioned to act through "successive, slight modifications," and Darwin envisioned a Cambrian evolution. However, Darwin knew that the fossil record demonstrated an explosion with a massive number of complex species appearing suddenly rather than gradually. Defiance of the evidence clearly indicates that Darwin let the scientific method be trumped by a preconceived theory. Bias ruled.

Darwin refers to "Cambrian" twenty-six times in *The Origin of Species,* writing that species "suddenly appear" in the Cambrian strata, "the lowest known fossiliferous rocks."[134] In a letter to Hooker, Darwin wrote that the problem was thought to be the same with plants: "Nothing is more extraordinary in the history of the Vegetable Kingdom, as it seems to me, than the apparently very sudden or abrupt development of the higher plants."[135]

To work around the Cambrian explosion problem, Darwin suggests that fossil evidence of life from an earlier period must be found below the "lowest known fossiliferous rocks," even though Darwin knew there was no evidence.

> Thus the words, which I wrote in 1859, about the existence of living beings long before the Cambrian period … Nevertheless, the difficulty of assigning any good reason for the absence of vast piles of strata rich in fossils beneath the Cambrian system is very great.[136]

With no known evidence for life earlier than the Cambrian strata, Darwin sees this is a problem: "The sudden manner in which several groups of species first appear in our European formations, the almost entire absence, as at present known, of formations rich in fossils beneath the Cambrian strata, are all undoubtedly of the most serious nature."[137]

To salvage the theory from the fossil fiasco, Darwin proposed a number of other explanations. Clearly, in 1859, explorations of the Earth were limited, and the evidence of the fossil record was likewise limited. Therefore, Darwin reasons, "on the theory, such strata must somewhere have been deposited at these ancient and utterly unknown epochs of the world's history."[138]

Now, after 150 years, knowledge about the fossil record far exceeds what was known in 1859. According to Alfred Sherwood Romer's widely circulated textbook *Vertebrate Paleontology*, knowledge of the fossil record is nearly complete; 97.7 percent of the known living orders have been found in fossilized form.[139] Estimating the percentage of living fossils recovered in one region of North America, George Gaylord Simpson concluded in 1953 that the record was almost complete.[140]

Echoing the same conclusion, evolutionary biologists James Valentine and Douglas Erwin, in 1987, wrote that the evidence of early life on Earth is complete: "The sections of Cambrian rocks that we do have (and we have many) are essentially as complete as sections of equivalent time duration from similar depositional environments" in even more recent rocks.[141]

To address the completeness and quality of the known fossil record from the Cambrian period to the present, extensive surveys have been performed throughout the twentieth century. In 2000, British geologists M. J. Benton and colleagues concluded: "Early parts of the fossil record are clearly incomplete, but they can be regarded as adequate to illustrate the broad patterns of the history of life."[142]

But whatever flaws may be in the fossil record, Darwin's problem has not been resolved. J. W. Valentine and D. H. Erwin, in 1987, concluded that the evidence for "the explosion is real; it is too big to be masked by flaws in the fossil record."[143]

At the time, Darwin argued that this abrupt explosion of species is only a false appearance in the record, but "if true would be fatal to my views."[144] Even in 1859, Darwin knew that the theory was not aligned with the evidence.

Coming to terms with Darwin's "falsely" plague, the eminent paleontologist George Gaylord Simpson, in his 1953 book *The Major Features of Evolution,* weighed in with the sudden appearance of new species in the fossil record:

> In spite of these examples, it remains true, as every paleontologist knows, that most new species, genera, and families, and that nearly all new categories above the level of families, appear in the record suddenly and are not led up to by known, gradual, completely continuous transitional sequences.[145]

The Cambrian explosion completely undermines Darwin's theory of evolution "by means of natural selection." Discoveries in the Cambrian strata stunningly oppose any concept of "successive, slight variation." Niles Eldredge pointed out in 1981 that paleontologists can only hold on to Darwin's theory by ignoring the evidence: "paleontologists have been insisting that their record is consistent with slow, steady, gradual evolution where I think that privately, they've known for over a hundred years that such is not the case."[146]

While Darwin was wrong on the gradual appearance of species, the "earliest phase of rapid change [remains] undiscovered." The reason is that there was no rapid change in the beginning. What Darwin did have right was that species have been disappearing into extinction since the Cambrian explosion. Normal D. Newell of Columbia University writes, "All through the fossil record, groups—both large and small—abruptly appear and disappear.... The earliest phase of rapid change usually is undiscovered."[147]

Oxford University zoologist Mark Ridley, an expert on the evolution of reproductive behavior, reflected on the lack of evolutionary lineages in the fossil record and the fact that Darwin "could not cite a single example" in *The Origin of Species*.[148]

In 1983, Douglas Futuyma, world-renowned American biologist and member of the National Academy of Sciences, conceded that the "potentially embarrassing features of the fossil record" is that there is "virtually no evidence of transition(s)" in the fossil record since the "majority of major groups appear suddenly in the rocks."[149]

What Douglas Futuyma might have thought to have been "potentially embarrassing" in 1983, was surefire embarrassment by 1991. In light of the glaring contradictions with Darwin's theory of "successive, slight variations," paleontologists have largely abandoned searching the fossil record for the missing transitional links. Ernst Mayr, Darwin's bulldog of the twentieth century, resigned to the obvious contradiction:

> Paleontologists had long been aware of a seeming contradiction between Darwin's postulate of gradualism ... and the actual findings of paleontology. Following phyletic lines through time seemed to reveal only minimal gradual changes but no clear evidence for any change of a species into a different genus or for the gradual origin of an evolutionary novelty. Anything truly novel always seemed to appear quite abruptly in the fossil record.[150]

Ernst Mayr

Stefan Bengtson, winner of the 1995 Charles Doolittle Walcott Medal award from the National Academy of Sciences for research in

Precambrian and Cambrian life, minced no words in disbursing a searing dark tribute to a litany of "creation of myths" orchestrated by paleontologists out of the "debris of death."[151]

Paleontologist Stephen Gould pointed out that since the 1950s, even after decades of research, "The problem of the Cambrian explosion has not receded, since more extensive labor has still failed to identify any creatures that might serve as a plausible immediate ancestor for the Cambrian faunas."[152]

Weighing in on the Cambrian explosion in 2006, Italian geneticist Giuseppe Sermonti cuts to the chase:

> The explosion of types is not simply an insignificant, unresolved charade in the Sphinx's book of riddles. It is exactly the opposite of what Darwin's gradualist mechanism predicted regarding the origin of animal forms. Yet evolutionists do not seem to have been unduly troubled by this. Their theory of adaptation to anything.[153]

Now and Ever

Not only was Darwin wrong about the gradual appearance of species during early life on Earth, Darwin was wrong about natural selection resulting in gradual evolutionary changes. Fossil evidence reveals that species have not changed but have remained stable. Darwin had envisioned "that natural selection is daily and hourly scrutinizing, throughout the world."[154]

Darwin envisioned that natural selection "leads to divergence of character and to much extinction of the less improved and intermediate forms of life" gradually through "successive, slight variations."[155] Despite the popularity of Darwin's theory, the fossil evidence does not demonstrate "divergence"—the essence of evolution. Not only did species abruptly appear in the early history of the Earth, like an explosion, but also species have remained virtually unchanged. Playing on words, Stephen Gould writes, "This is truly the 'age of bacteria' —as it was in the beginning, is now and ever shall be."[156]

While the fossil record demonstrates variation within species, there is no fossil record evidence to support the origin of species or any species-to-species evolutionary sequence. Gareth V. Nelson of the American Museum of Natural History, in 1971, wrote in the *Annals of the New York Academy of Sciences* that there is no evidence for species-to-species evolution in the fossil record, and it "is a mistake to believe that even one fossil species or fossil 'groups' can be demonstrated to have been ancestral to another."[157]

Jeffrey S. Levinton, chairman of the Department of Ecology and Evolution at the State University of New York at Stony Brook, wrote in *Scientific American,* "Just as automobiles are fundamentally modeled after the first four-wheel vehicles, all the evolutionary changes since the Cambrian period have been mere variations on those basic themes."[158]

The origin of species remains a mystery, and evolution now simply means variation within a species. In 1980, in his book *The Panda's Thumb,* Stephen Gould clearly pointed out how the evidence contradicts Darwin's concepts of gradualism.[159]

As the fossil evidence continues to demonstrate stasis rather than evolution, Niles Eldredge conceded in 1985, "Stasis was conveniently dropped as a feature of life's history to be reckoned with in evolutionary biology."[160]

Biology historian, Peter J. Bowler, a Fellow of the American Association for the Advancement of Science, concluded in 1984 that there is "no sign of an evolutionary trend" from the evidence discovered in the fossil record.[161]

While the gradual change that Darwin had envisioned was certainly reasonable from a philosophical perspective in 1859, Stephen Gould points out in a paper published in *Paleobiology* in 1985 that "upon closer inspection," Darwin's vision dissolves.[162]

In the aftermath of the "argument," the fossil evidence now presents Darwin's theory with a major dilemma. It is the price to be paid for abandoning the scientific method. After 150 years, the emerging fossil evidence continues to diverge from Darwin's theory. In the same paper, Gould bemoans the fact that the fossil evidence has not demonstrated "any clear vector of accumulating progress (evolution)" and now "represents our greatest dilemma for a study of pattern in life's history."[163]

The Fossil Exodus

The halls of paleontology have no more direct fossil evidence to support Darwin's theory today than was available 150 years ago. In support of evolution, Darwin drew conclusions based on philosophy and not evidence. The emerging ranks of twentieth century paleontologists largely followed suit. Under the guise of a scientific profession, paleontologists have let the philosophy of evolution dominate science. Niles Eldredge pointed out in 1986, "It has been the paleontologist, my own breed, who has been most responsible for letting ideas dominate reality."[164]

In the 1976 Presidential Address at the *Geological Association*, D. V. Ager dismayed of the fossil record, and went on the record to say, "It must be significant that nearly all the evolutionary stories I learned as a student ... have now been 'debunked.'"[165]

Chances that further digging will discover some yet unknown golden evolution nugget seems more remote each passing day. There is not one geological column known anywhere in the world that contains a simple-to-complex geological column as theorized by Darwin. The convening of new fossil explorations is no longer the considered central method of discovering evidence for evolution.

In fact, the fossil record is no longer thought to contain evidence of evolution. As Eldredge pointed out in 1996, "No wonder paleontologists shied away from evolution for so long. It seems never to [have] happen[ed]."[166]

In solving the problems between Darwin's theory and evidence in the fossil record, some investigators have suggested minimizing the importance of the fossil record. However, geologists, for obvious reasons, have countered this approach, arguing that fossils are the "only direct evidence of the early history of life." Weighing in on the argument, Geologist William Schopf, in 1994, wrote, "There is only one source of direct evidence of the early history of life—the Precambrian fossil record; speculation made in the absence of such evidence, even by widely acclaimed evolutionists, has commonly proved groundless."[167]

Colin Patterson, senior paleontologist of the British Museum of Natural History, in his 1999 book entitled *Evolution,* takes a different approach by arguing that inference of ancestry from the fossil record is simply too tricky of a business: "Fossils may tell us many things, but one thing they can never disclose is whether they were ancestors of anything else."[168]

The chief science writer for *Nature*, Henry Gee, in his 1999 book *In Search of Deep Time,* concludes that while theoretically the fossil record should determine ancestry, the problem is that the conclusions drawn from the evidence can only be at best indirect evidence "made after the fact," which can never be a "testable hypothesis."[169]

Wow

The chasm between the fossil record and Darwin's theory continues to escalate. Douglas Futuyma, president of the Society for the Study of Evolution and the American Society of Naturalists, as well as editor of *Evolution*, abandoned Darwin's theory, stating, "The supposition that evolution proceeds very slowly and gradually, and so should leave thousands of fossil intermediates of any species in its wake, has not been part of evolutionary theory for more than thirty years."[170]

The list goes on. Paleontologist Robert Carroll even concluded, "Paleontologists, in particular have found it difficult to accept that the slow, continuous, and progressive changes postulated by Darwin can adequately explain major reorganizations that have occurred between dominate groups of plants and animals."[171]

Evolutionist Niles Eldredge acknowledges that paleontology, his own profession, has "proffered a collective tacit acceptance of the story of gradual adaptive change, a story that strengthened and became even more entrenched as the synthesis took hold. We paleontologists have said that the history of life supports that interpretation, all the while really knowing that it does not."[172]

Famed evolutionist and paleoanthropologist Ian Tattersall suggests that to continue with Darwin's theory is not from "evidence itself," but from an "unconscious mindset." In 1985, molecular biologist Michael Denton concluded in the book *Evolution, a Theory in Crisis*, "Fossils have ... failed to yield the host of transitional form demanded by evolutionary theory" and the "absence of transitional forms is dramatically obvious."[173]

Payday has come and is now gone. The relentless search for Darwin's missing links has left the theory of evolution alienated from the fossil record. Robert Wesson, in his 1991 book *Beyond Natural Selection,* concludes that the "gaps in the record are real" and the species are static.[174]

In abandoning the scientific method, like Darwin before, Stephen Gould laments that paleontologists "have paid an enormous price for Darwin's argument. We fancy ourselves as the only true students of life's history, yet to preserve our favored account of evolution by natural selection we view our data as so bad that we almost never see the very process we profess to study."[175]

What if Darwin had speculated on evolution based on the evidence? What scientific evidence would have been available during the early twentieth century? The answer is none: only a philosophy.

Darwin knew that even the existing Cambrian fossils contradicted his theory. The problem was with the evidence. However in 1859, evolution was "in the air," and the theory could not be restrained by the evidence. Darwin avoided the truth and explained away the fossil record with the argument it is "a history of the world imperfectly kept.[176]

Arguments, while logical, cannot create reality. Only reproducible evidence can demonstrate a natural law. The fact that Darwin's theory contradicts the fossil record is evidence that Darwin's theory of evolution cannot be described as a work based on the scientific method.

After 150 years of running after the fossil evidence, one wonders what Darwin's verdict would be based on the fossil evidence. Actually, we do know. Darwin said that the "inexplicable" evidence was a "valid argument against the views [evolution] entertained."[177]

The key evidence for early life on Earth must be found in the fossil record, buried in the sedimentary rocks. Yet nothing from the Burgess Shale to *Archaeoraptor* has supported Darwin's theory of evolution. Clearly, evolution, through "successive, slight modifications" of natural selection, never happened.

Moving forward with the known evidence, evolutionary paleontologists, led by Eldredge and Gould, have abandoned gradual evolution and adopted the theory that new species emerged spontaneously and immediately at various times in the history of life. The theory has been termed "punctuated equilibrium," which contradicts Darwin's theory: "If numerous species, belonging to the same genera or families, have really started into life at once, the fact would be fatal to the theory of evolution through natural selection."[178]

Use of the fossil record to support the Darwinian theory of evolution is on the edge of no return. The notion that paleontology will ever once again promote fossils as the evidence of gradual evolutionary change seems highly inconceivable.

The theory of punctuated equilibrium was originally proposed by Ernest Mayr in 1954, but was not popularized until the theory was modified by Niles Eldredge and Stephen Gould and presented at the annual meeting of the Geological Society of America in 1971.

Darwin was keenly aware of the fossil record problems. Certainly, the fossil record did not keep Darwin's theory alive. But if not the fossil record, what scientific evidence amassed immense support for Darwin's theory later in the twentieth century? The next chapters will explore why the Darwin phenomenon has continued after 150 years running.

Chapter Ten
Molecular Biology

It is no valid objection that science as yet throws no light on the far higher problem of the essence or origin of life. Who can explain what is the essence of the attraction of gravity?

—Charles Darwin[1]

Fading interest in fossil evidence, starting after the results from the HMS *Challenger* were published in 1895, forced a change in direction to find the missing links in Darwin's Tree of Life.

While the phrase did not exist in Darwin's day, "molecular biology," by the late twentieth century, was commonly understood, from the Yale science classroom to the readers of *Time* magazine. In *The Origin of Species,* Darwin uses the term "molecule" only once, in reference to the work of geologist Count Keyserling that connected the concept of evolution and the action of molecules. Darwin notes that Keyserling had suggested that the "germs of existing species may have been chemically affected by circumambient molecules of a particular nature, and thus have given rise to new forms."[2]

While Darwin knew little of molecular biology, the concept was to become one of the driving scientific forces of the twentieth century. The Tree of Life was envisioned to follow the evolution of molecules, from simple to complex, as the organisms became more complex. Darwin lent credence to the concept of molecules interacting and giving rise to new forms. Entering into this unknown realm, Darwin uses the term "atoms" three times in *The Origin of Species*. Darwin actually ridicules theories of rapid evolutionary change: "Do they really believe that at innumerable periods in the Earth's history certain elemental atoms have been commanded suddenly to flash into living tissues?"[3]

Atoms

Without question, what Darwin knew of atoms and molecules pales in comparison to what has been discovered over the past 150 years. Darwin used the term "atom" twice in a discussion of erosion: "worn away atom by atom, until after being reduced in size, they can be rolled about by the waves, and then they are more quickly ground into pebbles, sand, or mud."[4] Today, "atom" would not be used in this context. The only other use of the term "atom" was in discussing the characteristics of the honeycomb, in which Darwin uses the term to describe the source of coloration.[5]

Darwin uses the term "chemical" four times, but "chemistry" was never used in *The Origin of Species*. While not pretending to understand the molecular origins of life, Darwin speculated that the origin of life was produced by the complex action of natural laws yet to be discovered.

> It is interesting to contemplate an entangled bank, clothed with many plants of many kinds, with birds singing on the bushes, with various insects flitting about, and with worms crawling through the damp Earth, and to reflect that these elaborately constructed forms, so different from each other, and dependent on each other in so complex a manner, have all been produced by laws acting around us.[6]

Darwin, driven to find these natural laws accounting for the origin of life, eventually concluded that it was "by means of natural selection." Since the laws of gravity discovered by Isaac Newton were according to fixed laws, Darwin reasoned that "whilst this planet has gone cycling on according to the fixed law of gravity, from so simple a beginning endless forms most beautiful and most wonderful have been, and are being, evolved."[7]

In the same way that Newton discovered the laws of gravity, Darwin envisioned that the laws governing the origin of life to be simple, yet likewise inexplicable: "It is no valid objection that science as yet throws no light on the far higher problem of the essence or origin of life. Who can explain what is the essence of the attraction of gravity?"[8]

Organic Molecules

By the early eighteenth century, the essence of life was beginning to be seen as the unique action of organic molecules. Accordingly, the start of the origin of life was the assimilation of simple organic molecules that evolved to become more complex. The molecular basis for life gained the first scientific measure of momentum in 1773, when French chemist Hilaire Marian Rouelle discovered urea, the first organic molecule.

In 1828, challenging the common notion that organic molecules are substantially different from inanimate matter, German chemist Friedrich Woehler sent shock waves across the Western world by synthesizing urea in the laboratory. Simple chemicals, potassium cyanate, and ammonium sulfate, reacted to form an organic molecule. A new door was now open. The possibility that molecules could evolve into man, "produced by laws acting around us," was beginning to enter center stage.[9]

As the excitement of finding the massive fossil evidence as predicted by Darwin continued to fade, interest in the chemical evidence of evolution was gaining momentum. The emerging and unprecedented technological advances in the early twentieth century set the stage to take on the new challenge—demonstrating evolution via the evolution of complex molecules.

In 1924, Russian biochemist Alexander Ivanovich Oparin theorized that life on Earth developed through the gradual and spontaneous chemical evolution of organic molecules from inorganic molecules in some primeval Earth environment.[10] Oparin was a supporter of Darwin's theory of natural selection, and his popularity soared in Russia. He became the hero of Socialist Labour in 1969, received the Lenin Prize in 1974, received the Lomonosov Gold Medal in 1979 "for outstanding achievements in biochemistry," and was awarded five Orders of Lenin.

British scientist John Burton S. Haldane, working independently from Oparin, developed a similar scheme on how life may have originated on Earth; the two theories have become known as the Oparin–Haldane theory.[11] Both independently proposed that the early Earth atmosphere was primarily strongly reducing, with combinations of methane, ammonia, carbon dioxide, and water—an atmosphere resembling interstellar gases.

Table IV
Atmosphere in Origin of Life: Oparin and Haldane Models

	Oparin	Haldane
Primitive atmosphere	Reducing; composed of methane, ammonia, hydrogen, and water vapor	Reducing; composed of carbon dioxide, ammonia, and water vapor
Source of carbon for life	Methane	Carbon dioxide
Site of prebiotic evolution	Atmosphere, then oceans	Atmosphere, then oceans
Mechanism	Spontaneous appearance of coacervates followed by evolution to cell-like state	Synthesis of increasingly complex organic molecules in presence of ultraviolet light

In essence, life originated from inorganic molecules that developed into life-force organic molecules. In theory, chemicals spontaneously interacting to form self-replicating molecules along with increasingly complexity, gave rise to life. Evidence for the interstellar gas theory was supported by the discovery of organic material in the Orgeuil meteorite, which fell in France in 1864.

The Oparin–Haldane theory was based on the probabilities of random actions of interstellar gases giving rise to form organic molecules. Haldane's greatest contributions were in a series of ten papers on the mathematical theory of natural selection, entitled *"A Mathematical Theory of Natural and Artificial Selection."* For the first time, Haldane proposed a mathematical model of changes in gene frequencies through the interaction of natural selection with mutation. His 1928 book entitled *On Being the Right Size* estimated human mutation rates. In 1932, Haldane summarized the work in *The Causes of Evolution,* re-establishing natural selection as the premier mechanism of evolution.

Interest in the theory was even weaved into the center stage of twentieth century culture. Walt Disney's 1940 film, *Fantasia,* opens with a scene of primordial Earth seething with volcanic action to Stravinsky's

1913 classic music score *Rite of Spring*. In the scene, red-hot lava flows out over the land and into the sea, creating clouds of steam, and while lightening flashes across the sky, the camera slowly pans down to the ocean where beneath the surface mysterious lights emerge, and suddenly single-celled animals are seen moving speedily across the screen. Walt Disney's narrator declares this "a coldly accurate reproduction of what science thinks went on in the first few billion years of this planet's existence."[102]

Disney's scenario, created from the Oparin–Haldane scheme, was accepted as "what science thinks" about the first steps of the origin of life. At the time, it seemed reasonable to suppose that the original atmosphere of the Earth resembled interstellar gases, which is strongly reducing.

Life's Building Blocks

Nobel Prize–winning chemist Harold Urey restated the Oparin–Haldane scheme nearly twenty-five years later in 1952.[12] Urey was awarded the Nobel Prize in Chemistry in 1934 for his work on isotopes. During World War II, Urey directed the Manhattan Project at Columbia University that lead to the development of the atomic bomb. Urey even became a diplomat, leading the American mission to England to establish a cooperative agreement on development of the atomic bomb in the autumn of 1941.

In 1952, Urey published the book entitled *The Planets: Their Origin and Development*, speculating, like the Oparin–Haldane theory, that the early Earth's atmosphere was probably composed of ammonia, methane, and hydrogen—a reducing atmosphere. Credited with coining the term "cosmochemistry," Urey developed a theory on the evolution of stars.

The theory was published without ever being tested. When challenged by his graduate student, Stanley Miller, they performed the now-famous Miller–Urey experiment. After assembling a closed glass apparatus in Urey's laboratory, Miller pumped out the air and replaced it with methane, ammonia, hydrogen, and water, creating a reducing atmosphere—without oxygen—a gas composition resembling the atmosphere of Jupiter. "By the end of the week," Miller reported the water "was deep red and turbid."[13]

Just as Urey had predicted, chemical analysis of the tar solution revealed several organic compounds, including glycine and alanine, the two simplest amino acids found in proteins—the building blocks of life.[14] The experiment generated enormous excitement throughout the scientific community, and quickly found its way into almost every high-school and college textbook as the Miller–Urey experiment.

Confirmation came from experiments using ultraviolet radiation, which exists in outer space. The Miller–Urey experiment became an icon of evolution. By the late 1960s, after the initial excitement, geochemists began exploring for evidence that early Earth actually had a reducing atmosphere, lacking oxygen, as suggested by Oparin–Haldane and Urey. *So, some organic molecules were synthesized in a reducing (or) atmosphere.*

The Oxygen Factor

Oxygen holds the key to the Oparin–Haldane and Urey theories. A reducing atmosphere is devoid of oxygen, but today the atmosphere of the Earth consists of approximately 20 percent oxygen, yielding an oxidizing, not a reducing, atmosphere. Oxygen is essential for life, and life could never have survived in such a reducing atmosphere.

Analogous to the way that automobile engines use oxygen to produce energy from gasoline, living organisms use oxygen to produce energy. Oxygen transfers energy. Fortunately, for the Miller–Urey experiment, oxygen was absent in accordance to the reducing atmosphere theory. The presence of oxygen would have been explosive. However, the central question is—was the Earth's original atmosphere strongly reducing as proposed, lacking oxygen? From a cosmochemistry perspective, the Earth's original atmosphere had the same composition as interstellar gas clouds.

Just a year after Urey published the interstellar gas theory in *The Planets: Their Origin and Development,* University of Chicago geochemist Harrison Brown observed that the Earth's atmosphere was at least a million times lower than interstellar gases. Brown concluded that either the Earth lost its original interstellar atmosphere or it never had a reducing atmosphere.[15] Early chinks were discovered in the origin-of-life chain.

lower?

In the 1960s, Princeton University geochemist Heinrich Holland and Carnegie Institution geophysicist Philip Abelson agreed with Harrison

Brown's conclusions.[16, 17] Working independently, Holland and Abelson concluded that the Earth's primitive atmosphere was not derived from reducing interstellar gases, but from oxidizing gases released by the Earth's own volcanoes. From the available evidence, the Earth's ancient and modern atmospheres are the same—oxidizing, not reducing.

In the reducing atmosphere, large amounts of methane would be expected in Earth's earliest rocks. However, the evidence does not support the view. As Philip H. Abelson, editor of *Science* magazine and author of articles published by the American Association for the Advancement of Science, explained in 1966: "What is the evidence for a primitive methane-ammonia atmosphere on Earth? The answer is that there is no evidence for it, but much against it."[18]

The controversy surrounding a reducing versus an oxidizing atmosphere raged from 1960 to 1980 as geologists began to examine sediments for the presence of oxygen early in Earth's history. Sediments rich in the highly oxidized red form of iron—iron oxide—reflect the presence of oxygen early in Earth's history. In 1979, Canadian geologists Erich Dimroth and Michael Kimberly concluded: "In general, we find no evidence in the sedimentary distributions of carbon, sulfur, uranium, or iron, that an oxygen-free atmosphere has existed at any time during the span of geological history recorded in well-preserved sedimentary rocks."[19]

In 1982, British geologists Harry Clemmey and Nick Badham wrote that the evidence showed "from the time of the earliest dated rocks at 3.7 billion years, Earth had an oxygenic atmosphere."[20]

In 1975, Belgium biochemist Marcel Florkin denounced the reducing atmosphere theory, stating, "the concept of a reducing primitive atmosphere has been abandoned," and the Miller–Urey experiment is "not now considered geologically adequate."[21]

Along with geological evidence, even biological evidence supports the existence of oxygen in Earth's early history. In 1975, British biologists J. Lumsden and D. O. Hall reported that the enzyme superoxidase dismutase, used by living cells to protect themselves from the damaging effects of oxygen, was present even before the advent of photosynthesis.[22]

By 1977, onetime reducing advocates Sidney Fox and Klaus Dose conceded that a reducing atmosphere did "not seem to be geologically realistic because the evidence indicates that … most of the free hydrogen probably had disappeared into outer space and what was left of methane and ammonia was oxidized."[23] According to Fox and Dose, not only did the Miller–Urey experiment start with the wrong gas mixture, but the theory was not geologically relevant: "The inference that Miller's synthesis does not have a geological relevance has become increasingly widespread."[24]

In 1983, Miller and colleagues, responding to criticism, were able to produce a small amount of the simplest amino acid glycine by sparking an atmosphere containing carbon monoxide and carbon dioxide instead of methane. But glycine was the only amino acid produced.[25] of N of O_2

John Horgan wrote in *Scientific American* in 1991 that an atmosphere of carbon dioxide, nitrogen, and water vapor "would not have been conducive to the synthesis of amino acids," as conceived by Oparin–Haldane and Miller–Urey.[26]

Since the late 1970s, the irrelevance of the Miller–Urey experiment has garnered near-consensus among geochemists. As Jon Cohen wrote in *Science* in 1995, many origin-of-life researchers now dismiss the 1953 experiment because "the early atmosphere looked nothing like the Miller–Urey simulation."[27]

The Orgeuil meteorite, which fell in France in 1864, supported the early Earth interstellar gas theory, including a reducing atmosphere. However, it was not until nearly 100 years later after the "discovery" of the meteorite, in 1963, that X-ray technology proved that the Orgeuil meteorite was a human invention—a fraud.[28] Interestingly, in 1864 the great debate in France was centered on the possibility of spontaneous generation of life from inorganic molecules. Then later that year, Louis Pasteur delivered his famous lecture at the Sorbonne, debunking the theory of spontaneous generation of life.

Evolutionary biochemist Franklin Harold, professor emeritus at Colorado State University, admitted as much in his 2001 book *The Way of the Cell*, when he wrote that the origin of life, despite the advances in biochemistry, continues to stand as a profound mystery, concluding that of "all the unsolved mysteries remaining in science, the most consequential may be the origin of life."[29] The 1998 issue of *National Geographic* highlights the issues surrounding the Miller–Urey experiment. Science writer Richard Monastersky distances scientists from the experiment by writing: "Many scientists now suspect that the early atmosphere was very different from what Miller first supposed."[30]

How an organic molecule, the essential component of life, spontaneously evolved from an inorganic molecule has eluded scientists for 150 years. Because of the absence of scientific evidence for the evolution of chemicals to life, in 2006 Harvard University commissioned a multimillion-dollar project entitled "The Origins of Life in the Universe Initiative." The purpose of the initiative is to resolve the origin of life mystery by taking an expansive interdisciplinary approach to answering the ultimate question—"how biological evolution can emerge from chemistry."[31]

Table V
Atmosphere Conditions: Oxidizing versus Reducing

Oxidizing Present Earth atmosphere	Reducing Oparin-Haldane
Nitrogen	Methane (carbon + hydrogen)
Oxygen	Ammonia
Carbon dioxide (carbon + oxygen)	Hydrogen
Water vapor (hydrogen + oxygen)	Water vapor (oxygen + hydrogen)

Molecular Tree of Life

While the battle raged over the composition of the Earth's early atmosphere, researchers were racing to trace evolution's pathway via molecular evolution. According to the Tree of Life theory, molecules became more complex as species became more complex. This molecular approach was becoming increasingly important since the fossil record was floundering to support evolution.

Urea was discovered in 1777 and was first synthesized in 1828, but The biological role of the protein remained unknown for 100 years. In 1932, German-born chemist Hans Krebs discovered the urea cycle at the University of Freiburg in Germany. Also known as the Krebs cycle, or the ornithine cycle, it was the first metabolic cycle discovered.

The discovery of a mechanism for a biological process demonstrated that science was getting closer to discovering Darwin's natural laws, now through molecular biology. In 1953, Krebs was awarded half of the Nobel Prize in Physiology for his "discovery of the citric acid cycle."

The word "protein" comes from a Greek word meaning "of primary importance." By the 1930s, research was under way to uncover the sequences of amino acids in proteins, the means of primary importance.

The linear sequence of amino acids in structuring a protein can be considered analogous to letters structuring a sentence. Early laboratory results demonstrated that proteins varied from species to species. In the early 1950s, investigators demonstrated that the insulin molecule was different among the cow, the pig, and the sheep and had distinct chemical compounds.[32] Molecular evidence was pointing to life evolving by "successive, slight" changes, molecule by molecule.

Hemoglobin

One of the first protein molecules examined to develop a molecular Tree of Life was the hemoglobin molecule. Hemoglobin is found in all animal kingdoms, from bacteria to man. One of the initial papers about the evolution of the molecular biology of hemoglobin was written by Rossi-Fanelli in 1955 and was published in *Nature*.[33]

Hemoglobin is an iron-containing protein responsible for oxygen transport in the red blood cells of the blood in both vertebrates and invertebrates. In theory, the closer the evolutionary ancestry between species, the more similar the proteins should be. For example, the hemoglobin of man would be more similar to that of a dog than of a fish, since the evolution of dog and man occurred closer together than that of fish and man.

Early molecular evidence was compatible with the theory of evolution. Initial laboratory analysis demonstrated that differences in the hemoglobin molecule between man and dog only varied less than 20 percent, compared to more than 50 percent between man and fish. Since the dog is closer to man than to the fish, in evolutionary terms, the implications were that molecular biology would be a powerful tool in tracing the pathways of evolution.[34]

The finding intensified the international research on developing the evolutionary Tree of Life using hemoglobin. On further investigation, however, scientists discovered that the tree branches are actually not connected. Finally, molecular biologist Richard Dickerson and colleagues concluded (1969) that the successive, slight changes in the presumed hemoglobin molecule Tree of Life simply do not exist. Referring to the evolution of hemoglobin, Dickerson admits that it "is hard to see a common line of descent snaking in so unsystematic a way through so many different phyla."[35]

Research into the evolution of hemoglobin continues to be active, however. A search of the National Center for Biotechnology Information Web database, using the search terms "hemoglobin AND evolution," retrieves more than two thousand articles.[36] While new possible mechanisms or a new investigative approach or likely scenario have been suggested, the goal of constructing a Tree of Life from hemoglobin, one of life's most important molecules, remains a mirage. In 2007, Ron Milo of Harvard University, concedes, "The contrasting mode of variation suggests that physiological changes in hemoglobin are connected to evolutionary changes in some nonrandom way that begs an explanation."[37]

The technological advances in every aspect of molecular biology during the later twentieth century began to shift interest from the composition of the protein to the sequence of amino acids.

Cytochrome C

The first complete amino acid sequence of a protein, insulin, was determined by British biochemist Frederic Sanger in 1955 at the University of Cambridge; Sanger won the Nobel Prize in Chemistry in 1958 for his pioneering work.

Determining the amino acid sequence of proteins from different species became the cornerstone for investigating the molecular basis for evolution. Emile Zuckerkandl and Nobel Prize winner Linus Pauling, in 1962 and 1965, respectively, further popularized the concept that evolution can be explored through the successive, slight changes in molecular biology.

Of the protein molecules studied at the time, cytochrome C, unlike insulin and hemoglobin, became a prime candidate for investigation due to its greater ubiquitous distribution throughout the animal kingdom. Cytochrome C is found from the least to the most complex of species. In cells, cytochrome C functions to facilitate the transfer of energy. Cytochrome C quickly emerged as the favorite molecule to trace the steps of evolution for the purposes of discovering the Tree of Life.

As an enzymatic protein, cytochrome C is uniquely composed of about the same number of about 100 amino acids throughout the animal kingdom with 19 amino acids occurring in exactly the same sequence. Differences in the identity and positions of the remaining amino acids became the center of interest in tracing the "successive, slight modifications."

Initial evidence on cytochrome C lent support to Darwin's theory; the percentage changes in cytochrome C first appeared to occur in concert with changes in hemoglobin. The hemoglobin variation between man and dog is 20 percent, and cytochrome C variation is 5 percent. The hemoglobin variation between man and carp is 50 percent, and cytochrome C variation is 13 percent.[38] These microanalysis parallels pointed to a connection between comparative gross anatomy and molecular biology.

Armed with the new field of sequencing, scientists began to accumulate a library of protein sequences. As the evidence began to accumulate, however, on closer examination there were no "successive, slight modifications" in protein sequences connecting species as anticipated. Like hemoglobin, the Tree of Life branches were not connected in any evolutionary sequence. Species on the molecular level are unique. In fact, cytochrome C was similar, even between distant species. Biochemist R. E. Dickerson explains that the "more one approaches the molecular level in the study of living things, the more similar they appear, and the less important the differences between, for instance, a clam and horse become."[39]

Reflecting on the emerging evidence in 1985, Australian molecular biologist Michael Denton connected with the growing disappointment in Darwin's theory:

As more protein sequences began to accumulate during the 1960s, it became increasingly apparent that the molecules were not going to provide any evidence of sequential arrangement in nature, but were rather going to reaffirm the traditional view that the system of nature conforms fundamentally to a highly ordered hierarchic scheme from which direct evidence for evolution is emphatically absent.[40]

In 1972, to demonstrate the percentage difference in protein sequences, Margaret Dayhoff, working as a pioneer in the emerging field of informatics along side Carl Sagan, published protein sequences in a matrix, which is now known as the Dayhoff Atlas of Protein Sequences and Structure. Commenting on the matrix, molecular biologist Michael Denton said, "The most striking feature of the matrix is that each identifiable subclass of sequences is isolated and distinct ... All the sequences of each subclass are equally isolated from the members of another group. Transitional or intermediate classes are completely missing from the matrix."[42]

Rather than displaying evolutionary sequences in proteins, the matrix demonstrates the unlinked uniqueness of each species: no protein sequence links have been discovered between species. The gradual evolution of cytochrome C was expected, but the matrix demonstrates no successive, slight evolutionary trends from species to species. The matrix demonstrates that each species is equal in distance in terms of protein sequences from their most likely ancestor—the bacteria. Astoundingly, Michael Denton noted that starting from the bacteria, "organisms as diverse as man, lamprey, fruit fly, wheat and yeast, all exhibit a sequence divergence of between sixty-four percent to sixty-seven percent from this particular bacterial cytochrome ... this must be considered one of the most astonishing findings of modern science."[43]

Denton is certainly not alone. Italian geneticist Giuseppe Sermonti, in 1987, came to the same conclusion that the differences in cytochrome C between species were compatible with the null hypothesis because there is no evidence to support any sequence pattern between species.[44]

This complete molecular biology disconnect from the most basic Darwinian evolutionary scheme of fish to amphibian to reptile to mammal has been stunning, shattering early excitement over tracing evolution through protein sequence changes. The gradual divergence one expects to see in an evolutionary sequence from cytochrome C is completely missing. Michael Denton explains that there "is not a trace at a molecular level of the traditional evolutionary series: cyclostome [lamprey] → fish → amphibian → reptile → mammal. Incredibly, man is as close to a lamprey as are fish!"[45]

Rather than compiling a table of sequences to quantify evolutionary sequences, Dayhoff's Atlas of Protein Sequences has demonstrated the uniqueness of species, not the evolution of species. Cytochrome C research continues, but a cytochrome C Tree of Life has never been constructed.

Like hemoglobin and cytochrome C, the molecular sequence in insulin between species does not follow any Tree of Life scheme either. Of the fifty-one amino acids, human insulin differs from cow insulin by three amino acids and from pig insulin by one amino acid. The significant finding is that while the number of amino acids is similar, changes in amino acid sequences does not follow any evolutionary sequence. *#'s similar, sequences not*

Insulin is in the relaxin-like family of molecules. In the most extensive review of the literature ever conducted, University of Melbourne, Australia, molecular biologists Tracey Wilkinson and colleagues concluded in 2005 that the relaxin-like family does not demonstrate an amino acid evolutionary sequence. Even the "distantly related species show high similarity."[46]

In this landmark evaluation, Wilkinson concludes that the sequence in the relaxin-like family does not simulate the Tree of Life (phylogenic tree) sequences as expected from the Darwinian theory of evolution: "None of the phylogenetic tree construction programs used was able to completely resolve the evolution of the relaxin-like peptide family."[47]

Molecular biologists studying the luteinizing-releasing hormones discovered species uniqueness, not evolutionary sequences. The luteinizing hormone-releasing hormone in amphibians and mammalians is identical but different from birds, reptiles, and certain fish, suggesting that mammals are more closely related to amphibians than birds.

The story gets more interesting. Since the luteinizing-releasing hormone is regulated by the gonadotropin-releasing hormone, Italian molecular biologists, led by Maria M. Di Fiore in 2000, were surprised to discover that the gonadotropin-releasing hormone in the chicken and the rat is identical. Expecting to discover the evolutionary sequence, they were lead to a contrary conclusion: "Now it is evident that this is not the case."[48]

Calcitonin is a ubiquitous hormone throughout the Animal Kingdom that regulates calcium metabolism. Ironically, human calcitonin is more similar to calcitonin in the salmon than in the pig—not an expected evolutionary finding. To investigate the evolutionary sequence, molecular biologists from Portugal and England collaborated in 2006 to identify the sequence of evolution of the secretin family of molecules, to which calcitonin belongs. However, like hemoglobin, cytochrome C, insulin, and the luteinizing hormone, the amino acid sequence of calcitonin, the Tree of Life branches are not connected in any evolutionary sequence. Species on the molecular level are unique. The international collaborative study was only able to conclude that the study gave a "better understanding" in that there is no evolutionary sequence for calcitonin.[49]

While the construction of a Tree of Life would be expected to start at the root of the tree, ironically, after over fifty years of research in modern molecular biology, there has been no consensus on the root of the tree—life's original cell. Professor of evolutionary biology in the Department of Zoology at the University of Oxford, Thomas Cavalier-Smith pined: "Despite great advances in clarifying the family Tree of Life, it is still not agreed where its root is or what properties the most ancient cells possessed—the most difficult problems in phylogeny."[50]

Reflecting on Darwin's Tree of Life, Italian geneticist Giuseppe Sermonti concluded in 2005 that the "genealogical tree that shows forms gradually varying and diverging—the 'fact' for which Darwinism proposed its revolutionary explanation—is nowhere to be found."[51]

On a molecular level, like the fossil record, the elusive missing links remain missing. The molecular biology of each species is unique. And the evidence is overwhelming. Michael Denton concluded: "Thousands of different sequences, protein and nucleic acid, have now been compared in hundreds of species but never has any sequence been found to be in any sense the lineal descendants or ancestor of any other sequence."[52]

What has been discovered is that the more extensive the investigation, the more perplexing any evolutionary molecular sequence. In an overview of the evolution of protein complexes, molecular biologists Jose B. Pereira-Leal, Emmanuel D. Levy, and Sarah A. Teichmann of the MRC Laboratory of Molecular Biology at Cambridge University in England conclude by stating that the evidence leaves more questions and challenges than answers: "These insights into the evolution of protein complexes still leave us with challenges in terms of understanding the modular nature of cellular networks."[53]

Perhaps the uniqueness of life is not determined by molecular biology. Looking at the evidence, Sermonti writes that the message emerging from the data indicates that molecular biology does not determine the species: "From a biochemical standpoint the horse and the horsefly are essentially the same."[54]

In 1977, echoing the same sentiment, Francois Jacob, a founding father of biochemical genetics, wrote, "Biochemical changes do not seem to be the main driving force in the diversification of living organisms."[55]

Species appear to be distinct entities and not sequential or transitional. On the molecular level, species are unique. While there are more than 800 species of frogs, all of which look superficially the same, there is a greater variation of molecular structure between them than there is between the bat and the blue whale. While there are at least twenty-six species of the protozoan *Tetrahymena*, all of which are nearly identical in structure, there are enormous differences between their homologous proteins. The same is true of the more than 2,000 species of fruit flies.

Today, not a single cohesive Tree of Life has been universally accepted using molecular biology to demonstrate the successive, slight changes in the sequence of a single molecule as expected from Darwin's theory of natural selection. More importantly, an organism is not composed of just a single molecule. The simplest organism is composed of no less than 600 protein molecules.

Random Chance

One of the biggest questions in molecular biology is—what is the chance that a single protein molecule could have actually been formed by mere chance?

In principle, probabilities smaller than 1 over 1,050 are thought of as having a zero probability. Since an average-sized protein molecule is composed of 288 amino acids with 12 different types of amino acids, this protein can be arranged in 10^{300} different ways, which is 10 followed by 300 zeros. Since 10^{300} far exceeds 1,050, the probability of the formation of only one protein molecule by random chance is zero. Molecular biologist Harold Blum concludes that from the mathematical perspective, probability of a protein autonomously assimilating by chance is zero: "The spontaneous formation of a polypeptide of the size of the smallest known proteins seems beyond all probability."[56]

If the random chance of forming a single molecule is less than zero, the probability of the random forming the simplest organism is even more improbable. Since the smallest bacteria known, *Mycoplasma hominis* H39 contains 600 types of proteins, to determine the probability of forming the simplest organism, the calculation for one protein would have to be repeated for each of these 600 different types of proteins. The result staggers even the concept of impossibility. Chandra Wickramasinghe, professor of applied mathematics and astronomy, University College Cardiff, Wales, concluded:

> The likelihood of the spontaneous formation of life from inanimate matter is one to a number with 40,000 noughts after it ... It is big enough to bury Darwin and the whole theory of evolution. There was no primeval soup, neither on this planet nor on any other, and if the beginnings of life were not random, they must therefore have been the product of purposeful intelligence.[57]

Not only does the biological activity of an amino acid depend on the sequence of amino acids, the function is dependent on the folding structure of the protein.[58] The three-dimensional structure is essential to the biological activity of proteins. The advent of the mad cow disease highlights the importance of spatial configuration. None of the

calculations on the sequence of amino acids accounts for possible variations in the structure of the protein. In the book *Chance and Necessity*,
Nobel Prize winner Jacques Monod, speculating on the probability
of the origin of life, concluded, "its *a priori* probability was virtually
zero."[59]

The more that is known, the more complex the puzzle grows.
Biochemist Leslie Orgel, at the University of California, San Diego,
concluded that life never could have originated "by chemical means."[60]
In the book *Climbing Mount Improbable* (1996), even ardent molecular
biologist Richard Dawkins, in reviewing the probability of random
chance resulting in the origin of life, concluded:

> So the sort of lucky event we are looking at could be so
> wildly improbable that the chances of its happening,
> somewhere in the universe, could be as low as one in a
> billion billion billion in any one year.[61]

On the topic of the origin of life, or "abiogenesis," Wikipedia.org, in
2008, posted, "There is no truly 'standard model' of the origin of life."[62]
While the early twentieth century evolutionary mathematical models
of Haldane and Oparin initially appeared promising, as technology
advanced onto the frontiers of knowledge, now with reality coming into
focus, scientists have become increasingly skeptical at the prospect of life
arising from inorganic materials by random chance alone.

Molecular Clocks

While ideas come and go, some bad ideas just never seem to leave.
To estimate the pace of evolution, in 1962 molecular biologist Emile
Zuckerkandl and Nobel Prize winner Linus Pauling were working at
Caltech on hemoglobin evolution and expressed the idea of "molecular
anthropology" as a new discipline. The idea was later termed the molecular clock theory.[63] The purpose of the molecular clock is to estimate the
rate of evolution for individual molecules. In 1962, molecular sequence
problems were just emerging.

Zuckerkandl and Pauling postulated that in a protein, each amino
acid randomly changes at a constant rate. If the estimated time for
divergence between species and the number of amino acid changes since

that time can be determined from the fossil record, the rate of change can be calculated. This rate of molecular change (time per amino acid change) has been called the molecular clock.

The molecular clock for the hemoglobin molecule can be calculated simply by knowing the time of divergence and the number of amino acid changes. Since horse and man are thought to have diverged about 100 million years ago and the alpha chain differs between horse and man by about twenty amino acids, the calculated molecular clock for hemoglobin is one amino acid change over five million years (100,000,000 years/20 amino acids = 5,000,000 years/amino acid change.) This same approach can used to determine the molecular clock for genes by comparing times of divergence and the number of changes in the DNA.

The method for determining the molecular clock seems simple enough. Determining the number of amino acid changes can be done in a molecular biology laboratory. Because of scant fossil record evidence to support evolution, estimating times of divergence from the fossil record can become a dicey matter. Experiments for divergence in the fossil record cannot be performed in a laboratory.

Not only is the fossil record an issue, but Zuckerkandl and Pauling could hardly expect what was to be discovered next—when more than one molecule is examined, different clocks appear to be running for each molecule; that is, each molecule's clock runs at a different speed.

This problem of different clocks was first discovered during the study of hemoglobin and cytochrome C in man and the carp. The level of hemoglobin in man differs from the level of hemoglobin in carp by 50 percent, while the level of cytochrome C in man differs from the level of cytochrome C in carp by only 13 percent.[64] Hemoglobin and cytochrome C were expected to have evolved at the same rate, but this is not supported by evidence.

This means that hemoglobin and cytochrome C have intrinsically different molecular clocks. Compounding the problem further is that even the simplest of species is composed of hundreds of molecules. Molecular biologist Denton concluded that if the Zuckerkandl and Pauling theory is correct, "then it is necessary to propose not just two clocks but one for each of the several hundred protein families, each ticking at its own unique rate."[65]

The question is can molecules evolve independently at different rates in a species? Since each molecule interacts with a vast number of molecules in an organism symbiotically, how can one molecule independently evolve? To support the theory of evolution, different rates of evolution must have an explanation.

As the molecular data began to accumulate during the early 1990s, it became increasingly apparent that the theory was intrinsically even more problematic when examining evolution from the context of the entire organism and the fossil record.[66] At the core of Darwinian evolution are the successive, slight changes in molecules. However, how different molecules can evolve at different rates in the same organism remains an enigma.

While there was hope that these different rates of evolution were a fluke or that there would be some work-around to the problem, no known workaround has yet emerged. The fact that different molecules appear to be evolving at different rates is now an unavoidable reality, if evolution ever did develop through successive, slight variations.

The popularity of molecular clocks waned in the early 1990s. There has been a resurgence of interest during the past ten years, however.[67] Conceding that molecular components of a species must have evolved at different rates, molecular biologists have developed workarounds. One workaround proposed that the different molecular clocks might eventually average out based on a yet undiscovered natural law.[68] At present though, research into these molecular clock workarounds are only "interesting" and "promising," at best. As molecular evolutionist Naoyuki Takahata explains: "These examples show some interesting and promising ways to connect the molecular clock to the study of species-specific life history traits and spontaneous mutation."[69]

Information from the molecular clock was once thought to be one of the most useful tools in establishing evolutionary biology. How the evolution of each molecule can run by a different molecular clock in the same organism continues to undermine a cohesive theory of molecular evolution.

The pursuit to resolve the clock issue has reemerged onto center stage because the rate of molecular change is foundational to evolution.[70] If the molecular mechanisms of evolution cannot be traced, the only logical conclusion is that molecular biology has played no role in evolution.

A popular theory forwarded to explain differences in the molecular clock is the metabolic rate hypothesis. The goal was to integrate the different metabolic rates of an organism into a single molecular clock. Using metabolic rate measurements and DNA sequence data for more than 300 metazoan species for twelve different genes, scientists at the Centre for the Study of Evolution at the University of Sussex in the United Kingdom disproved this theory in 2007 by concluding that "we find no evidence that mass-specific metabolic rate drives [molecular clocks]."[71]

In taking a different approach, by exploring the relationship between body size and differences in the molecular clock, the same research group at the University of Sussex in 2006 concluded that body size has no influence on molecular clocks (substitution rates): "We find no evidence of any influence of body size on invertebrate substitution rates."[72] Evolutionary biologists Megan Woolfi and Lindell Bromham, from the Centre for the Study of Evolution at the University of Sussex, after studying seventy independent species between islands and mainland taxa, including vertebrates, invertebrates, and plants from nineteen different island groups, concluded: "Overall substitution rates do not differ significantly."[73]

Zuckerkandl and Pauling originally postulated that divergence is determined by the fossil record. In taking a different approach, investigators are estimating times of divergence from laboratory data, rather than from the fossil record. Commenting on the results of this approach in 2007, Naoyuki Takahata, of The Graduate University for Advanced Studies in Japan, wrote in the journal *Genetics,* "It is now clear that any kind of molecular clock ticks erratically, but it is nevertheless widely used [unforunately] for estimating species divergence times."[74] Undermining the scientific method, the molecular clock then becomes the dependent variable while the time of divergence becomes an independent variable—clever move.

Recognizing the chaos and "conspicuously discordant results" with the use of molecular clocks, biologist Kevin J. Peterson of Dartmouth College and geologist Nicholas Butterfield of the University of

Cambridge emphasize the importance of testing the fossil record as a dependent and not an independent variable.[75]

How Zuckerkandl and Pauling's simple postulate has become so complicated begs the question, are molecular clocks real? Professor of evolutionary biology Thomas Cavalier-Smith of the University of Oxford in England concluded in a paper entitled *Cell Evolution and Earth History: Stasis and Revolution* that the answer is no: "Evolution is not evenly paced and there are no real molecular clocks."[76]

Available evidence bodes negatively for the usefulness of molecular clocks in establishing any shape for the Tree of Life. What was originally thought to become a cornerstone for molecular evolution is now irreconcilable with evolution and created chaos in evolutionary thought.

> The difficulties associated with attempting to explain
> how a family of homologous proteins could have evolved
> at constant rates has created chaos in evolutionary
> thought.[77]

Rather than supporting the theory of evolution, the molecular clock evidence and the sequence data actually undermine the theory of evolution through "successive, slight" variations in molecular biology. Just as hope in the fossil record, the origin of life, and the sequence of amino acids dissipated, the hope that molecular clocks will become an evidential, evolutionary cornerstone is vaporizing. In 2005, geneticist Giuseppe Sermonti wrote: "Once the universal 'molecular clock' was shelved, biochemists ceased to question (in any case dubious) datings proposed by paleontologists."[78]

After 150 years of research, scientists have not been able to trace the evolution of any molecule to man—or even the rate of the evolutionary process. Had evolution delivered the origins of life, the concept of molecular clocks would have been brilliant.

The "successive" concept holds the key to Darwin's theory; the term is used 109 times in *The Origin of Species*. It is foundational to the principle of natural selection. Darwin wrote that natural selection could only act through successive changes, never rapidly:

On the theory of natural selection, we can clearly understand why she should not; for natural selection acts only by taking advantage of slight successive variations; she can never take a great and sudden leap, but must advance by short and sure, though slow steps.[79]

Molecular biologists beginning in the early twentieth century had expected to trace the organization of inorganic to organic molecules as well as the successive molecular changes as the species evolved. Clearly, however, the convergence of molecular evidence does not support the theory. Darwin concluded that if the evidence does not support "numerous, successive, slight modifications, my theory would absolutely break down."[80]

Stopping a running train is no small matter. With the remaining momentum, the most recent trend has been to investigate the concept that life did not start by arranging molecules into proteins, but rather into nucleic acids.

Mad Cow Disease

The first signs of the mad cow disease emerged in 1986, when British cattle begin to suffer from a condition similar to scrapie in sheep. Due to the behavior of the sick cows, the condition was nicknamed "mad cow disease." At the time, the cause of the disease was unknown. A British advisory committee in the early 1990s issued a statement that the cattle would be a "dead-end host." In May 1995, Stephen Churchill, at the age of nineteen, became the first victim attributed to mad cow disease. By the turn of the century, mad cow disease had spread from England to North America and Japan.

Mad cow disease is an infection that causes a neurodegenerative disease known as Bovine Spongiform Encephalopathy (BSE). The infectious agent in BSE is simply a misfolded ingested protein called a prion. There is no known cure for the fatal disease.

At issue is the fact that mad cow disease challenges the central dogma of evolutionary biology that DNA is the controller of life. Scientists had known for some time that proteins can control the proteins independent from the control of DNA. From the misfolding of the spatial configuration of one protein, actions of surrounding proteins can be changed. Therefore, not only is the sequence of amino acids critical, so is the "spatial configuration that folds them into the proper association."[81] The spatial configuration of a molecule also determines cellular information. This unexpected phenomenon has been termed "heredity by contact."[82]

The "heredity by contact" hypothesis was first postulated by Stanley Prusiner, American neurologist and biochemist, and winner of the 1997 Nobel Prize in Physiology or Medicine. After the structure of DNA and the genetic code, the "heredity by contact" effect is now considered by many to be the most important event in biology in the past half century.

On tempering this "heredity by contact" view, geneticist Sermonti explains that what "has happened is not the multiplication of a pathogen, and not a true case of molecular heredity, but transmission of a deformity in the molecule arising out of contact with another deformed molecule—a 'rotten apple' effect."[83]

The RNA World

Ever since the glaring oxidizing-atmosphere error could no longer be ignored, scientists have largely abandoned proteins as the first molecular building blocks of life, and nucleic acids have taken center stage. While DNA is composed of nucleic acids, it is dependent on specific proteins for replication. Therefore, DNA cannot be considered as the forerunners of proteins. Today, the focus of research on the origin of life has changed from proteins to the role of RNA.

The initial role of RNA in molecular biology began to emerge in 1939. RNA is now recognized as having the greatest potential to have been the first molecular building block of life.[84] In 1968, two of the Nobel Prize in Physiology or Medicine awards were awarded for pioneering work on RNA. In 1976, Belgium scientists, led by Walter Fiers at the University of Ghent, determined the first complete nucleotide sequence of an RNA virus genome.[85]

RNA is much simpler than proteins. While proteins are composed of at least twenty amino acids, RNA is composed of only four nucleic acids, although it does require the attachment of a sugar and phosphate group.

Unlike the double-stranded DNA, the RNA molecule is single-stranded and much shorter than DNA. In 1982, Thomas Cech and Sidney Altman demonstrated that not only are RNA molecules carriers of genetic information, but they also can have catalytic functions and can participate in cellular reactions—like a protein. In 1989, Thomas Cech and Sidney Altman shared the Nobel Prize in Chemistry for their work on RNA.

In 1986, American physicist, biochemist, and molecular biology pioneer Walter Gilbert was the first to propose the term "RNA world hypothesis" for the origin of life. Gilbert suggested that because RNA can synthesize itself in the absence of proteins, RNA may have originated on the early Earth before proteins or DNA; this is known as the RNA world. According to the RNA world hypothesis, the RNA molecule later evolved into DNA and protein molecules. While the DNA molecule evolved into a data storage role, the protein molecules evolved into a catalytic role.

In 1959, Spanish Catalan biochemist Joan Oró began to synthesize adenine, a key component of RNA and DNA, from hydrogen cyanide, similar to a Miller–Urey experiment.

Like the Miller–Urey experiment though, the lack of geological evidence for hydrogen cyanide in the fossil record is missing. Another problem with hydrogen cyanide is that at room temperature, it becomes a gas toxic to cellular metabolism. During the German Nazi regime in the mid-twentieth century, hydrogen cyanide was used as an agent for mass murder.

To date, not one laboratory experiment with realistic early Earth elements and conditions has produced a single nucleic acid. Scripps Research Institute biochemist Gerald Joyce states that the "most reasonable interpretation is that life did not start with RNA."[86] The origin of life is so difficult a problem that German researcher Klaus Dose stated in 1988 that the RNA theory is "a scheme of ignorance. Without fundamentally new insights in evolutionary processes … this ignorance is likely to persist."[87]

In 1998, Leslie Orgel, senior research fellow and research professor at the Salk Institute for Biological Studies, where he directed the Chemical Evolution Laboratory, acknowledged that "we are very far from knowing whodunit"[88] or what were the early environmental conditions on the Earth.

Nearly twenty years later, the role of RNA in the origin of life remains elusive, if not improbable. In 2007, commenting in *Proceedings of the National Academy of Sciences of the United States of America* on a paper by Phillipp Baaske and Eugene V. Koonin, senior investigator, National Center for Biotechnology Information, National Library of Medicine, National Institutes of Health, stated that while more is known about RNA, the evolutionary role of RNA has severe difficulties and "still is a hypothetical entity; … the evolutionary path to the translation systems remains essentially uncharted."[89, 90]

After 150 years running, how life spontaneously began, to which Darwin alluded, remains a mystery. The deputy editor of the journal *Nature,* and science writer, Nicholas Wade, reported in the *New York Times* in June 2000 "everything about the origin of life on the Earth is a mystery, and it seems the more that is known, the more acute the puzzle gets."[91]

The lack of molecular evidence to support a natural view for the origin of life has Darwin's theory scrambling. After years of investigation, molecular biologist Michael Behe concluded in *Darwin's Black Box* that it "was once expected that the basis of life would be exceedingly simple. That expectation has been smashed."[92]

Molecular biology has only emphasized the enormity of the gaps between competing theories on the origin of life and evidence. Even as early as 1988, Klaus Dose, at the Interdisciplinary Science Reviews, reflected on the difficulty of understanding origins of life:

> More than 30 years of experimentation on the origin of life in the fields of chemical and molecular evolution have led to a better perception of the immensity of the problem of the origin of life on Earth rather than to its solution. At present, all discussions on principle theories and experiments in the field either end in stalemate or in a confession of ignorance.[93]

Molecular biology, once commissioned as the vanguard of evolution late in the twentieth century, now stands with branches in hand but powerless to place the branches on Darwin's Tree of Life. Evolutionary molecular biology stands at a crossroad. Molecular biologist Behe reflects:

> Molecular evolution is not based on scientific authority. There is no publication in the scientific literature—in prestigious journals, specialty journals, or books— that describe how molecular evolution of any real complex, biochemical system either did occur or even might have occurred. There are assertions that such evolution occurred, but absolutely none are supported by pertinent experiments or calculations. Since no one knows molecular evolution by direct experience and since there is absolutely no authority on which to claim knowledge ... the assertion of Darwinian molecular evolution is merely bluster.[94]

Evolutionary molecular biologist W. Ford Doolittle of Dalhousie University in Nova Scotia concluded in 1999 that molecular biology has "failed to find the 'true tree,' not because their methods are inadequate or because they have chosen the wrong genes, but because the history of life cannot properly be represented as a tree."[95]

The Tree of Life through successive, slight changes is a philosophy not supported by scientific evidence. Scientific evidence does not support the theory that the origins of life were the products of natural laws. In 1998, evolutionary molecular biologist Carl Woese pined that "incongruities can be seen everywhere in the universal tree" of life.[96]

By the 1980s, research in molecular biology was poised as the "method superior to paleontology" in defining the natural mechanisms for evolution. However, as early as 1986, molecular biologist Christian Schwabe of the University of Iowa recognized that "It seems disconcerting that many exceptions exist to the orderly progression of species as determined by molecular homologies; so many in fact that I think the exception, the quirks, may carry the more important message."[97]

Where the fossil record had failed, it was hoped that molecular biology would bridge the evidence gaps in Darwin's theory. But instead of bridging the gaps, the gaps have become even bigger. Molecular biologist Michael Denton explains: "Instead of revealing a multitude of transitional forms through which the evolution of the cell might have occurred, molecular biology has served only to emphasize the enormity of the gaps."[98]

French molecular biologist and Nobel Prize winner Jacques Lucien Monod concluded that identifying the origins of life "is not so much a problem as a veritable enigma."[99] Echoing the same sentiment, biochemist and Nobel Prize winner Francis Crick, in the book *Life Itself*, concludes "that in some sense, the origin of life appears at the moment to be almost a miracle."[100]

Giuseppe Sermonti, in his 2005 book *Why a Fly is not a Horse,* wrote that a molecular basis for evolution does not exist since "the molecular mechanisms invoked to explain evolution are all fundamentally either degenerative or conservative."[101] Sermonti concedes that if "we want to solve the problem underlying every origin of species in molecular terms, we have to admit that for the moment the answer is not forthcoming."[102]

Failure to identify any plausible natural law for the molecular origin of life continues unresolved. Life, rather than originating sequentially, can be best characterized as distinct, unique, and not reproducible. However, Darwin never said that molecular biology was the key to his theory. For Darwin, embryology held the key evidence.

Chapter Eleven
Embryology

So again it is probable, from what we know of the embryos of mammals, birds, fishes, and reptiles, that these animals are the modified descendants of some ancient progenitor.

—*Charles Darwin*[1]

Of all the facts in *The Origin of Species*, embryology held the most important facts to support the theory of evolution through natural selection, according to Darwin. In a letter to Asa Gray in September 1860, Darwin wrote that "embryology is to me by far the strongest single class of facts in favor" of the theory.[2]

Three years earlier, in September 1857, Darwin had written to Asa Gray that embryology was just one of the supporting facts: "Why I think that species have really changed, depends on general facts in the affinities, embryology, rudimentary organs, geological history, and geographical distribution of organic beings."[3] Then, just two months before the release of the first edition of *The Origin of Species* in September 1859, Darwin wrote to Charles Lyell, "Embryology in Chapter VIII is one of my strongest points I think."[4]

What had been most impressive were the similarities in structures between species: the five fingers in the hand of a man, the five "fingers" in the wing of a bat, and similarities in the embryo between species. Darwin was fascinated by embryology. Writing in his autobiography, Darwin recalls: "Hardly any point gave me so much satisfaction when I was at work on the *Origin*, as the explanation of the wide difference in many classes between the embryo and the adult animal."[5]

Similarity

Darwin's premise was that the similarity between the structure and the embryo of animal and man was primary proof that man evolved from animals. These similarities in structure were called homology. In evolutionary biology, homology refers to any similarity between characters that is due to their shared ancestry. In *The Descent of Man,* published in 1871, Darwin writes in the first chapter that homology provides the "ample and conclusive evidence in favour of the principle of gradual evolution."[6]

Homology is derived from the ancient Greek "to agree." Darwin defines homology as that "relation between [similar] parts [like the hand], which results from their development from corresponding embryonic parts."[7] Darwin's classic example of homology is the hand.

Noticing that the hand between species is essentially the same, Darwin explains: "What can be more curious than that [of] the hand of a man, formed for grasping, that of a mole for digging, the leg of the horse, the paddle of the porpoise, and the wing of the bat, [if] all be constructed on the same pattern, and should include similar bones, in the same relative positions?"[8] Evidence in homology and embryology at the time was based on unmeasured observations.

Since the hand of man and the wing (hand) of the bat appear to be homologous structures, Darwin logically concluded that both must have evolved from the same embryological source—from the same original ancestor. Taking the logic further, the hand of the man and the wing of the bat must have been derived from the same embryological tissue during development, evolving from some original ancestor, "some lower form." Using this line of logic, Darwin linked homologous structures with embryological development:

> We have seen in the first chapter that the homological structure of man, his embryological development and the rudiments which he still retains, all declare in the plainest manner that he is descended from some lower form.[9]

Haeckel

This embryological approach to evolution was not original to Darwin. The connection between homology and embryology was originally developed by German biologist Fritz Müller (1821–1897) and German biologist Ernest Haeckel (1834–1919). As contemporaries,

Darwin gave them credit in his autobiography: "Within late years several reviewers have given full credit of the idea to Fritz Müller and Haeckel, who undoubtedly have worked it out ... much more fully, and in some respects more correctly, than I did."[10]

Actually, Müller and Haeckel built on the work of German biologist Karl Ernst von Baer (1792–1876), who is now recognized as the founder of modern embryology. In 1826, Baer was the first to discover the mammalian ovum. In 1827, he completed research entitled "Ovi Mammalium et Hominis genesi" at Saint-Petersburg's Academy of Science, which established that mammals develop from eggs. Baer's work in the nineteenth century set the stage for Müller, Haeckel, and Darwin. Baer, using the technology of the nineteenth century, noted similarities in the embryo among different species and formulated what has become known in the field of embryology as "Baer's law."

As *The Origin of Species* continued to gain attention, Haeckel extended Baer's law and popularized the controversial "recapitulation theory." Haeckel was the one who actively promoted that a species' embryological development (ontogeny) traces the species' entire evolutionary development (phylogeny). This means that in the case of man, the embryo transforms from a single cell, to a tadpole, to a fish, to an amphibian, to a monkey, and finally to man. In other words, at the different stages of development, the embryo is actually a series of ancestor species. The sequences of the embryo retrace the steps of evolution.

Haeckel coined this process with the now-famous phrase "ontogeny recapitulates phylogeny." In other words, the human embryo retraces—repeats (recapitulates)—all stages representing its ancestors, from the single cell stage to man. In theory then, seeing the human embryo grow would be like watching a silent movie of our ancestral history and presumably evolution in action.

Darwin was not an embryologist, and instead relied on the work of others. In *The Origin of Species,* Darwin gave credit to Haeckel: "Professor Haeckel in his "Generelle Morphologie" and in [other] works has recently brought his great knowledge and abilities to bear on what he calls phylogeny, or the lines of descent of all organic beings. In drawing up the several series he trusts chiefly to embryological characters [to establish evolutionary sequences]."[11]

Haeckel became a legendary figure of the nineteenth century by discovering, describing, and naming thousands of new species and coining many of the common terms in use today, including "phylum," "phylogeny," and "ecology." In the United States, the 13,418-foot summit on the eastern slopes of Sierra Nevada of California overlooking the Evolution Basin is named in his honor, Mount Haeckel.

By drawing embryos, Haeckel attempted to show that embryos of all species are virtually identical from the earliest stages and that embryos develop through their evolutionary ancestors. Haeckel called his theory the "biogenetic law," which is summarized by the now-famous phrase "ontogeny recapitulates phylogeny." Haeckel's theory was compatible with Darwin's theory of descent from a common ancestor. In *The Origin of Species,* Darwin wrote, "So again it is probable, from what we know of the embryos of mammals, birds, fishes, and reptiles, that these animals are the modified descendants of some ancient progenitor."[1]

Haeckel made Darwin's abstract concepts more concrete, and Haeckel's "ontogeny recapitulates phylogeny" became a gold mine for Darwin. As a member of the intellectual elite, Darwin gained a powerful, influential comrade. Eventually, Haeckel's concepts of embryology became "second to none in importance" to Darwin's theory. In *The Origin of Species,* Darwin wrote:

> Thus, as it seems to me, the leading facts in embryology, which are second to none in importance, are explained on the principles of variation in the many descendants from some ... ancient progenitor.[12]

Biogenetic Law

In the opening paragraph of the "Development and Embryology" section of *The Origin of Species,* Darwin envisions that this "is one of the most important subjects in the whole round of natural history."[13]

The biogenetic law—ontogeny recapitulates phylogeny—became widely popular with the intellectual elite. Herbert Spencer, a prominent classical liberal English philosopher, supported and collaborated with Haeckel and Darwin. Eventually, Darwin incorporated the phrase "survival of the fittest," which had been popularized by Spencer in the fifth edition of *The Origin of Species.* The influential American geologist, Jean Louis Rodolphe Agassiz, at Harvard University, sided with Haeckel's biogenetic law. Darwin wrote in *The Origin of Species* that Haeckel's law "accords admirably well with our theory."[14]

Darwin knew that Haeckel's theory required collaborating evidence from the fossil record, even though there was only a small chance of finding the evidence. In *The Origin of Species,* Darwin envisions "the embryo ... as a sort of picture, preserved by nature, of the former and less modified condition of the species." Darwin continues: "This view may be true, and yet may never be capable of proof ... until beds rich in fossils are discovered far beneath the lowest Cambrian strata—a discovery of which the chance is small."[15]

Haeckel was motivated. In looking for the fossil remains of the first man, Haeckel theorized that evidence of human evolution should be found in the Dutch East Indies, now known as Indonesia. Eventually one of Haeckel's students, Eugene Dubois, accepted the challenge to search for the first man and went to explore the Dutch East Indies. Eventually, in 1891, Dubois arranged mammalian bones, naming the composite "Java man."

Haeckel's perspective was powerful. At the time, many influential critics had completely rejected Baer's initial premise that the embryos of all species were similar. Darwin knew that proof of the theory was tentative pending either direct or indirect collaborating evidence from the fossil record.

In 1894, embryologist Adam Sedgwick even recognized that Baer's theory of similarity was bogus and "not in accordance with the facts of development."[16] Driving the point of dissimilarity of embryos between species into clarity, Sedgwick observed that if "Baer's law has any meaning at all, surely it must imply that animals so closely allied as the fowl and duck embryo would be indistinguishable in the early stages of development ... yet I can distinguish a fowl and a duck embryo on the second day."[17]

Sedgwick completely contradicts Baer, Darwin, and Haeckel, by recognizing that each species is noticeably distinct from the beginning. Sedgwick highlighted that the embryos among various species demonstrate differences, not similarities, and wrote, "a species is distinct and distinguishable from its allies from the very earliest stages all through development."[18]

Taking on the challenge, Haeckel passionately wielded the new biogenetic law against the critics to a point that was even offensive to Darwin. In a unique display of true character, recognizing that Haeckel was becoming too adamantly offensive, Darwin wrote to Haeckel in 1867, asking him to tone down the rhetoric and become more tolerant:

> I have long observed that much severity leads the reader to take the side of the attacked person.... As you will surely play a great part in science, let me as an older man earnestly beg you to reflect on what I have ventured to say. I know that it is easy to preach and if I had the power of writing with severity I dare say I [should] triumph in turning poor devils inside out and exposing all their imbecility. Nevertheless, I am convinced that this power does no good, [and] only causes pain. I may add that as we daily see men arriving at opposite conclusions from the same premises, it seems to me doubtful policy to speak too positively on any complex subject however much a man may feel convinced of the truth of his own conclusions.[19]

One of Haeckel's critics from the beginning included Baer, the Father of Embryology. Baer and Darwin's cautious premonition has proved to be sound. For more than a century, scientists have known that Haeckel's law and the evidence from his drawings were fundamentally flawed—and fraudulent. Eminent evolutionist Stephen Gould concluded in the March 2000 issue of *Natural History* that Haeckel's drawings, now famous, were characterized by "inaccuracies and outright falsification" and that "Haeckel had exaggerated the similarities by idealizations and omissions. He also, in some cases—in a procedure that can only be called fraudulent—simply copied the same figure over and over again."[20]

Yet, Haeckel became a legend in the late eighteenth century. After earning a doctorate in medicine degree in 1857, Haeckel discovered that compassion for suffering patients was less attractive than pursuing a career in the developing field of natural sciences.

As such, Haeckel continued his education at the University of Jena for three years, eventually earning a doctorate in zoology. He later became a professor of comparative anatomy at the University of Jena. Haeckel remained at the university for forty-seven years, from 1862 to 1909. Haeckel is credited for naming thousands of newly discovered species between 1859 and 1887, though not one species were thought to be one of Darwin's missing links.

Haeckel was in the echelons of the academic elite during the eighteenth century. From 1866 to 1867, he traveled from Germany to the Canary Islands. En route, Haeckel arranged to meet Charles Darwin, Thomas Huxley, and Charles Lyell. The flamboyant Haeckel is even credited with coining the phrase "First World War." In 1914, Haeckel was quoted in the *Indianapolis Star* on September 20 as saying: "There is no doubt that the course and character of the feared 'European War' ... will become the first world war in the full sense of the word."[21]

Besides numerous scientific memoirs and illustrations, Haeckel created an extensive library of forty-two works totaling nearly 13,000 pages. Today however, Haeckel is famous for all the wrong reasons. Haeckel's infamous embryo drawings published in 1874 that quickly made their way into biology text books are now known to be a fabrication of his imagination—not from a scientific discovery.

While Darwin did not include Haeckel's drawings in *The Origin of Species* or in the *Descent of Man* since their publication predated Haeckel's most infamous drawing in 1874, Haeckel's drawing has been used as evidence to support Darwin's theory of evolution in biology textbooks since 1901. Darwin's works, however, plainly reflect Haeckel's influence. Writing in the *Descent of Man*: "we can understand how it has come to pass that man and all other vertebrate animals have been constructed on the same general model, why they pass through the same early stages of development, and why they retain certain rudiments in common." Darwin continues: "Consequently we ought frankly to admit their community of descent."[22]

While Darwin never specifically commented on the credibility of the drawings, natural scientists have not stayed silent. In 1894, Adam Sedgwick argued against Haeckel: "There is no stage of development in which the unaided naked eye would fail to distinguish between them with ease.... I need only say with regard to them that a species is distinct and distinguishable from its allies from the very earliest stages all through the development."[23]

While Haeckel's legacy, including the Java man, is lamentable, Sedgwick's legacy continues into the twenty-first century. One of Sedgwick's students at Cambridge University, William Bateson, was eventually credited for coining the term "genetics" in a letter written to Sedgwick in 1905.

Haeckel's "recapitulation" problem originated with his line of reasoning. Sedgwick recognized that Haeckel's theory was founded by deductive, rather than inductive, reasoning. Like Darwin, Haeckel had blatantly abandoned the scientific method. In 1909, Sedgwick explained: "The recapitulation theory originated as a deduction from the evolutionary theory and as a deduction it still remains."[24]

Early twentieth century American embryologist Frank Lillie, of the University of Chicago, argued against Haeckel's law of recapitulation. Lillie found no scientific evidence to support Haeckel's contention that the embryos of different species look similar. In 1908, Lillie wrote in his textbook *The Development of the Chick* that "it never happens that the embryo of any definite species resembles in its entirety the adult of a lower species, nor even the embryo of a lower species; its organization is specified at all stages from the (egg) on, so that it is possible without any difficulty to recognize the order of animals to which a given embryo belongs."[25]

Walter Garstang of Oxford University, one of the first to study the functional biology of marine invertebrate larvae, argued against Haeckel's law. To Garstang, Haeckel's law was "demonstratively unsound." Garstang's best-known works on marine larvae were written in the form of poems that were published after his death in the book entitled *Larval Forms and Other Zoological Verses.* The book describes and illustrates the controversies in evolutionary biology of the time. In using poetic license, Garstang argues against Haeckel's law using a house or cottage metaphor: "A house is not a cottage with an extra story on top. A house represents [a] higher grade in the evolution of a residence, but the whole building is altered—foundations, timbers, and roof—even if the bricks are the same."[26]

By 1956, the renowned paleontologist and embryologist, Conrad Waddington, disillusioned by Haeckel's theory, left the field of embryology to explore the role of genetics in evolution. Commenting on the role of embryology in evolution, Waddington declared that this "type of analogical thinking which leads to theories that development is based on the recapitulation of ancestral stages, or the like, no longer seems at all convincing or even interesting to biologists."[27]

The train of Haeckel's critics continued to gain momentum in the late twentieth century. By 1958, eminent British embryologist Gavin de Beer published three editions of a book on embryology criticizing Haeckel's "ontogeny recapitulates phylogeny" theory: "Recapitulation .... i.e., the pressing back of adult ancestral stages into early stages of development of descendants, does not take place."[28] De Beer concluded in 1958 that Haeckel's recapitulation law was "a mental straight jacket" that has "thwarted and delayed" embryological research, and lamented that "the theory of recapitulation has had a great and, while it lasted, regrettable influence on the progress of embryology." [29, 30]

Haeckel's train of critics hooked up a long line of boxcars. Paul R. Ehrlich of Stanford University and author of the famous overpopulation book, *The Population Bomb*, wrote in 1963 that Haeckel's law now only has a leading role in mythology—not in science.

> This generalization was originally called the biogenetic law by Haeckel and is often stated as 'ontogeny recapitulates phylogeny.' This crude interpretation of embryological sequences will not stand close examination, however. Its shortcomings have been almost universally pointed out by modern authors, but the idea still has a prominent place in biological mythology.[31]

As director of the British Museum of Natural History and president of the Linnean Society, de Beer revisited Haeckel's biogenetic law, stating that the "enthusiasm of the German zoologist Ernst Haeckel, however, led to an erroneous and unfortunate exaggeration of the information."[32]

American evolutionary paleontologist George Gaylord Simpson, curator of the American Museum of Natural History and Museum of Comparative Anatomy at Harvard University, was quick to the chase in denouncing Haeckel's biogenetic law, writing in 1965 that it "is now firmly established that ontogeny does not repeat phylogeny."[33]

Though lacking evidence, Haeckel's biogenetic law is still debated in certain circles, probably because Darwin said embryology was "by far the strongest single class of facts," and because of the absence of any better evolutionary mechanism. In an article published in *Science* magazine in 1969, Walter Bock, professor of evolutionary biology at Columbia University, bemoans how "the biogenetic law has become so deeply rooted in biological thought that it cannot be weeded out in spite of its having been demonstrated to be wrong by numerous subsequent scholars."[34]

German embryologist Erich Blechschmidt of the University of Guttingen regarded Haeckel's "Great Biogenetic Law" as one of the most egregious errors in the history of biology. In his book *The Beginnings of Human Life*, no words were minced in repudiating Haeckel's fraudulent forgeries: "The so-called basic law of biogenetics is wrong. No buts or ifs can mitigate this fact. It is not even a tiny bit correct or correct in a different form. It is totally wrong."[35]

The train continued to garner speed, shifting into fifth gear as it rolled into the latter part of the twentieth century. In 1976, Dartmouth College embryologist William Ballard concluded that Haeckel's theory and drawings were fabrications, demonstrated "only by semantic tricks and subjective selection of evidence," and only by "bending the facts of nature."[36]

With the concept of evolution traveling in the fast fifth gear, the warnings of the nineteenth century scientists were left fighting through the dust to see the light.

Drawings by Design

Haeckel's most famous embryo drawings were published in 1874, long after the publication of the first edition of *The Origin of Species* in 1859. While Darwin's interest was in the theory, he no doubt was influenced by Haeckel's earlier drawings. In the sixth edition of *The Origin of Species*, published in 1972, Darwin refers to Haeckel's drawings: "Professor Haeckel, in his 'Generelle Morphologie' and in [other] works, has recently brought his great knowledge ... [regarding] the lines of descent of all organic beings. In drawing up the several series..."[37]

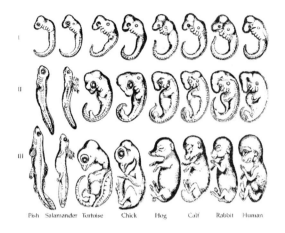

Fish Salamander Tortoise Chick Hog Calf Rabbit Human

While never using the phrase "ontogeny recapitulates phylogeny," Darwin clearly endorsed Haeckel's evidence: "As the embryo often shows us more or less plainly the structure of the less modified and ancient progenitor of the group, we can see why ancient and extinct forms so often resemble in their adult state the embryos of existing species of the same class."[38]

The now-infamous 1874 embryo drawing was certainly not Haeckel's first. Haeckel had previously been drawing embryos for the purpose of educating the public on the development of the embryo. In 1868, Haeckel published his drawings in two books—*Natural History of Creation* (*Naturliche Schopfungsgeschichte*) and *Uber die Enstehung and den Stammbaum des Menschengeschlechts*.

In reviewing Haeckel's books, Ludwig Rutimeyer, professor of zoology and comparative anatomy at the University of Basel, was quick to point out the problems with Haeckel's drawings and published these arguments against Haeckel in the *Archives of Anthropology* in 1868. Rutimeyer revealed that the dog embryo and human embryo shown on page 240 of Haeckel's book are completely identical, and that Haeckel had used the same drawing (woodcut) to portray a dog, a chicken, and a tortoise. Haeckel had only changed the label of the drawings. Rutimeyer concluded:

> Haeckel claims these works to be both easy for the scientific layman to follow, and scientific and scholarly. No one will quarrel with the first evaluation of the author, but the second quality is not one that he seriously can claim. These are works clothed in medieval formalistic garb. There is considerable manufacturing of

scientific evidence perpetrated. Yet the author has been very careful not to let the reader become aware of this state of affairs.[39]

Incredibly, nearly 150 years later, Haeckel's deeply rooted theory has remained a hot topic. In 1987, after years of investigation into other species, Canadian embryologist Richard Elinson of the University of Toronto reiterated Rutimeyer's 1868 observation: the early developmental patterns in frogs, chicks, and mice are not similar but are distinctly different. The evidence clearly undermines Haeckel's law. Elinson wrote that the embryology of these species "[is] radically different in such fundamental properties in egg size, fertilization mechanisms, cleavage pattern, and (gastrulation) movements."[40]

In 1997, in an effort to develop a scientific consensus on Haeckel's biogenetic law, Michael Richardson, an embryologist at St. George's Hospital Medical School, London, convened an international team of experts to rule on Haeckel's law. The team convened to examine and compare Haeckel's embryos to photographs of actual embryos from all seven classes of vertebrates.

The collaborative team collected embryos of fifty vertebrates from thirty-nine different species, including marsupials from Australia, tree-frogs from Puerto Rico, snakes from France, and an alligator embryo from England. Just as the previous scientific investigators discovered, rather than finding similarities, the scientific team found vast differences in the embryos of the different species. In fact, the species were so radically different that the team reported in the *Anatomy and Embryology* journal that Haeckel's drawings were actually unrelated to the real specimens—they were only theoretical drawings.[41]

In an interview of Richardson by Nigel Hawkes published in *The Times* (London), Richardson's team concluded that Haeckel was "an embryonic liar": "This is one of the worst cases of scientific fraud. It is shocking to find that somebody one thought was a great scientist was deliberately misleading. It makes me angry ... What he [Haeckel] did was to take a human embryo and copy it, pretending that the salamander and the pig and all the others looked the same at the same stage of development. They don't ... These are fakes."[42]

The team discovered that Haeckel not only changed the drawings by adding, omitting, and changing features, but he also forged the scale of the drawings. In a review of the team's findings by evolutionary geneticist Elizabeth Pennisi in an article entitled "Haeckel's Embryos: Fraud Rediscovered," published in *Science* in 1997, Pennisi stated, "[Haeckel] also fudged the scale to exaggerate similarities among species, even when there were 10-fold differences in size. Haeckel further blurred differences by neglecting to name the species in most cases, as if one representative was accurate for an entire group of animals."[43]

The team concluded that Haeckel drew a human embryo and played with the labeling. Richardson is quoted in *The Times* (London): "What he did was to take a human embryo and copy it, pretending that the salamander and the pig and all the others looked the same at the same stage of development."[44]

The theory—not the evidence—drove Haeckel. When the evidence did not fit the theory, Haeckel changed the evidence. In a 2001 article published in *Nature*, Richardson discovered that, "When we compare [Haeckel's] drawings of a young echidna embryo with the original, we find that he removed the limbs ... This cut was selective, applying only to the young stage. It was also systematic because he did it to other species in the picture. Its intent is to make the young embryos look more alike than they do in real life."[45]

Elizabeth Pennisi highlights the fact that Haeckel's drawings may be one of biology's most dreadful pages in history by quoting Richardson: "It looks like it's turning out to be one of the most famous fakes in biology."[46]

Richardson's team noticed that Haeckel, in creating the fraud—to support the "ontogeny recapitulates phylogeny" paradigm—had to change the relative sizes of the embryos as well as delete and fabricate features. Sadly, Haeckel's drawings have been insidiously woven into the reference books of western academia, even *Gray's Anatomy*. A review in the *New Scientist* concluded by stating: "Although Haeckel confessed to drawing from memory and was convicted of fraud at the University of Jena, the drawings persist. 'That's the real mystery,' says Richardson."[47]

Actually, the Richardson team's discovery of the fraud was not a first. In the same year the drawings were published in 1874, professor Wilhelm His of Oxford University declared Haeckel's drawings to be fraudulent. Under academic pressure, Haeckel eventually wrote a quasi confession, but unfortunately the confession became lost in the excitement of the growing popular interest in evolution after publication of the book *Darwin and After Darwin,* in 1901. Pennisi wrote in *Science,* "Haeckel's confession got lost after his drawings were subsequently used in a 1901 book called *Darwin and After Darwin* and reproduced widely in English language biology texts."[48]

Not until more than thirty-five years later, after the release of the drawings in 1874, that Haeckel's confession was actually released to the public. In January 1909, the confession was published as a letter in the *Münchener Allegemeine Zeitung,* an international weekly publication for the sciences, arts, and technology. In the letter, Haeckel clearly states that the drawings were contrived by "comparative synthesis" and not by accurate reproduction. Without the fraud, the expected evolutionary embryonic sequences had obvious gaps. Haeckel concedes, "a small portion of my embryo pictures (possibly 6 or 8 in a hundred) are really 'falsified'."[49]

Even more sadly, Haeckel's only defense was that fraudulent practices were an accepted practice even by some of the "most esteemed biologists" of the day. Haeckel wrote, "After this compromising confession of 'forgery' I should be obliged to consider myself condemned and annihilated if I had not the consolation of seeing side by side with me in the prisoner's dock hundreds of fellow culprits, among them many of the most trusted observers and most esteemed biologists. The great majority of all the diagrams in the best biological textbooks, treatises, and journals would incur in the same degree the charge of 'forgery,' for all of them are inexact, and are more or less doctored, schematised, and constructed."[50] Indeed the scientific method had been abandoned not only by Darwin and Haeckel, but also by a large segment of the profession of biology.

No wonder even by 1988 Keith Thomson, professor of biology at Yale University, summed up the status of Haeckel's biogenetic law, declaring, "Surely the biogenetic law is as dead as a doornail."[51]

Homology—Same, but Different

For Darwin, the similarity in the five-fingered hand was strong evidence to support Haeckel's biogenetic law.[52] Similarities, otherwise known as homologies, were thought to reflect common ancestries. Haeckel and Darwin argued that these similarities began in the embryo. As the embryo develops, each structure replays evolution and serves as a record of the past condition of the species with the structures originating "from corresponding embryonic parts." [53, 54]

Little did Darwin know that by rating the importance of embryology as "second to none," Darwin unwittingly undermined the theory. By 1932, with mounting scientific evidence, biologist Sir Arthur Keith, in the book *The Human Body,* concluded: "Embryology provides no support whatsoever for the evolutionary hypothesis."[55]

By 1956, in the introduction to Darwin's *Origin of Species,* even the eminent Canadian biologist W. R. Thompson clearly stated: "The 'Biogenetic Law' as a proof of evolution is valueless."[56] Darwin would have been devastated.

One of the most influential embryological scientists of the mid-twentieth century was Gavin de Beer. In 1954, de Beer was knighted by the throne, and awarded the Darwin Medal of the Royal Society in 1957. As the evidence continued to mount, though, even this most ardent Darwin supporter could no longer avoid the facts. Starting in 1958, de Beer began to publish known contradictions between the evidence and Haeckel's theory. De Beer noted that similar parts, like the five-fingered hand, simply do not develop from similar embryonic sites.[59]

In the book *Homology: An Unresolved Problem* (1971), de Beer compared the five-fingered hands of the amphibian, lizard, and man. De Beer discovered that while in the amphibian the hand originates from embryonic trunk segments 2, 3, 4, and 5, the amphibian's common ancestor, the five-fingered hand of the lizard originates from completely different embryonic sites, from trunk segments 6, 7, 8, and 9. From the evidence De Beer concluded: there is no embryonic sequential change from the amphibian to lizard.[60]

Since both the amphibian and the lizard are common ancestors to man, according to Haeckel's theory, the embryonic sites for the hand in man should be similar. When the embryonic sites of the five-fingered hand in man were examined, however, de Beer discovered again that the hand originated from different embryonic sites—trunk segments 13, 14, 15, 16, 17, and 18. The theory, "ontogeny recapitulates phylogeny" was not compatible with the evidence. Not only were the sites of origin different, the numbers of sites were different.[61]

Table VII
Embryonic Sites of Origin

Vertebrate	Trunk Segment Sites of Origin
Amphibian (Neut)	2, 3, 4, 5
Lizard	6, 7, 8, 9
Man	13, 14, 15, 16, 17, 18

De Beer discovered that not only was the development of the hand incompatible with Haeckel's theory, he found problems in the development of the amniotic sac. The amniotic sac that surrounds the developing embryo in reptiles, birds, and mammals is considered a homologous structure. However, the origin of the amniotic sac in mammals is completely different from that of reptiles and birds. De Beer discovered that the "correspondence between homologous structures cannot be pressed back to similarity of position of the cells in the embryo, or of the parts of the egg out of which the structures are ultimately composed."[62]

De Beer also discovered that evidence obtained from a developing eye of a frog contradicts Haeckel's biogenetic law. There are two species of frog, *Rana fusca* and *Rana esculent,* and the eyes of both species appear identical. In *Rana fusca,* the optic cap induces the epidermis to differentiate into a lens. De Beer discovered that if the optic cap is removed, no lens will develop, while in *Rana esculent,* if the optic cap is removed, the lens develops completely. De Beer concluded that even though the two species' eyes appear homologous, they do not originate from the same site in the embryo, contradicting Haeckel's "ontogeny recapitulates phylogeny" theory.[63]

One of Haeckel's famous fabrications was that the human embryo, while going through the fish stage, developed fishlike gills. This fabrication survived a long life without evidence. However, by 1965, William Beck and George Gaylord Simpson, in the widely used high school biology textbook *Life: An Introduction to Biology,* stated that the "human embryo does not have any differentiated gill tissue, and the gill-like pouches do not have open gill slits as in fishes. Fins are lacking. The tail is not at all like any fish's tail. Indeed, the resemblance to an adult fish is vague and superficial."[64]

Strongly held core values are difficult to change. By 1985, the picture was becoming clearer. Even developmental biologist Pere Alberch began to concede. In the article "Problems with the Interpretation of Developmental Sequences," published in the journal *Systematic Zoology* in 1985, Alberch states that it is "the rule rather than the exception" that "homologous structures form from distinctly dissimilar initial states."[66]

In 1985, stepping back and looking at the big picture, molecular biologist Michael Denton concluded in the pivotal book *Evolution: A Theory in Crisis* that the "demise of any sort of straightforward explanation for homology, one of the major pillars of evolutionary theory, has become so weakened that its value as evidence is greatly diminished."[67]

Phasing out Haeckel's law has been a process. By the late twentieth century, the link between embryology and homology became only a history lesson of "good reasoning, bad science." In the words of biologist Richard Hinchliffe, published in 1990 in the book entitled *Organizational Constraints on the Dynamics of Evolution*, "Embryology does not contribute to comparative morphology by providing evidence of limb homology."[68]

In 1999, writing in the journal *New Scientist,* evolutionary biologist Ken McNamara summarized Haeckel's 150-year legacy: "[Haeckel] called this the biogenetic law, and the idea became popularly known as recapitulation. In fact, Haeckel's strict law was soon shown to be incorrect. For instance, the early human embryo never has functioning gills like a fish, and never passes through stages that look like an adult reptile or monkey."[69]

Clearly, embryology alone is no longer "second to none" in importance to Darwin's theory, or Darwin's theory is dead. However, interest in Darwin's theory is not dead.

The Evo-Devo Deal

By the early 1990s after more than 150 years of investigation, a consistent natural law accounting for evolution had become more elusive than ever. The fossil record, molecular biology, nor embryology alone could trace the steps of evolution, and these disciplines were working together. With reality knocking at the door, what emerged as a priority was the need to integrate the disciplines and facilitate the convergence of evidence.

To facilitate the convergence of evidence, a new discipline developed called evolutionary developmental biology, nicknamed evo-devo. By integrating evidence from paleontologists to geneticists, the goal was to develop a unified theory and discover the elusive natural laws of evolution.

Ironically, the model for this new scientific perspective developed out of Haeckel's "ontogeny recapitulates phylogeny" theory. In the 2007 book *From Embryology to Evo-Devo: A History of Developmental Evolution*, biology historians Manfred D. Laubichler and Jane Maienschein acknowledges that even though Haeckel's theory has not withstood the test of time, the intuitive component of the theory remains the best-known model to study the mechanisms of evolution:

> Ernst Haeckel's biogenetic law, 'ontogeny recapitulates phylogeny,' still represents the canonical formulation of this relationship. The fact that even though it has long been disproven, at least in its radical form, the biogenetic law still discussed in textbooks is, at the very least, a testament to its intuitive appeal.[70]

To launch this new perspective, two new journals were founded: *Evolution & Development,* and *Molecular and Developmental Evolution,* an independent section of the *Journal of Experimental Zoology.* Even *Archiv für Entwicklungsmechanik der Organismen,* founded in 1890 and the oldest journal in the field of experimental embryology, was renamed *Genes, Development, and Evolution.*

To support evo-devo, the National Science Foundation has established a specific panel devoted to the evolutionary developmental biology. The Society for Integrative and Comparative Biology (formerly the American Society of Zoologists) now has specific sections for evolutionary developmental biology.

Evolutionary biologist Brian K. Hall wrote in *Scientific American* in 2005 "it could be argued that evo-devo was born when Darwin concluded that the study of embryos would provide the best evidence for evolution."[71]

One of the first challenges of evo-devo was to identify a series of species to serve as a model to integrate the evidence. While evo-devo researchers have tested a range of species from sea anemones to dung beetles, no such species has emerged as a potential model. Evolutionary biologists Ronald Jenner and Matthew Wills, from the Department of Biology and Biochemistry at the University of Bath, England, are quoted in *Science Daily* as stating that it "is fair to say that, since its inception, some workers feel that evo-devo hasn't quite lived up to its early expectations."[72]

Like the fossils, molecular biology, and now embryology, the convergence of evidence has so far failed to reveal the consistent sequential changes theorized by Darwin. After more than fifteen years of research, now skepticism abounds regarding whether evo-devo will ever deliver an integrated mechanism for evolution. Biology historians Manfred D. Laubichler and Jane Maienschein concede that there are fundamental problems with this integrated perspective:

> On the one hand, current evolutionary developmental biology includes more than just developmental and evolutionary biology; on the other hand, it is still unclear whether the Evo-Devo focus can succeed in providing new perspectives.[73]

Today the goal of discovering macroevolution from species to species has given way to the study of changes within a species—microevolution.[74] Like any entity going out of business, convergence of diverse disciplines has sparked raging debates over even the definition of evo-devo. At the end of the proverbial day, this embryological evo-devo perspective has once again not revealed any natural law that can account for evolution.

Today, the phenomenon of homology, which Darwin thought so perfectly accounted for descent from a common ancestor, survives only as a myth in dusty library books. Perhaps Darwin was on track when he unknowingly wrote about the edge of science:

> He will be forced to admit that these great and sudden transformations have left no trace of their action on the embryo. To admit all this is, as it seems to me, to enter into the realms of miracle, and to leave those of Science.[75]

Chapter Twelve
The Rise of Genetics

Towards the end of the work I gave my well abused hypothesis of Pangenesis. An unverified hypothesis is of little or no value.
—*Charles Darwin*[1]

The rise and fall of genetics as the explanation of evolution over the past 150 years is without question one of the most astounding stories of modern science. Never in the history of science has there been a rival pursuit comparable to the study of genetics.

Times have changed since 1859, however. In the first edition of *The Origin of Species,* Darwin never used the terms genetics, genetic, or genes. By the sixth edition, Darwin used genetic twice, but never genetics, genes, or gene. The reason is the field of modern genetics was unknown to Darwin. By the turn of the twenty-first century however, genetics had become the cornerstone of evolution.

New Variations

The study of genetic has addressed two of Darwin's greatest theoretical dilemmas: the origin of new variations and the inheritance of new variations by the next generation. In 1859, Darwin proposed "use and disuse" along with pangenesis to explain the origin of new variations and the "blending" method of inheritance. As Darwin was not yet certain of these explanations, he wrote in *The Origin of Species,* "Our ignorance of the laws of variation is profound."[2]

The origins of new variations are central to Darwin's theory. For natural selection to operate there must be a source of new variations. Without new variations, natural selection is essentially rendered useless. Darwin wrote, "natural selection can do nothing until favourable individual differences or variations occur."[3]

Use and Disuse

In search of an explanation for the origin of new variations, Darwin subscribed to the logic of Greek philosophers Hippocrates and Aristotle and the earlier French naturalists Comte de Buffon and Jean-Baptiste Lamarck. The logic proposed that new variations originate through the process of "use it or lose it." In the book *Philosophie Zoologique,* published in 1809, Lamarck proposed two natural laws for evolution. The first law that Lamarck proposed was that "a more frequent and continuous use of any organ gradually strengthens, develops, and enlarges that organ ... while the permanent disuse of any organ ... progressively diminishes its functional capacity, until it finally disappears."[4]

In the second law, Lamarck envisioned that these new variations are preserved through the action of reproduction.[5] Eventually ascribing to Lamarck's version of evolution, Darwin wrote:

> Lamarck was the first man whose conclusions on the subject excited much attention. This justly celebrated naturalist first published his views in 1801 ... he first did the eminent service of arousing attention to the probability of all changes in the organic, as well as in the inorganic world, being the result of law, and not of miraculous interposition.[6]

In theory, Darwin envisioned new variations arise through the process of use and disuse. These new variations are then acted on by natural selection bring about the evolution of new species. Darwin wrote, "We should keep in mind, as I have before insisted, that the inherited effects of the increased use of parts, and perhaps of their disuse, will be strengthened by natural selection."[7]

This concept of use and disuse was later challenged by German biologist August Weismann in 1883. In the search for the actual origin of new variations, Darwin proposed an even more fundamental problem—pangenesis.

Pangenesis Challenge by Pasteur

Pangenesis became Darwin's approach for explaining how new variations actually originate. Pangenesis was similar in concept to the Aristotelian school of "spontaneous generation," popularized by Lamarck, in which new life-forms appear spontaneously.

In the mid-nineteenth century, the topic of spontaneous generation was hot and controversial. Spontaneous generation was first challenged in 1668 by the Italian Francesco Redi and later by Italian biologist Lazzaro Spallanzani in 1768. In 1861, Louis Pasteur performed a series of experiments that finally and decisively undermined the foundations of spontaneous generation by supporting the cell theory.

Darwin derived the term "pangenesis" from Greek fertility deity "Pan," and the term "genesis," meaning "the coming into being"— origin. Darwin outlined the theory of pangenesis in his 1868 work *The Variation of Animals and Plants Under Domestication.* Pangenesis is not mentioned in *The Origin of Species,* even in the sixth edition, which was published in 1872. As reflected in a letter to John Jenner Weir in 1868, Darwin knew that the theory was sketchy, at best: "You will find pangenesis stiff reading, and I fear [you] will shake your head in disapproval."[8]

In the pangenesis theory as defined by Darwin, the entire body contains atoms or units, which are capable of being programmed by the experiences of life and are capable of reproduction.[9] These atoms or units are called "gemmules." According to Darwin, these gemmules become the "sexual elements" of reproduction that determine the constitution of the next generation.[10]

Pangenesis was central to Darwin's theory. In a letter to Charles Lyell in 1867, Darwin wrote, "I am inclined to think that if it be admitted as a probable hypothesis it will be a somewhat important step in biology."[11] While the transfer of accumulated experiences of life to the next generation was central to Darwin's theory, the concept of pangenesis was more of a dream than a discovery. In an 1867 letter to Asa Gray, Darwin wrote:

> The chapter on what I call Pangenesis will be called a
> mad dream, and I shall be pretty well satisfied if you

think it a dream worth publishing; but at the bottom of my own mind I think it contains a great truth.[12]

One of the major problems for Darwin was discovering how new variations could actually originate. Since Darwin was desperate for a source of new variations, the concept of gemmules seemed to be a logical way to accumulate and deliver new information to the next generation. Italian geneticist Giuseppe Sermonti explains how Darwin envisioned the offspring acquiring new variations from the cumulative experience of the parents:

> According to pangenesis, the entire organism generates the offspring. Only in this way could Darwin explain the evolution of species—i.e., as a decanting of the vicissitudes of the parents' lives into the offspring. For Darwin, evolution was the cumulative experience of the world's organisms over time.[13]

Darwin's theory was based only on logic and not scientific evidence. There is no scientific evidence to support pangenesis. In hedging the stakes with the pangenesis theory, in *The Variation of Animals and Plants Under Domestication,* Darwin entitles the chapter twenty-seven "Provisional Hypothesis of Pangenesis." Provisional it was, and provisional it remains.

Pangenesis was simply speculative theorizing, hoping that it would be shown to be true some day. In a letter to Fritz Müller in 1870, Darwin wrote, "Pangenesis will turn out true some day!"[14] In 1867, Darwin confided to Joseph D. Hooker that pangenesis was purely speculative: "I should very much like to hear what you think of 'Pangenesis,' though I fear it will appear to *every one* far too speculative."[15] With a little play on words, Darwin wrote a letter to Hooker expressing concern over the "provisional hypothesis" of pangenesis, stating, "I fear Pangenesis is stillborn."[16]

By the late nineteenth century, as the eclipsing of Darwinian evolution seemed imminent, the search was on to discover the origin of new variations. Without an origin for new variations, Darwin's theory was destined to be another philosophical relic. While Darwin was working on concepts of blending and pangenesis, a monk, using the scientific method, made a discovery that would revolutionize Darwinism into neo-Darwinism.

Mendel Challenges Blending

Darwin subscribed to the popular belief of "blending inheritance." In the same way that black and white paint can be blended into a gray color, "blending inheritance" resulted from the mating of two different species. How this concept was thought to play out is seen in a letter to Joseph Hooker, in which Darwin wrote:

> If you cross two very distinct races, you may make (not that I believe such has often been made) a third and new intermediate race. But if you cross two exceedingly close races or two slightly different individuals of the same race, then in fact you annul and obliterate the differences.[17]

Ironically, blending inheritance in effect actually reduces variation. Just as two colors blended together become one, two species "blended together" become one—an intermediate species.

Unfortunately, little did Darwin know that even before the publication of the fourth edition of *The Origin of Species* in 1866, Gregor Mendel had presented the now-famous paper entitled "Experiments on Plant Hybridization," laying the foundations of modern genetics.

To promote blending inheritance, Darwin presented a diagram in *The Origin of Species,* with the explanation that "on the principle of inheritance, all the forms descended, for instance from A, would have something in common. In a tree, we can distinguish this or that branch, though at the actual fork the two unite and blend together."[18] Darwin reasoned that in order to maintain species, related species must be kept separate to avoid blending of two species into one species.[19]

Even more damaging to the evolutionary theory with blending inheritance is the fact that neither the black nor the white color can be recovered from the color gray. Clearly, Darwin was unaware that Mendel had discovered that variations are separate and independent. With Mendelian genetics, the black and white in the gray can be recovered because they still maintain their distinctiveness as red and white variations.

The concept of blending inheritance was completely incompatible with evolution. With blending inheritance, once a trait is lost, the trait is no longer recoverable, similar to the way color once mixed cannot be recovered. Unrecoverable variations shrink—not expand—the spectrum of gene variations.

While Darwin was using logic to devise theories in the Down house library, Mendel was observing, measuring, and experimenting with pea plant in the field. Using the scientific method, Mendel discovered two fundamental laws of inheritance that were to demolish Darwin's theory of blending inheritance and pangenesis.

In 1865, Mendel presented his findings at the Natural History Society of Brünn in Moravia. The following year, Mendel's "Experiments on Plant Hybridization" paper was published in the *Proceedings of the Natural History Society of Brünn*.

Now known as the "Father of Modern Genetics," Mendel is credited with discovering two laws of inheritance: the law of separation and the law of independence. Given a white and a black expression in the first generation, combining white and black in the second generation was observed to result in a gray expression. In the third generation, because the inheritance is separate and independent, the white and black were once again expressed. Characteristics remain independent and separate through the generations.

Mendel was known to have read Darwin, but there is no evidence that Darwin was even

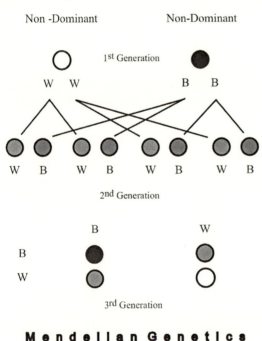

Non -Dominant   Non-Dominant

1st Generation

W   W   B   B

W B   W B   W B   W B

2nd Generation

B   W

B
W

3rd Generation

Mendelian Genetics

aware of Mendel. Mendel's findings were largely ignored because the theory of blending inheritance had been widely accepted and promoted by scientists, including Darwin. Geneticist Sermonti notes, "What really happened was that Mendelism ruled out almost all of the forces that Darwin had invoked to explain evolution."[20]

While neither Darwin's pangenesis hypothesis nor Mendel's laws of inheritance garnered much attention, challenges to Lamarckian concepts of "use and disuse" and "spontaneous generation" continued.

In 1869, just ten years after the publication of *The Origin of Species,* the first step in identifying what is now known as DNA was discovered. In Switzerland working with white blood cells, Friedrich Miescher at Felix Hoppe-Seyler's laboratory at the University of Tübingen discovered a substance he called "nuclein." Eventually, Miescher's student, Richard Altmann, renamed "nuclein" nucleic acid.

Between 1869 and the early twentieth century, a succession of discoveries began revealing how nucleic acid held the keys to inheritance. Chromosomes were then later discovered in 1875 by German biologist Oskar Hertwig while he was studying reproduction in sea urchins. Along with the new discoveries came mounting challenges for the theories of pangenesis and blending inheritance.

Weismann Barrier

German biologist August Weismann, at the University of Freiburg, launched the first scientific evidence directly challenging Darwin's theory. Now known as the "Weisman Barrier," in 1883, Weismann cut off the tails of mice from twenty-one generations. Seeing that the twenty-second generation still had tails, Weismann concluded that the evidence contradicted Darwin's theory of use and disuse. Weismann concluded that despite obvious reasons for change in the mice, "continuity" was observed, not new variations.[21]

The search was on to find the origin of new variations—the engine of evolution. Darwin's theory of use and disuse was finished. Ernst Mayr called Weismann "the second most notable evolutionary theorist of the nineteenth century, after Charles Darwin." Fortunately, the emerging technologies were revealing the molecular biology of inheritance through nucleic acid—deoxyribonucleic acid (DNA).

The Rediscovery

In 1885, German biochemist Albrecht Kossel discovered the chemistry of one of Friedrich Miescher's "nuclein" compounds, now known as "adenine." Writing, "it is highly probable that four nucleic acids exist," Kossel eventually discovered all four DNA nucleic acids by 1894. These nucleic acids were named adenine, guanine, thymine, and cytosine. A fifth nucleic acid, uracil, was discovered later in 1900. For his work on determining the composition of Miescher's nuclein, Kossel was awarded a Nobel Prize in 1910.

German botanist, Carl Correns, in taking the same approach as Mendel, studied inheritance in the hawkweed plant. In January 1900, Correns published his results, citing both Darwin and Mendel. In his paper entitled "G. Mendel's Law Concerning the Behavior of the Progeny of Racial Hybrids," Correns is credited for rediscovering and reinstating Mendel's laws of segregation and independence.

At the same time that Correns was studying the hawkweed plant, Dutch botanist Hugo de Vries was experimenting with hybridizing the evening primrose plant. De Vries also confirmed Mendel's laws of separation and independence. De Vries published the results of his research in the French journal *Comtes Rendus de l'Académie des Sciences* later in 1900. Modifying Darwin's pangenesis hypothesis, de Vries coined the term "pangenes" to indicate that inheritance is transferred by particles, which are now known as DNA. Recognizing Darwin's limitation, De Vries wrote the now-famous phrase: "Natural selection may explain the survival of the fittest, but it cannot explain the arrival of the fittest."[22]

By the turn of the twentieth century, the mechanism of Darwin's theory was going through a crisis. Not only were the vast paleontological expeditions failing to reveal the "inconceivably great" number of missing links in the fossil record, but also the blending theory was not lining up with the biological evidence for inheritance. Stepping out to contradict Darwin's theory of natural selection through successive, slight changes, de Vries proposed that species actually arose in a single event. According to de Vries, a mutation was at the origin of this event.

Even though the concept of genetic mutations was unknown to de Vries in 1900, the mutation theory was eventually to emerge as the chief contender for revising the mechanism of Darwin's theory by eliminating the theories of blending inheritance, pangenesis, use and disuse, and spontaneous generation. What was emerging was a new form of Darwinism—neo-Darwinism. Research into the theory of evolution, once overshadowed by its proponents, Lamarck and Darwin, was set for a new approach. Sermonti notes: "Mendel remained unnoticed for 35 years, as long as Darwinism was in the ascendant, and his laws came to light only when evolutionism was going through a crisis late in the early 1900s."[23]

Contrary to Darwin's theory of blending, Correns, de Vries, and Austrian Erich von Tschermak, using the scientific method, reestablished Mendel's laws that new variations do not arise, but are static, permanent, and indifferent to the environment. The principles of Mendel's discovery in 1866 were established long before the gene's physical entity was discovered.

Mendel's rediscovery was not universally accepted and especially not by Darwin's old guard. In 1908, Russel Wallace, commenting on the rediscovery of Mendel's laws, wrote, "As playing any essential part in the scheme of organic development, the phenomena seem to me to be of the very slightest importance."[24]

The rediscovery of Mendel's laws, however, opened a new, yet unknown frontier. English biologist William Bateson became the first person known to use the term "genetics" in a personal letter to Alan Sedgwick in 1905. Scientists eagerly launched into the mission of discovering the role genetics plays in evolution. In 1909, Danish botanist Wilhelm Johanssen coined the word "gene" for the hereditary unit found on a chromosome, but how evolution was linked to Mendel's laws was still unknown.

The rediscovery of Mendel's laws eclipsed Darwin's version of inheritance, placing the mechanisms of evolution into a tailspin. However, in 1942, Julian Huxley published *Evolution: The Modern Synthesis,* and Ernst Mayr published *Systematics and The Origin of Species,* linking genetic mutations to the origin of new variations. The link formulated a foundation for a new Darwinism.

Names given to this new Darwinism include "modern synthesis," "synthetic theory," "modern evolutionary synthesis," and neo-Darwinism. Genetic mutations solved Darwin's largest unresolved problems—the origin and inheritance of new variations.[25] The problem of the origin of variation was of no small concern to Darwin. How variations arise was never resolved. Darwin concedes, "we are far too ignorant to speculate on the relative importance of the several known and unknown causes of variation."[26] Again, Darwin would not even "pretend to assign any reason why this or that part has varied."[27]

Genetic mutations as the origin of variation complemented natural selection perfectly. Natural selection needed new variations, but natural selection could not produce variations. Darwin had written that "natural selection, or the survival of the fittest, does not necessarily include progressive development—it only takes advantage of such variations as [they] arise and are beneficial to each creature under its complex relations of life."[28]

The works of Huxley and Mayr formalized a blending of Darwin and Mendel's work. At the core of the proposed mechanism for evolution, the elusive new variations were thought to arise from random genetic mutations. With new variations originating through mutation, natural selection could act and lead to the evolution of new species. Darwin explains, "natural selection acts solely by accumulating slight, successive, favourable variations, it can produce no great or sudden modifications."[29]

What had been emerging as the unifying concept for Huxley and Mayr was the work of Thomas Hunt Morgan. Starting in 1908, Morgan began exposing fruit flies to radiation and observing how radiation-induced mutations caused morphological changes in the fruit fly.

Extrapolating on de Vries concepts of mutation, Thomas Morgan published his theory in his 1915 book *The Mechanism of Mendelian Inheritance.* Morgan was eventually awarded a Nobel Prize in 1933 for identifying chromosomes as the vector of inheritance. Conrad Hall Waddington, reflecting on the importance of Morgan's work, wrote: "Morgan's theory of the chromosome represents a great leap of imagination comparable with Galileo or Newton."

Many questioned the relevance of the finding, including Thomas Morgan. Commenting on the paradox between the ubiquitous nature of similar genes yet with differential expression: "At first sight it may seem paradoxical that a guinea pig that can develop areas of black hair should have white areas of hair if, as [is] the case, the cells of both areas carry the same genes."[30]

Through the 1930s, the number of geneticists who accepted Morgan's theory as giving rise to new variations for evolution continued to increase. In 1937, American geneticist Theodosius Dobzhansky subscribed to the fundamental tenet of neo-Darwinism:

> Mutations and chromosomal changes ... constantly and unremittingly supply the raw materials for evolution.[31]

Dobzhansky was one of the engineers of the modern evolutionary synthesis, which united Mendelian genetics with evolution. Neo-Darwinism incorporates natural selection and genetic mutations. One of the pioneers of neo-Darwinism, Dobzhansky envisioned that evolution originates through changes in DNA.[32]

DNA on Center Stage

In a series of key experiments, microbiologists George Beadle and Edward Tatum demonstrated the biochemical role of DNA by exposing the bread mold *Neurospora crassa* to radiation. Radiation-induced mutations resulted in changes in specific enzymes. In a paper published in 1941, Beadle and Tatum proposed a direct link between gene mutations and changes in enzymatic reactions, which lead to the "one gene, one enzyme" hypothesis.[33] In 1958, Beadle and Tatum were awarded the Nobel Prize for developing the concept that one gene specifies for one enzyme rather than multiple enzymes.

In clarifying the central role of DNA in inheritance, molecular biologists Oswald Avery, Colin McLeod, and Maclyn McCarty performed experiments with different strains of bacteria, using genetic material, to transform one strain of bacteria into another. The bacterial species *Pneumococcus* can exist in two strains, the "S" (smooth) and the "R" (rough). The smooth strain is covered with a polysaccharide capsule that protects the bacteria from the host immune system, while the rough strain does not have a protective coating. When the smooth strain's genetic material was transferred into the rough strain, the rough strain was transformed into the smooth strain. These results, which demonstrated that genetic material controls even the strain of bacteria, were published in the *Journal of Experimental Medicine* in 1944.[34] The evidence seemed clear—not only does DNA play a key role in determining the biochemistry of the organism, it plays a key role in inheritance.

Inroads into the mechanisms of inheritance hinted that new variations in DNA were the driving mechanism in the evolution of life. Evolution through neo-Darwinism emerged as a fact. American paleontologist George Gaylord Simpson, the most influential paleontologist of the twentieth century, became a major player in creating a fusion of Darwinian natural selection and Mendelian genetics, as highlighted in the widely popular book *Tempo and Mode in Evolution,* published in 1944.[35]

By 1945, though the connection between mutation and DNA still had not been established, mounting scientific evidence continued to establish a direct link between DNA and biological control. In a series of experiments in 1952, American biochemists Alfred Day Hershey and Martha Chase discovered that nucleic acids, not proteins, are the genetic materials and means of inheritance.[36]

The Hershey—Chase experiment paved the way for the next step—discovering the structure of DNA, seemingly the code of life. Excitement over the prospect of discovering the structure of the code of life was electrifying.

Early in 1953, using X-ray crystallography, Nobel Prize winner Linus Pauling proposed a triple helix model of DNA. Taking the same type of approach, James D. Watson and Francis Crick, English scientists at King's College London, proposed their own structure of DNA, the double helix, which was published later in 1953.[37]

The discovery was monumental. Now that the actual structure of genetic information had been discovered, the mysterious events of evolution were open for further molecular investigation. Since it was agreed that genetics held the key to discovering evolution's Tree of Life, Watson and Click's article became symbolic of a transition between two ages—the "classical age" of biology, and the "new" age of molecular biology.

Eventually, in 1962, the trio of Watson, Crick, and King's College colleague Maurice Wilkins was awarded the Nobel Prize for discovering the molecular structure of nucleic acids. Since Watson and Crick's discovery in 1953, investigations into evolutionary biology have moved from the fossil fields to the laboratory benches, as changes in DNA became the leading field of investigation.

By the late 1950s, neo-Darwinism—evolution through mutation and natural selection—had become a standard subject in biology textbooks. In the 1957 biology textbook *An Introduction to Biology*, Gaylord Simpson, of the American Museum of Natural History, wrote, "Mutations are the ultimate raw materials for evolution."[38]

Central Dogma

In 1958, with the structure of DNA known, Francis Crick declared the "central dogma of molecular biology"—evolution acts through DNA mutations and natural selection—now more popularly known as neo-Darwinism. In theory, the dogma maintains that DNA contains life's genetic information based on the sequences of nucleotides in DNA. DNA alone regulates the physical processes of life through molecular synthesis.[39] Sermonti, in 2005, wrote:

> One of the fundamental principles of molecular biology (now enshrined as the 'Central Dogma') assigned to DNA the role of absolute governor of life and inheritance for the cell, and consequently from the organism.[40]

In 1961, French biochemists Jacques Monod and François Jacob further discovered that DNA regulates cell metabolism by directing the biosynthesis of enzymes, but through a carrier molecule—not by DNA alone. Specifically, Monod and Jacob reported finding a carrier step between the DNA and the synthesis of proteins. The carrier step involves a messenger—messenger ribonucleic acid—known as mRNA. Once encoded by DNA, mRNA carries "information" encoded in the cell for protein synthesis. For this discovery, Monod and Jacob, along with André Lwoff, were awarded the Nobel Prize in 1965. Jacob wrote, "everything is written in the nucleic acid message."[41]

The central dogma, which maintains that DNA contains the information that governs life, places DNA center stage as the controller of evolution. In DNA, acquired or induced chance mutations from the environment create the new variations. Natural selection, then, acts to preserve the new variations. The accumulation of new variations eventually leads to the evolution of a new species.

By 1970, the central dogma of neo-Darwinism was well established in academic circles. In 1971, Jacques Monod, who was one of the three men awarded the Nobel Prize in Physiology or Medicine in 1965, declared in the book *Le Hasard et la nécessité (Chance and Necessity)*: "The mechanisms of Darwinism [are] at last securely founded."[42]

Crick advanced the "central dogma of molecular biology" in a *Nature* paper published in 1970, stating that not only is inherited genetic information stored in DNA, but also once information is transferred into proteins, the information cannot be transferred back to DNA. Crick wrote that the "central dogma of molecular biology deals with the detailed residue-by-residue transfer of sequential information. It states that such information cannot be transferred back from protein to either protein or nucleic acid."[43]

In 1983, Douglas J. Futuyma, from the State University of New York at Stony Brook, became an ardent supporter of Crick's central dogma: "By far the most important way in which chance influences evolution is the process of mutation. Mutation is, ultimately, the source of new genetic variations, and without genetic variation, there cannot be genetic change. Mutation is therefore necessary for evolution."[44]

The emerging new technologies were opening the laws of nature. Life is simply like a set of building blocks—as predicable as Newton's laws of gravity. As Sermonti observed, "It seemed as though life could be disassembled and reassembled like a child's blocks. Some people then placed their faith in the omnipotence of biology and the prospect—it seemed only a matter of time—of being able to put life together and change it in a test tube."[45]

Four-Winged Fruit Fly

With the central dogma widely accepted, the next scientific frontier was to trace the actions of genetic mutation and natural selection in the formation of a new species from a DNA perspective.

Earlier in the twentieth century, Thomas Morgan, working with fruit flies, was one of the first to demonstrate how mutations can change a species. His results were published in the 1915 book *The*

Mechanism of Mendelian Inheritance. Even though DNA had not yet been discovered, Morgan was able to mutate two-winged fruit flies into four-winged fruit flies using radiation. The four-winged fruit fly was widely heralded as the earliest evidence that the first step in evolution was a mutation.

Later in the century, in 1978, geneticist Edward Lewis of the California Institute of Technology reproduced Morgan's mutations while studying the fruit fly mutation using the new genetic technology. Expanding on the Beadle–Tatum theory of "one gene, one enzyme," Lewis demonstrated that the four-winged fruit fly was actually produced by three mutations in the single large gene called Ultrabithorax.[46] Lewis was awarded a Nobel Prize in 1995 for his work on the fruit fly.

Even though four-winged fruit flies could be produced with radiation-induced mutations, the question was, were the mutations actually beneficial? What was the advantage of the mutation? Ironically, the four-winged fruit fly is not as well suited for flight as the two-winged fruit fly. Commenting on fruit flies' flying abilities, Ernst Mayr, in 1963, wrote that the mutated four-winged fruit flies "are such evident freaks that these monsters can be designated only as 'hopeless.' They are so utterly unbalanced that they would not have the slightest chance of escaping elimination."[47]

The four-winged fruit fly is a useful window on the genetics of development, but it provides no evidence that mutations are advantageous or actually supply the raw materials for evolution. Sermonti wrote, "One spur to research on mutations was the hope that an accumulation of these might lead to a new species. But this never happened."[48]

Sequence to Complexity?

At the core of evolution is the concept of sequential changes. The Wikipedia.org states, "evolution is the process of change in the inherited traits of a population of organisms from one generation to the next."[49] The "change" in evolution is from the simple to the more complex—molecules to man. Darwin wrote that as "natural selection acts solely by accumulating slight, successive, favourable variations, it can produce no great or sudden modifications."[50]

In examining the genetics of the fruit fly, there should be "slight, successive" genetic changes from one generation to the next and from species to species. In the family of fruit flies, known as *Drosophila*, there are more than fourteen hundred species. According to the theory, there should be "slight, successive" changes in the DNA nucleotides between species.

However, the evidence demonstrates that there are no "slight, successive" changes in nucleotide base pairs from species to species. In the *Drosophila* species, the number of nucleotide base pairs ranges from 127 to 800 million. Efforts to construct a Tree of Life through "slight, successive" changes in nucleotide base pairs from species to species have been unsuccessful. Each species of *Drosophila* appears to remain distinct and unique. Table VIII lists the number of estimated genome sizes as measured by the number of nucleotide base pairs in several different *Drosophila* genomes.

Table VIII
Nucleotides within Species

Drosophila Species	Nucleotide Pairs (in millions)
D. americana	300
D. arizonensis	225
D. eohydei	234
D. funebris	255
D. hydei	202
D. melanogaster	180
D. miranda	300
D. nasutoides	800
D. neohydei	192
D. simulans	127.5
D. virilis	345

Breeding different *Drosophila* species together leads to infertile offspring, indicating that the species are distinct and cannot be blended. Not only are the species distinct, but also research indicates that the species have not evolved but have remained stable. In 1977, in the book *Evolution of Living Organisms,* Pierre-Paul Grassé, president of the French Academie des Sciences, states: "The fruitfly [*Drosophila melanogaster*], the favorite pet insect of the geneticists, whose geographical, biotopical, urban, and rural genotypes are now known inside out, seems not to have changed since the remotest times."[51]

Similar to the unexpected non-successive genetic changes seen with *Drosophila* in the number of base pairs from species to species, there appears to be no "slight, successive" changes in the number of nucleotide base pairs from fish to mammals. While fish and man have approximately three to five billion nucleotide base pairs, the amphibian, which is considered an intermediate in evolutionary terms, has 10 to 100 billion nucleotide base pairs, demonstrating a disconnection between increased complexity and evolution. Sermonti wrote, "Displaying the table of the genetic code as though it demonstrated the unveiling of life's interlocking puzzle is a mistaken enterprise."[52]

Table IX
Nucleotides between Species

Species	Nucleotide Pairs (in billions)[55]
Fish	0.3 to 3
Birds	2
Worms	2
Reptiles	2
Mammals	2
Amphibians	10–100
Flowering Plants	500

The same problem emerges when searching for the Tree of Life from the "slight, successive" changes on the number of genes. Ironically, the fruit fly has approximately the same number of genes as humans. These observations drove Sermonti to reflect: "It was thought that a gene count might offer a better index of organismal complexity.... And where did this all lead? To the conclusion that biochemical complexity has little to offer in explaining evolution."[54]

Table X
Genes between Species

Species	Gene Count (in thousands)[55]
Bacteria	3–5
Yeast	6
Cenorhabditis elegans (worm)	20
Humans	25–30
Drosophilia (Fruit fly)	26

What was once a closed case in the mid-twentieth century, Crick's central dogma is now being challenged by emerging new genetic discoveries. The stage is now set for a new revolution in neo-Darwinism. Sermonti suggests that the "arbitrary nature of the genetic code imparts to the whole of life a general character of mystery and indetermination."[56]

Even in counting chromosomal pairs, no corresponding increase in complexity exists. The donkey has an estimated thirty-one chromosomal pairs, while humans have only twenty-three; additionally, the fern plant has five times more chromosomes than humans. There is a disconnection between the number of chromosomes and evolution. Sermonti notes: "Between chromosome number and evolution of species, it was immediately clear that no clear relationship exists."[57]

Table XI
Chromosome Pairs between Species

Species	Chromosome Pairs[55]
Ascaris (Roundworm)	1
Drosophilia (Fruit Fly)	4
Zebras	16
Soft Wheat	21
Humans	23
Donkeys	31
Horses	31
Fern	150

One of the biggest problems for the central dogma centers on the origin of DNA from an inert Earth. In order for mutations to occur, there first must be an ordered series of nucleotides to mutate. The most commonly recognized simplest organism, excluding viruses, is *Mycoplasma genitalium,* which has about 580,000 nucleotide base pairs.

This brings up the question, how did the DNA of *Mycoplasma genitalium* originally organize by chance? Spontaneous organization of 580,000 nucleotide base pairs statistically eliminates nature as an original cause and evokes Lamarck and Darwin's concepts of "spontaneous generation"—a theory long rejected by Pasteur.

Of all the range of issues surrounding evolution, the only universally accepted concept among scientists is that the molecular origins of life has no universal central dogma. Reflecting on the limits of science, Sermonti concludes: "Science has taken on the great wager … and lost."[59]

Chapter Thirteen
Evidence

I cannot see how the belief that all organic beings, including man, have been genetically derived from some simple being, instead of having been separately created.
—Charles Darwin, in a letter to Mrs. Boole, December 14, 1866.[1]

In 1859 when the first edition of *The Origin of Species* was published, Mendel was still six years away from publishing his sentinel paper, "Experiments on Plant Hybridization." While Darwin only tentatively gave credence to the popular concept of spontaneous generation, he envisioned that "natural selection has been the most important, but not the exclusive, means of modification."[2]

Since natural selection was one of the most important means, what direct evidence did Darwin document in drawing that conclusion? Lengthy arguments for evidence that plants and animals struggle for survival fill the pages of *The Origin of Species*. Specifically, Darwin argued along the same lines as Lamarck, that the long neck of the giraffe developed from the struggle to reach the leaves higher in the trees.

This evidence, though, was only indirect evidence. Darwin explains: "In order to make it clear how, as I believe, natural selection acts, I must beg permission to give one or two imaginary illustrations."[3] This is the first sentence in the section entitled "Illustrations of the Action of Natural Selection, or the Survival of the Fittest." The fact is Darwin did not have any direct evidence. It was not until 1898 that something approaching direct evidence emerged.

After a severe snowstorm in Providence, Rhode Island, Brown University biologist Hermon Bumpus found a large number of English sparrows close to death. Bumpus captured and transferred 136 of the sparrows back to his laboratory, where nearly half of them died.

In measuring the living and the dead, Bumpus found that the survivors tended to be shorter and lighter males. Based on this evidence,

Bumpus concluded that the storm had taken a greater toll on sparrows that deviated [from] the most from the "ideal type," and claimed that the pattern of differential survival was due to natural selection.[4]

Bumpus later became director of the American Museum of Natural History, serving from 1906 to 1907. For decades, Bumpus' work was the closest biologists had come to observing natural selection directly. Thereafter, in the late 1950s, British physician and geneticist Bernard Kettlewell's European peppered moth, *Biston betularia*, phenomenon became the classic textbook example of natural selection.

Kettlewell's Peppered Moths

In England, following the rise of industrialization and driven by the use of coal, the natural color of the surrounding countryside began to darken. Soot from the burning coal began covering vegetation. The previously dominate light-colored peppered moths once found on light-colored tree trunks began decreasing, while the number of dark-colored moths began increasing on the darkened tree trunks in the areas surrounding the coal-operated industries.

This phenomenon became known as "industrial melanism," but its causes remained a matter of speculation until the early 1950s, when Bernard Kettlewell designed a field experiment to test the actions of natural selection. Kettlewell's experiment was simple. After breeding populations of light- and dark-peppered moths in his laboratory, Kettlewell marked the moths' wings with a drop of paint so that they could be traced.

Kettlewell released the marked light and dark moths in two types of wooded areas in England—polluted and nonpolluted. The polluted and darkened wooded area was near the highly industrialized city of Birmingham, the powerhouse of the Industrial Revolution. The other area Kettlewell selected was an unpolluted wooded area on the coast of Southwest England near the town of Dorset.

Setting traps to catch the moths, Kettlewell reported that the moths matching the color of the tree trunks were more likely to survive. On the night following the release of the dark moths around Birmingham, Kettlewell set traps for the moths. Of the 447 dark-colored moths that were marked and released, 123 (28 percent) were trapped, while only 18 (13 percent) of the 137 light-colored moths that were marked and released were trapped. To explain the results, Kettlewell concluded that predatory birds ate light-colored moths when they became more conspicuous on pollution-darkened tree trunks, leaving the dark-colored variety to survive and reproduce. As a classic example of the survival of the fittest, natural selection, Kettlewell wrote that the "birds act as selective agents, as postulated by evolutionary theory."[5]

Kettlewood repeated the same experiment in the wooded areas surrounding Dorset while being filmed by Niko Tinbergen, who later shared the 1973 Nobel Prize in Physiology and Medicine with two other men. Of the 496 light-colored moths marked and released, only 62 (13 percent) were trapped, and only 30 (6 percent) of the 473 dark-colored moths were trapped.[6]

Kettlewell's direct evidence for natural selection was a first, and the study was quickly incorporated in introductory biology textbooks, complete with pictures of moths on tree trunks. Kettlewood concluded that the change from light to dark-peppered moths during the Industrial Revolution is "the most striking evolutionary change ever actually witnessed in any organism."[7]

The results of Kettlewell's experiments provided what appeared to be substantial evidence of natural selection acting on variation in a population and actually changing a species. Kettlewell later published an article in *Scientific American* in 1959 entitled "Darwin's Missing Evidence."[8]

Biologists since the early 1980s, however, began uncovering serious issues with Kettlewell's story. Later in the 1950s, antipollution legislation addressing the darkening effects of industrialization was enacted, and the darkened countryside began to lighten. Field studies that began in the 1960s confirmed that the proportion of light-peppered moths increased as pollution decreased. In 1975, British geneticist P. M. Sheppard called the phenomenon "the most spectacular evolutionary change ever witnessed and recorded by man, with the possible exception of some examples of pesticide resistance."[9]

The presumed evolutionary process was simply a swing of the pendulum in the proportion of light- and dark-peppered moths. The light-colored moths did not evolve into dark-colored moths. Fortunately, the light-colored moths remained genetically distinct and did not become extinct as predicted by Darwin: "The theory of natural selection is grounded on the belief that each new variety, and ultimately each new species, is produced and maintained by having some advantage over those with which it comes into competition; and the consequent extinction of less-favoured forms almost inevitably follows."[10]

During the 1970s, investigators attempted to retrace Kettlewell's experiment. Unexpectedly, Liverpool biologist Jim Bishop was not able to reproduce Kettlewell's premise that light-peppered moths predominate in unpolluted areas.[11] In 1975, D. R. Lees and E. R. Creed studied peppered moths in the little-polluted areas of East Anglia where the light-colored moths were considered better camouflaged than the dark-colored moths. Lees and Creed found that the dark-colored moths predominated, accounting for 80 percent of the moths. The investigators concluded: "Either the predation experiments and tests of conspicuousness to humans are misleading, or some factor or factors in addition to selective predation are responsible for maintaining the high melanic frequencies."[12]

Factors Kettlewell did not control for in his experiments included studying the moths during the night and photographing the moths on tree trunks. In most of Kettlewood's experiments, moths were released and observed during the day, but peppered moths are night-flyers, normally resting in places before dawn and not on tree trunks. In 1984, Finnish zoologist Kauri Mikkola, discrediting textbook pictures of moths on tree trucks, stated in the *Biological Journal of the Linnean Society* that "the normal resting place of the Peppered Moth is beneath small, more or less horizontal branches (but not on narrow twigs), probably high up in the canopies, and the species probably only exceptionally rests on tree trunks."[13] In the same year, Italian biologists Giuseppe Sermonti and Paola Catastini criticized Kettlewood's daytime releases, declaring, "the evidence Darwin lacked, Kettlewood also lacked."[14]

In revisiting the issue in 1999, Japanese biologist Atuhiro Sibatani likewise concluded in the European journal *Rivista di Biologia* that the "story of industrial melanism must be shelved at least for the time being, as a paradigm of neo-Darwinian evolution."[15] American biologist Theodore Sargent and colleagues in 1998 contended in *Evolutionary Biology* "that there is little persuasive evidence, in the form of rigorous and replicated observations, to support this explanation at the present time."[16]

Commenting on Kettlewell's moth example of natural selection at work, University of Chicago evolutionary biologist Jerry Coyne, in 1998, wrote in *Nature,* "From time to time, evolutionists re-examine a classical experimental study and find, to their horror, that it is flawed or downright wrong."[17]

Coyne further acknowledges that the fact that peppered moths do not rest on tree trunks "alone invalidates Kettlewood's release-and-recapture experiments, as moths were released by placing them directly onto tree trunks."[18] After reviewing Kettlewood's original papers, Coyne concluded that Kettlewood's "prize horse of evolution in our stable of examples" of evolution "is in bad shape, and while not yet ready for the glue factory, needs serious attention."[19]

Facing the reality of current knowledge, Coyne painted a bigger picture: "We must stop pretending we understand the course of natural selection."[20] Coyne's dismay and disillusionment recalled the childhood memory of discovering the truth about Santa Claus.[21] In 2005, Italian geneticist Sermonti encapsulated Kettlewell's experiment in the book *Fly is not a Horse* and took the stance that the "fairy tale of the peppered moth is plausible, but untrue."[22]

Once a showcase example of natural selection in action, Kettlewell's peppered moths stand like Darwin's example, an "imaginary illustration." To the question, what is the genetic evidence for evolution of the peppered moth according to the central dogma?—the answer is none. Even if the existence of mutations and the actions of natural selection were discovered in Kettlewood's minutiae of evolution, which interests science most, how moths came to be moths in the first place remains an enigma.

Darwin's Finches

The finches from the Galápagos Islands are some of the most studied and pivotal birds in the world. They played a major role in the acceptance of evolution by natural selection later in the twenty-first century. The Galápagos Islands consist of an oceanic archipelago with sixteen principal islands located on the equator, approximately 600 miles west of Ecuador. While the Galápagos Islands today have become a beautiful destination point, Darwin thought the islands were disdainful:

Nothing could be less inviting than the first appearance. A broken field of black basaltic lava, thrown into the most rugged waves, and crossed by great fissures, is everywhere covered by stunted, sunburnt brushwood, which shows little signs of life. The dry and parched surface, being heated by the noonday sun, gave to the air a close and sultry feeling, like that from a stove: we fancied even that the bushes smelt unpleasantly.[23]

While Darwin mentioned the "Galápagos Archipelago" seventeen times and "finches" three times in *The Origin of Species*, he never used the terms "Galápagos" and "finches" together in the same sentence or even in the same paragraph. Darwin did not use the finches in the *Origin* as evidence for evolution.

During Darwin's nearly five-week stay in the Galápagos Islands, he collected thirty-one finch specimens, which were not consistently tagged to identify the island of capture.[24] Darwin later documents the mixing of the finches from the different islands in the second edition of his *Journal of Researches*: "Unfortunately, most of the specimens of the finch tribe were mingled together."[25]

Only after returning to England did British ornithologist John Gould begin to arrange the finches by island. Frank Steinheimer of the Museum für Naturkunde, Berlin, speculates that Darwin, at the time, might have seen the finches as evidence of evolution. Steinheimer suspects that the myth of "Darwin and his finches" has been shown to be just that—speculation. We now know that in the field, Darwin did not appreciate the close relationship between the different species of finches on the Galápagos Islands."[26]

Of the information that Darwin wrote on the tags, most was fragmentary and incomplete. Eventually, identification of the island of collection was reconstruction based on the more carefully labeled collections of his shipmates. Today, only one original tag written by Darwin remains. Frank Sulloway, from the University of California, Berkley, notes: "Darwin did not begin to separate his ornithological collections by island while he was in the Galápagos Archipelago. Rather, whatever information he later provided in this connection was largely derived, after the *Beagle* voyage, from the carefully labeled collections of three other *Beagle* shipmates."[27]

After the finch collection had been arranged by John Gould, and seeing the gradations in the laboratory, Darwin tentatively eluded to evolution in the second edition of the *Journal of Researches,* published in 1845: "Seeing this gradation and diversity in one small, intimately related group of birds, one might really fancy that from an original paucity of birds in this archipelago, one species had been taken and modified for different ends."[28]

According to F. D. Steinheimer, Gould was principally responsible for turning Darwin's collection of finches into evidence for the theory of evolution. Steinheimer suggests "That it was mainly John Gould who made Darwin's collection and notes into a significant contribution to ornithology. The Galápagos finches subsequently proved to be more interesting than any other specimens in Darwin's collection."[29] Later, after Gould's arrangement, Darwin commented on the gradation of the beak sizes: "The most curious fact is the perfect gradations in the size of the beaks of the different species of finches."[30]

In 1837, Darwin donated the thirty-one finches he collected in September 1835 and his other ornithological specimens to the Zoological Society of London for display. When the Zoological Society decided to close operations in 1855, the collection was offered for sale to the British Museum. Unfortunately, in the transaction, only nineteen of Darwin's original finches were relocated to the British Museum collection.[31]

It was not until the rise of neo-Darwinism in the 1930s that the Galápagos finches rose to legendary status. While the finches were first called "Darwin's finches" by Percy Lowe in 1936, British ornithologist David Lack is credited for popularizing the name and concept a decade later.

In the 1947 book *Darwin's Finches,* Lack, expanding on Darwin's correlation between variations in the size of the finch beaks and the different food sources, made the case that the beaks were the result of natural selection. According to Sulloway, "Darwin was increasingly given credit after 1947 for finches he never saw and for observations and insights about them he never made."[32]

In 1973, taking on the challenge of observing evolution in action, the husband-and-wife team of Peter and Rosemary Grant went to the

Galápagos Islands to observe and study Darwin's finches. The Grant's tracked thousands of individual finches across several generations over a period of years, showing how individual species, especially the beaks, adapted to environmental changes.

The timing of the Grants arrival coincided with the development of a drought condition. Taking the queue from Darwin, the Grants began measuring the overall size of the finches and the size of the beaks. During the drought conditions through 1977, the population of finches declined dramatically, to about 15 percent of the former population. The total body size of the finches increased, and the average size of the beak increased about 5 percent—or approximately half a millimeter larger—about the thickness of a human hair.

The evidence was hailed as the most dramatic example of natural selection in the wild. Darwin's finches became more popular than ever. In 1994, journalist Jonathan Weiner published a book entitled *The Beak of the Finch: A Story of Evolution in Our Time,* which highlighted the Grants' research. The book was awarded a Pulitzer Prize in 1995. Weiner called the evidence collected by the Grants "an icon" of evolution—"the best and most detailed demonstration to date of the power of Darwin's process."[33]

Then the rains came. While Darwin's finches became an icon, what got lost in the excitement was the evidence from the El Niño winter of 1982–83, which brought heavy rains to the Galápagos Islands, over ten times the normal amount of rainfall. With the rain and a plentiful food supply, the average beak size returned to the same smaller size, as measured before the drought. Reporting on the event in *Nature,* Peter Grant, with graduate student Lisle Gibbs, explains how the reversal in the climate reversed the direction of selection: "Large adult size is favored when food is scarce because the supply of small and soft seeds is depleted first, and only those birds with large bills can crack open the remaining large and hard seeds. In contrast, small adult size is favored in years following very wet conditions, possibly because the food supply is dominated by small soft seeds."[34]

In essence, the evolution observed during the drought of the mid-1970s reversed with the rains of the early 1980s. In 1991, Grant published an article in *Scientific American* that the evidence was "oscillating back and forth."[35] The question is can evolution be just an oscillation? What appeared as evolution in the drought later reversed when the rains returned. The lesson learned was that the changes in the size of the beak was inherited expression acting within a species, but had nothing to do with the emerging evolution of a new species.

The litmus test for identifying a new and distinct species is a sustainable interbreeding population. One of the more widely accepted definitions of a species was written by the champion of evolutionary theory, biologist Ernst Mayr. While at the American Museum of Natural History, Mayr stated, "species are groups of actually or potentially interbreeding natural populations, which are reproductively isolated from other such groups."[36] Interbreeding within a population defines the group as a species.

In the book *The Beak of the Finch*, Weiner suggests that what the Grants previously thought to be different species was actually one species. In 1993, the Grants published their interbreeding concerns in the *Proceedings of the Royal Society of London* and tempered the issue by inferring that interbreeding "[led] to the fusion of the species into one population."[37] In essence, the finches were the same species.

Not only were the finches in question successfully mating, the offspring were eventually the most fertile that the Grants recorded during their twenty years on the islands. Four chicks of this mating produced no less than forty-six grandchildren. In an article written for *Science* in 1992, the Grants appropriately conceded that this "calls into question their designation as species."[38]

The Grants documented many other pairings of "different species" of finch. Like Lack, the Grants eventually dubbed them as hybrids, and not a different species. In the same 1992 *Science* article, the Grants further explains that the "three populations of ground finches on Genovesa would similarly be reduced to one species.... At the extreme, six species would be recognized in place of the current fourteen, and additional study might necessitate yet further reductions."[39]

Indeed, further investigations were needed since the evidence raised questions about whether Darwin's finches were simply not just one species. What is clear is that the apparent pattern of change is the convergence and not the divergence of species. This pattern actually represents a reversal of Darwin's concept of the Tree of Life.

In the article entitled "Convergent Evolution of Darwin's Finches Caused by Introgressive Hybridization and Selection," published in *Evolution* in 2004, Peter Grant noted that the "species without post-mating barriers to gene exchange can alternate between convergence and divergence when environmental conditions oscillate."[40] Grant's conclusion is compatible to Sermonti's contention that natural selection only conserves a species within a norm.

Since Darwin's finches can exchange genetic information, which results in either convergence or divergence, it puts into question whether the finches are actually a product of evolution. Since interbreeding cannot exist between different species, Darwin's finches are one species, according to Ernst Mayr's definition.

To determine the molecular difference between the species, Akie Sato of the Max-Planck-Institut für Biologie in Germany lead a research team to study mtDNA, cytochrome C, and the control region. The purpose was to retrace the "successive, slight" molecular steps between the species and create a molecular Tree of Life for the Galápagos finches. The team published their findings in the *Proceedings of the National Academy of Science* in 1999, in a paper entitled "Phylogeny of Darwin's finches as revealed by mtDNA sequences." The research team reported that the "traditional classification of ground finches into six species and tree finches into five species is not reflected in the molecular data. In these two groups, ancestral polymorphisms have not, as yet, been sorted out among the cross-hybridizing species."[41]

In reexamining the evidence, Sato headed another research team to study the same molecular markers. The results pointed to the same conclusions—the molecular evidence from the finches cannot be arranged into a Tree of Life scheme. This means that there was no evidence that Darwin's finches evolved from one species into a number of species. In the article entitled "On the Origin of Darwin's Finches," published in the journal *Molecular Biology and Evolution* in 2001, Sato concluded that in "the absence of a detailed and statistically well supported phylogeny of the genus Tiaris, we are currently unable to reconstruct their morphological evolution and distinguish between these possibilities."[42]

In essence, there is no reproductive or molecular evidence that the finch population evolved from one species into a number of different species. The evidence actually points to the finches being one species that can adapt to environmental changes but not evolve into different species. Like Kettlewood's moths, Darwin's finches as icons of evolution are entering extinction.

Bacterial Resistance

As the evidence against Kettlewell's moths and Darwin's finches was gaining momentum later in the twentieth century, these issues were being overshadowed by advances in genetics. By the late 1950s, the concept of evolution through genetic mutations was taking center stage. American paleontologist George Gay Simpson wrote in 1957, "mutations are the ultimate raw materials for evolution."[43]

By the early 1980s, Crick's central dogma was thought to hold the key to evolution. Douglas J. Futuyma wrote in his book *Science on Trial* that evolution stands at the mercy of mutations. Mutations form the central dogma of evolution and are by "far the most important way in which chance influences evolution ... Mutation is, ultimately, the source of new genetic variations, and without genetic variation there cannot be genetic change. Mutation is therefore necessary for evolution."[44]

The U.S. 2000 National Medal of Science recipient Peter Raven, along with George Johnson, in their widely circulated public school textbook, *Biology,* explains that the concept of evolution stands at the mercy of mutations: "All evolution begins with alterations in the genetic message.... Genetic change through mutation and recombination provides the raw materials for evolution."[45]

To prove that Darwin was not wrong, David Quammen, in the leading November 2004 *National Geographic* article entitled "Was Darwin Wrong?" used antibiotic resistance through mutation as proof. Quammen explains that the bacteria by natural selection "acquire resistance to drugs that should kill them. They evolve. There's no better or more immediate evidence supporting the Darwinian theory than this process of forced transformation among our inimical germs."[46]

Underlying the role of mutations lies the question, are mutations just nature's way of adapting to the environment and not evidence of evolution? In 1943, published in a paper entitled "Mutations of Bacteria from Virus Sensitivity to Virus Resistance," microbiologist Salvador Luria, biophysicist Max Delbrück, and bacteriologist and geneticist

Alfred Hershey discovered that mutations occur at a constant rate.[47] In 1969, they were awarded the Nobel Prize in Physiology and Medicine "for their discoveries concerning the replication mechanism and genetic structure of virus."

With constant mutations, comes the next question, is the resistance a product of new mutations? Experimenting with the antibiotic streptomycin at the University of Wisconsin-Madison in the early 1950s, Joshua Lederberg, along with his graduate student Norton Zinder, demonstrated that bacteria never previously exposed to streptomycin was already resistant to the antibiotic.[48]

Later in the twentieth century, scientists at the University of Alberta revived bacteria from members of the historic "Franklin Expedition" who mysteriously perished in the Arctic nearly one-hundred and fifty years ago in 1845. The scientists were surprised to discover that the bacteria recovered from the intestines of the explorers had the same level of antibiotic resistance as modern bacteria. In a 1990 headline story in the Canadian *Sunday Herald* entitled "Ancient Bacteria Revived," Ed Struzik reported that not "only are the six strains of bacteria almost certainly the oldest ever revived … three of them also happen to be resistant to antibiotics. In this case, the antibiotics clindamycin and cefoxitin, both of which were developed more than a century after the men died, were among those used."[49] The recovered bacteria had resistance even before encountering the antibiotics. Resistance in these organisms was not a product of new mutations in response to antibiotic exposure. Resistance was preexistent.

More recently, preexistent resistance in bacteria has also been observed in viruses and insects. After studying the AIDS virus in 2000, theoretical biologists Ruy M. Ribeiro and Sebastian Bonhoeffer, from the Wellcome Trust Centre for the Epidemiology of Infectious Diseases, published an article in the *Proceedings of the National Academy of Sciences of the United States of America* about their observation of the same phenomena in viruses. Resistance was preexistent and was not acquired by mutations. Ribeiro and Bonhoeffer concluded, "the key to drug resistance lies in the diversity of the viral population at the start of therapy."[50]

In examining viral models to determine whether the antimicrobial resistance was acquired during treatment through mutation or was preexistent, Ribeiro and Bonhoeffer concluded that the resistance was "most likely caused by the preexistence of resistant mutants."[51] Resistance did not originate from environmental pressures.

The resistance was not new, nor was resistance acquired through any novel mutation mechanism not previously inherent in the organism. Even in insects, resistance existed prior to the introduction of insecticides. In 1978, in a *Scientific American* article entitled "The Mechanisms of Evolution," evolutionary geneticist Francisco Ayala wrote:

> Insect resistance to a pesticide was first reported in 1947 for the housefly [Musca domestica] with respect to DDT [synthetic pesticide]. Since then, resistance to one or more pesticides has been reported in at least 225 species of insects and other arthropods. The genetic variants required for resistance to the most diverse kinds of pesticides were apparently present in every one of the populations exposed to these man-made compounds.[52]

While bacterial resistance through the mutation model is a logical mechanism for evolution, the reality is the bacteria have remained bacteria, the virus has remained a virus, and the fly has remained a fly. Preexistent genetic variants determine the range of mutations. In 1977, Pierre-Paul Grassé, president of the French Academy of Sciences, observed, "bacteria, the study of which has formed a great part of the foundation of genetics and molecular biology ... stabilized a billion years ago."[53]

The question is whether the mutations are the "raw material for evolution" or nature's means for the microbes to adapt to the environment. Reflecting on the interpretation of mutations, Grassé wondered, "What is the use of their unceasing mutations if they do not change?" Grassé concludes, "the mutations of bacteria and viruses are merely hereditary fluctuations around a median position; a swing to the right, a swing to the left, but no final evolutionary effect."[54]

Acknowledging that while novel mutations do occur, molecular biologist Soren Lovtrup, of the University of Goterborg in Sweden, writes, "micromutations do occur, but the theory that these alone can account for evolutionary change is either falsified, or else it is an unfalsifiable, hence metaphysical theory." Lovtrup continues by lamenting the core of evolution's central dogma:

> I suppose that nobody will deny that it is a great misfortune if an entire branch of science becomes addicted to a false theory. But this is what has happened in biology.... I believe that one day the Darwinian myth will be ranked the greatest deceit in the history of science. When this happens many people will pose the question: How did this ever happen?"[55]

Are the mutations directed or undirected? At the core of the central dogma is the accumulation of novel mutations. For Darwin, evolution was a directed process. In *The Origin of Species,* Darwin wrote, "I was so convinced that not even a stripe of colour appears from what is commonly called chance."[56] Acquiring directed variations is the kingpin in Darwin's theory. Darwin envisioned that "all spontaneous variations in the right direction will thus be preserved."[57]

Since the origin of new variations is mutation dependent, does the evidence favor directed or undirected (random) mutations? In an article published in *Nature* in 1988, Harvard University molecular biologists John Cairns, Julie Overbaugh, and Stephan Miller reconfirmed the Luria–Delbrück experiment, demonstrating the undirected nature of bacterial mutations: "As the result of studies of bacterial variation, it is now widely believed that mutations arise continuously and without any consideration for their utility."[58]

To further investigate whether mutations are directed or random, the Cairns team suggested a series of experiments to detect directional mutations. In 2002, Susan Slechta of the University of Utah and Jing Liu of the Swedish Institute for Infectious Disease Control published a research paper using Cairn's team's suggestions and discovered no evidence for directional mutations.[59]

Continuing the search for directional mutations, researchers Monica Sala and Simon Wain-Hobson from the Pasteur Institute in France examined eighty-five sets of proteins from viruses that are known to infect bacteria, plants, and mammals. The results, published in a 2000 paper entitled "Are RNA Viruses Adapting or Merely Changing?"

revealed that even though viruses mutate rapidly, the mutations occur randomly and are not directed to adapting to the environment.[60]

To the question, can microorganisms change over time? The answer is yes. To the next question, are they "purposefully evolving"? The answer is no. Genetic mutations responsible for antibiotic resistance in bacteria do not arise from the need of an organism to develop such resistance. As evolutionist, Douglas Futuyma explains, "the adaptive 'needs' of the species do not increase the likelihood that an adaptive mutation will occur; mutations are not directed toward the adaptive needs of the moment.... Mutations have causes, but the species' need to adapt isn't one of them."[61]

From the current scientific evidence, two key factors emerge in reference to mutations. First, a range of potential favorable mutations is an inherent characteristic of an organism, and second, these potential mutations occur in a random fashion unrelated to the environmental conditions. This current evidence contradicts Darwin's theory of the directed and purposeful evolution of species—that forms the Tree of Life.

Beyond Mutations

While bacteria can become resistant to antibiotics through preexistent genetic variants, there are other known mechanisms of resistance beyond mutation.

In a process known as conjugation, bacteria exchange genetic material that is analogous to copulation. This genetic material, known as a plasmid, begins to replicate and then codes for the production of enzymes that can defend attacks against the cell, including antibiotics. Plasmids can confer antibiotic resistance.

Microorganisms can also incorporate foreign genetic material into their own genetic material in two ways: transformation (transposition) or transduction. In transformation, sections of DNA from the surrounding environment diffuse directly into the cell and become incorporated in DNA of the microorganism. During transduction, the foreign DNA is transported into the cell by means of a virus. The net result is acquired resistance without mutation.

The Mutation Advantage?

While antibiotic resistance through mutations within preexisting genetic variability can be advantageous, newly acquired mutations beyond the range of preexisting genetic variability only decreases the viability and virulence—the survival advantage.

Even Ernst Mayr observed in 1942, "it is a considerable strain on one's credulity to assume that finely balanced systems such as certain sense organs (the eye of vertebrates or the bird's feather) could be improved by random mutations."[62]

While Crick's central dogma was gaining momentum, American geneticist and Nobel Prize winner Hermann J. Muller observed, "the great majority of mutations, certainly well over 99 percent, are harmful in some way, as is to be expected of the effects of accidental occurrences."[63]

Mutations are known to be overwhelmingly deleterious. C .P. Martin, an evolutionist, wrote in the *American Scientist* in 1953 that "mutations are really assaults on the organism's central being, its basic capacity to be a living thing."[64]

Evolutionary geneticist Theodosius Dobzhansky, who was awarded a Nobel Prize "for the discovery that mutations can be induced by X-rays," observed that most mutations are not advantageous. In the 1955 book *Evolution, Genetics and Man*, Dobzhansky wrote, "most mutants which arise in any organism are more or less disadvantageous to their possessors."[65]

By the 1970s, in light of the significance of mutations in the scheme of evolution, Motoo Kimura noted in *The John Hopkins Medical* journal that "from the standpoint of population genetics, positive Darwinian selection represents a process whereby advantageous mutations spread through the species. Considering their great importance in evolution, it is perhaps surprising that well-established cases are so scarce."[66] Twenty-five years later, in 2001, the leading evolutionary biologist at Harvard University, Ernst Mayr, conceded in the book *What Evolution Is* that "the occurrence of new beneficial mutations is rather rare."[67]

Molecular biologists at the University of Texas Institute for Cellular and Molecular Biology came to the same conclusion, stating that while "mutation is the basis of adaptation ... most mutations are detrimental, and elevating mutation rates will impair a population's fitness in the short term."[68] When Charles Baer was a postdoctoral research associate in evolutionary genetics at the University of Florida, he wrote in 2008 in the journal *PLoS Biology*: "There is overwhelming evidence that the great majority of mutations with detectable effects are harmful."[69]

To protect against the damaging effects of mutations to the gene pool, nature reverses mutations through a DNA mutation repair process. It has been estimated that in humans as many as one million individual molecular mutations per cell per day undergo the DNA repair process.[70]

In their 2002 book *Acquiring Genomes: A Theory of the Origins of Species*, evolutionists Lynn Margulis, from the University of Massachusetts Amherst, and Dorion Sagan, the son of Carl Sagan, explain that although "many ways to induce mutations are known ... none leads to new organisms. Mutation accumulation does not lead to new species or even to new organs or new tissues ... even professional evolutionary biologists are hard put to find mutations, experimentally induced or spontaneous, that lead in a positive way to evolutionary change."[71]

In summarizing their conclusions, Margulis and Sagan note that mutations, rather than being the source of evolution, "tend to induce sickness, death, or deficiencies. No evidence in the vast literature of hereditary change shows unambiguous evidence that random mutation itself, even with geographical isolation of populations, leads to speciation."[72]

Speculating on the magnitude of mutations required by the central dogma for the process of evolution, Pierre-Paul Grasse writes:

> "The opportune appearance of mutations permitting animals and plants to meet their needs seems hard to believe. Yet the Darwinian theory is even more demanding: a single plant, a single animal would require thousands and thousands of lucky, appropriate events. Thus, miracles would become the rule: events with an infinitesimal probability could not fail to occur ... There is no law against day dreaming, but science must not indulge in it.[73]

Bacterial resistance through new mutations followed by natural selection illustrates a theoretical mechanism for evolution, but to date there is no scientific evidence for the origin of new microbial species through this mechanism. As the emerging evidence continues to undermine Crick's central dogma, what may be at stake is not the issue of mutations but a more fundamental issue—the role of DNA. Sermonti suggests the answer may be illustrated in the butterfly.

The Butterfly

With stunning colored wings and symbolism, butterflies have been collected and admired through the ages. In ancient Greece, the word for butterfly is ψύχη (*psychē*), primarily meaning "soul" and "mind." In the Chinese culture, butterflies flying in pairs are a symbol of love. Butterflies were etched in Egyptian hieroglyphics 3,500 years ago. Chinese Taoist philosopher Zhuangzi, who lived around the fourth century BC, had a dream where he became a butterfly flying without a care about humanity. Awaking from the dream, Zhuangzi thought to himself, "Was I before a man who dreamt about being a butterfly, or am I now a butterfly who dreams about being a man?"

In the twenty-first century, while studies on butterfly wing light reflection have led to developing more efficient light-emitting diodes, studies on the butterfly DNA is shedding light on the role of DNA and evolution.

Butterfly symbolism is derived from the butterfly's unique life cycle. The first larval caterpillar stage transitions into an inactive pupal stage. Inside the pupa, digestive juices destroy the larva's body, leaving only a few cells, which develop into the beautiful colorful winged butterfly stage. Most baffling, the butterfly metamorphosis occurs with the same DNA. Italian geneticist Sermonti points out that "examples of highly divergent forms possessing one and the same DNA are so conspicuous and so numerous that the marvel is that they have attracted so little attention. As a symbol of morphological diversity emerging from genetic identity, we can take the caterpillar and the butterfly."[74]

Even more astounding in the case of the butterfly, Sermonti notes, "what we call metamorphosis is not really a change in form. Once the pupa or chrysalis stage is reached, the caterpillar starts emptying itself: its organs dissolve, and its outer covering is shed. Only certain groups of cells, called marginal disks, remain vital. From these cells develop all the structures of adult."[74]

The larva of the butterfly does not just change form, but actually dissolves and rebuilds into the structure of a butterfly—a new life-form, not a transformation. From the same DNA arise completely different organisms. According to Sermonti, the same DNA, then, can play different roles: "DNA may lend itself to such diverse forms, but it is not the DNA that imposes the blueprint, nor is it the hormones that do the organizing."[75]

The presence of the same DNA in different life-forms has been given the term "genomic equivalence." The question is, if the genes are the same, why are the life-forms so different?

The surprising answer is that life-forms are not DNA dependent. This means that control of the cell is beyond the DNA, or "epigenetic."

Brian Goodwin, Canadian developmental biologist and key founder of theoretical biology, focuses on the methods of mathematics and physics to understand processes in biology. In taking this perspective, Goodwin concludes that while "genes are responsible for determining which molecules an organism can produce, the molecular composition of organisms does not, in general, determine their form."[76]

In 1990, H. Frederik Nijhout of the Department of Biology at Duke University, a critic of Crick's central dogma, came to the conclusion that "the only strictly correct view of the function of genes is that they supply cells, and ultimately organisms, with chemical materials."[77]

Investigations to correlate life-forms with genes have failed to demonstrate the expected correspondence between life-form changes and changes in the gene that constitute the "stuff of evolution," according to the central dogma. Even though the DNA of humans and chimpanzees is approximately 99 percent identical, correspondence between the two is worlds apart. Evolution does not follow DNA or DNA mutations. According to molecular biologist Rudolf Raff and Thomas Kaufman at the Indiana Molecular Biology Institute, evolution by DNA mutations "is largely uncoupled from morphological evolution."[78]

While in the butterfly the same DNA results in completely different life-forms, just the opposite is known to occur. The eye of the octopus is nearly similar to the human eye, yet there are no similarities in DNA.

Octopus Eye

Darwin was fascinated with the development of the eye, yet the eye presented a challenge to Darwin's theory of natural selection. Darwin wrote in *The Origin of Species,* "To suppose that the eye with all its inimitable contrivances for adjusting the focus to different distances, for admitting different amounts of light, and for the correction of spherical and chromatic aberration, could have been formed by natural selection, seems, I freely confess, absurd in the highest degree."[79]

Darwin was at a loss to explain how the eye of any species formed through natural selection, and much less able to explain why the octopus eye is similar to the vertebrate eye. There are differences between the octopus and vertebrate eye; however, the fact remains that while the final form is similar, the DNA of the octopus and vertebrate is completely different and challenges the view traditionally about the role of DNA.

The similarity between the two has been interpreted as an example of convergent evolution in which a similar form and function develops in two unrelated organisms—phylogenetic lines. The well-known evolutionist Frank Salisbury at Utah State University, commenting on the likelihood of convergence, concluded:

> Even something as complex as the eye has appeared several times: for example, in the squid, the vertebrates, and the arthropods. It's bad enough accounting for the origin of such things once, but the thought of producing them several times according to the modern synthetic theory makes my head swim.[80]

Even if convergence were true, the fact is that DNA plays no role even in the discussion of convergent evolution. These facts led Italian geneticist Giuseppe to conclude, "DNA is not the starting place."[81]

Ironically, similar structures, termed "homologous" structures, are not necessarily controlled by similar DNA. Biologists have known this since the early 1970s. In 1971, British evolutionary embryologist Gavin de Beer wrote:

> Because homology implies community of descent from
> … a common ancestor, it might be thought that genetics
> would provide the key to the problem of homology.
> This is where the worst shock of all is encountered …
> [because] characters controlled by identical genes are not
> necessarily homologous … [and] homologous structures
> need not be controlled by identical genes.… It is now
> clear that the pride with which it was assumed that the
> inheritance of homologous structures from a common
> ancestor explained homology was misplaced.[82]

The neo-Darwinian theory subscribes to the concept that information is contained in the DNA, and because genes are transferred from generation to generation, then similar genes must be from a common ancestor. Logically, the purpose of similar genes is to give rise to homologous structures. However, this is not the case universally. Similar structures can exist without similar DNA, and thus questions arise regarding the validity of the common ancestor theory as measured by DNA. De Beer concluded, "the inheritance of homologous structures from a common ancestor … cannot be ascribed to identity of genes."[83]

Nearly twenty years later, evolutionary molecular biologist Gregory Wray from Duke University came to the same conclusion that "this association between a regulatory gene and several non-homologous structures seem to be the rule rather than the exception."[84]

What once seemed impossible is now possible—while the same DNA can produce widely different forms, widely different DNA can produce similar forms. According to Italian geneticist Sermonti: "We may conclude that with the same DNA the most disparate forms (as in caterpillars and butterflies) can be made, while with disparate DNA, forms can be made that are almost the same … with … equivalent organs (as with the eyes of the human and octopus)."[85]

In the end, DNA can be irrelevant in determining the final form of life: "With a single information molecule all kinds of beings can be made, and with the most disparate kinds of information molecules the same end-result can be obtained."[86]

DNA acts as an independent agent, and at other times, DNA acts as a dependent agent. DNA alone does not determine life-forms.

The Platypus

This duck-billed oddity is like a mammal, a bird, and a reptile all in one species. When the platypus, nicknamed the "watermole," was first discovered in 1797 by early European settlers near the Hawkesbury River, outside Sydney, it triggered a lasting controversy. The perplexed local governor sent specimens back to Mother England for study.

The "watermole" was equally mystifying in England. One zoologist suggested it was a "freak imposture" sold to gullible seamen by Chinese taxidermists. Suspecting the fraud, they tried to pry the "duck's bill" off of the pelt, leaving marks on the bill that are still visible today and on display at the British Museum in London.[87]

In 1802, an English scientist confirmed that the creature was neither freak nor fraud, and he labeled the specimen "platypus" because of its flat bill and gave it the scientific name *Ornithorhynchus anatinus*. Since then, the platypus has stood as an iconic conundrum in natural history.

Not only do these furry animals actually lay eggs like a bird, the young feed on breast milk like a mammal and make venom like a snake. Although the platypus has been an evolutionary conundrum, the structure of the platypus genome has now been deciphered. In comparing the platypus genome with genomes of the human, mouse, dog, opossum, and chicken, researchers found that the platypus shares 82 percent of its genes with these animals.[88]

For egg production, the platypus genome matches for the ZPAX genes that had previously been found only in birds, amphibians, and fish, and it shares with the chicken a gene for a type of egg-yolk protein called "vitellogenin." For breast milk production, the platypus has genes for the family of milk proteins called caseins, which map together in a cluster matching humans. Lastly, the male platypus has spurs on its hind legs loaded with lethal venom, rising from duplicate reptilian-like genes.[89]

The platypus exemplifies how similar genes in different species produce the same function. The question is how does genetic identification clarify the evolutionary place of the platypus? The answer is it does not. According to Richard Gibb, Director of the Human Genome

Sequencing Center at Baylor College of Medicine in Texas, concluded, "there is nothing quite as enigmatic as a platypus. You have got these reptilian repeat patterns and these more recently evolved milk genes and independent evolution of the venom. It all points to how idiosyncratic evolution is."[90]

Darwin proposed in *The Origin of Species* that as new species arise through "successive, slight changes," there is corresponding extinction. Darwin wrote "extinction and natural selection go hand in hand." The evidence however points to the conservation, not the extinction, of variations.[116]

A team led by Gregory Hannon of Cold Spring Harbor Laboratory in New York sequenced microRNAs, which regulate gene expression, from six platypus tissues, and also found a mix of reptile and mammal similarities, concluding that we "have microRNAs that are shared with chickens and not mammals as well as ones that are shared with mammals, but not chickens."[91]

Classifying the platypus into an evolutionary paradigm has been a challenge. In 1992, Australian biologist, Michael Archer wrote, "Indeed, evolutionary scientists are baffled about the ancestry of the platypus."[92]

As baffling as the platypus was in 1992, it is even more so today, despite the availability of genomic sequencing. Francis S. Collins, Director of the National Human Genome Research Institute, concedes: "At first glance, the platypus appears as if it was the result of an evolutionary accident. But as weird as this animal looks, its genome sequence is priceless for understanding how mammalian biological processes evolved."[93]

What is missing from the "priceless for understanding" is how the platypus fits into the Tree of Life sequence as Darwin envisioned. It is no wonder that Darwin never mentions the platypus in any of his writings, including *The Origin of Species*. In the light of the central dogma, the genetics is now known to play an array of conflicting roles. In the butterfly, similar genes are associated with different forms and functions. In the octopus, different genetics results in similar forms and functions. In the platypus, the same genes in different species are associated with the same functions.

Genes are independent, not sequential. As Richard Gibb points out, it "all points to how idiosyncratic evolution is."[94]

One Gene, One Enzyme?

In the early neo-Darwinism era, a series of key experiments by microbiologists George Beadle and Edward Tatum demonstrated the biochemical role of DNA. Their experiments led to the proposed association between gene mutations and changes in enzymatic reactions, which ultimately led to the "one gene, one enzyme" hypothesis.[95] For their work, Beadle and Tatum were awarded the Nobel Prize in Physiology or Medicine in 1958 for developing the concept that one gene specifies for one enzyme rather than multiple enzymes.

The "one gene, one enzyme" hypothesis was revolutionary for the 1940s, but the emerging evidence by the late twentieth century challenged the simplicity of the hypothesis. It is now known that a single gene can have multiple effects, in which case it is called a pleiotropic gene.

In the *Drosophila* fruit fly, a single gene controls not only the color of the eye but also the shape of the female sex organs. In the house mouse, a single gene not only controls the mouse's coat color but also its body size. In the domestic fowl, a single gene controls both the formation of the crest of its feathers as well as the development of the skull.[96]

Single genes control multiple effects that would not be considered to have a connection with each other. In the chicken, a single gene mutation causes (1) underdevelopment of the wings, (2) no claws on the feet, (3) underdevelopment of downy feathers, (4) disappearance of lungs and the air sac, and (5) underdevelopment of the ureter, with no kidney formation.[97] A classic example of a pleiotropic gene mutation is the human disease phenylketonuria, or "PKU." A single mutation in this gene causes mental retardation and reduced hair and skin pigmentation.

The simple concept of "one gene, one enzyme" that played a pivotal role in the development of neo-Darwinism has not stood the test of time. A team lead by Richard Lifton of the Yale University School of Medicine reported in *Science* in October 2004 that a single mutation in a mitochondria gene causes the development of a constellation of symptoms: hypertension, a high concentration of blood cholesterol, and lower than normal concentrations of magnesium. This mutation altered one base—a thymidine was changed for a cytosine—in the gene for a mitochondria transfer RNA (tRNA), which carries amino acids to the ribosome for protein synthesis.[98]

Serious problems have emerged in connecting mutations to any series of evolutionary steps. Geneticist John Endler at the University of California wrote in 1988 that although "much is known about mutation, it is still largely a 'black box' relative to evolution. Novel biochemical functions seem to be rare in evolution, and the basis for their origin is virtually unknown."[99]

Jerry Coyne, of the Department of Ecology and Evolution at the University of Chicago, arrives at the unanticipated verdict: "We conclude—unexpectantly—that there is little evidence for the neo-Darwinian view: its theoretical foundations and experimental evidence supporting it are weak."[100]

Sickle-Cell Anemia

The *Secret of Life* series aired in 2001 on PBS, the Public Broadcasting Station, in the episode "Accidents of Creation," opens with sunny beach sands bustling with fishing boats and fishermen, with the narrative story line beginning:

> For thousands of years, coastal West Africans have fished the sea to make a living. But 4,000 years ago, a deadly disease arrived here, bringing death and misery. Its name: malaria. That people live and fish here today demonstrates the power of evolution—power felt in the widespread effect of just one mutation in a single gene.[101]

Of all the mutations in humans, the sickle-cell mutation is the one most commonly used to demonstrate the "power of evolution." The question is what is the mutational advantage of the sickle-cell mutation?

In 1949, a research team at the California Institute of Technology headed by Linus Pauling was the first to demonstrate that sickle-cell disease occurs because of one abnormal amino acid in the hemoglobin molecule.[102] Sickle-cell anemia was the first genetic disease linked to a mutation, a milestone in the history of molecular biology.

In sickle-cell anemia, a point genetic mutation causes the β-globin chain of hemoglobin to replace glutamic acid with another amino acid, valine. Under valine substitution in low oxygen conditions, the hemoglobin chain polymerizes and distorts red blood cells into the shape of a sickle, and the red blood cells become rigid, decreasing their ability to pass through capillaries and leading to painful vascular occlusion and ischemia.

The cardinal advantage of sickle-cell anemia is an improved chance of surviving a malarial infection. Malaria is a unicellular parasite transmitted by the *Anopheles* mosquito. While most individuals stricken with malaria, even those without the sickle-cell trait, survive after an illness of two to three weeks, in Africa the disease kills one child in twenty before the age of five. Yet in contrast, 50 percent of African children stricken with sickle-cell disease die within their first twelve months of life, a mortality rate greater than a malarial infection.

Even the National Institutes of Health *Fact Sheet on Sickle Cell Disease* for children in the United States reports as recently as 1970 that the average patient with sickle-cell disease died in childhood. Approximately ten percent of children with sickle-cell disease suffer fatal or debilitating strokes.[103] Children living through sickle-cell anemia to adulthood can expect the lifelong risk of periodic, painful attacks with an average life expectancy of approximately forty-five years.

While the sickle-cell mutation can provide a slight advantage to those stricken with malaria, the mutation otherwise has a profound negative effect on the quality of life. Mutations are not associated with healthy outcomes. The U.S. National Library of Medicine's online Genetics Home Reference Handbook states: "Mutations have very serious effects, [and] they are incompatible with life."[104]

Mutations in humans have not been advantageous. At least 4,000 diseases have been linked to specific diseases, including Huntington's disease, neurofibromatosis, Marfan syndrome, hereditary nonpolyposis colorectal cancer, cystic fibrosis, Tay-Sachs disease, spinal muscular atrophy, hypophosphatemia, Aicardi syndrome, hemophilia A, Duchenne muscular dystrophy, color blindness, muscular dystrophy,

Leber's hereditary optic neuropathy, Down syndrome, and Klinefelter's syndrome. Mutations result in pain and suffering.

Contradicting the PBS series' conclusion regarding the advantages of the sickle-cell mutation, Sermonti, in 2005, concluded:

> To say that blind mutations are the driving principle of the world, and to rely on the rare fortunate mistake, is a poor resource, quite apart from the fact that transgressions of the kind needed by Darwinian evolution have never been documented.[105]

Perspective

Skepticism over the advantages of mutations was largely ignored, though not absent, during the late twentieth century. As early as 1980, H. S. Lipson, the eminent British physicist and evolutionist, authored an article in *Physics Bulletin* titled "A Physicist Looks at Evolution," that sparked quite a controversy when he questioned the foundations of evolution:

> In fact, evolution became in a sense a scientific religion; almost all scientists have accepted it and many are prepared to 'bend' their observations to fit with it.... I have always been slightly suspicious of the theory of evolution because of its ability to account for any property of living beings. I have therefore tried to see whether biological discoveries over the last thirty years or so fit in with Darwin's theory. I do not think that they do. To my mind, the theory does not stand up at all. [106]

Not one known genetic mutation in humans has been shown to be beneficial—only detrimental. The late twentieth century was flooded with excitement over the prospect of genetic engineering curing genetic mutations. Since then, however, not a single genetic engineering manipulation has resulted in a single cure. Sermonti concluded, "none of the manipulations making up genetic engineering has succeeded in taking off, and that nature has effectively defended its frontiers."[107]

Now in the early twenty-first century, the tide is beginning to turn. While research into genetics has been revealing, the field of genetics has not discovered any natural law to account for Darwin's evolutionary process. How genetics and genetic mutations provide the origin of variation required of evolution is now more problematic than ever and impossible to ignore. Under the title "Evolutionary Development Biology," the writers in Wikipedia.org conclude the species variation (phenotypic variation) cannot be explained by genetic variation: "Currently, it is well understood how genetic mutation occurs. However, developmental mechanisms are not understood sufficiently to explain which kinds of phenotypic variation can arise in each generation from variation at the genetic level."[108]

Niles Eldredge concurred in 2006, stating in *New Scientist* that over "the past 50 years, advances in molecular and developmental biology have outstripped a truly integrated synthesis of evolutionary theory."[109]

The foundation and pivotal centerpiece of evolution is based on the origin of variation. Without any new variations, there can be no evolution. Darwin wrote, "natural selection can do nothing until favourable individual differences or variations occur."[110] In determining the origin of variation, not much has changed in the past 150 years. Darwin recognized that our "ignorance of the laws of variation is profound. Not in one case out of a hundred can we pretend to assign any reason why this or that part has varied."[111]

The rise of mutations as the origin of variations in the neo-Darwinian era emerged as the driving force of evolution. It became the central dogma of evolution in 1958 and saved Darwin's theory from extinction. Now fifty years later, mutations have emerged as a problem rather than an asset. The concept of evolution is now more at risk than ever. Perhaps American humorist Mark Twain's comments on science, in the book *Life on the Mississippi,* should have been heeded earlier: "There is something fascinating about science. One gets such wholesale returns of conjecture out of such a trifling investment of fact."[112]

Conjecture is not a science. It is no wonder that Darwin's bulldog of the twentieth century, Ernst Mayr, recanted on the scientific nature of biology: "Biology, even though it has all the other legitimate properties of a science, still is not a science like the physical sciences."[113]

The laws of nature cannot be determined simply by popularity. British medical doctor Edward Jenner (1749–1823) was scorned when he suggested infecting people with a less virulent strain of smallpox to gain immunity. Afterward, he lived as a man whose reputation had been crushed. Yet today the vaccine, along with the World Health Organization, has largely eradicated smallpox. The Austrian physician Ignaz Semmelweis (1818–1865), noticing a high mortality rate among surgical patients, suggested that the deaths resulted from surgeons washing neither their hands nor their instruments between patients. While Semmelweis was ridiculed during his lifetime, today washing hand and washing surgical instrument stands as a cornerstone of patient safety in hospitals.

Popularity can be dead wrong. The only proven method for determining the laws of nature is though the use of the scientific method. As we have seen, Darwin clearly abandoned the scientific method for a "point of view." Darwin called *The Origin of Species* "one long argument from the beginning to the end."[114]

While an argument can develop a philosophy, only the evidence can discover a natural law. Science can use evidence, but science has little use for simply an argument. For Darwin, "the argument" framed the paradigm of the theory. In the end, it is no wonder that the natural laws of evolution through "successive, slight modifications" have not been found. Darwin recognized, however, that the theory eventually could not survive without the evidence. Without the evidence, the validity of the theory would be void:

> If it could be demonstrated that any complex organ existed, which could not possibly have been formed by numerous, successive, slight modifications, my theory would absolutely break down.[115]

Just as Isaac Newton made it a mission to discover the laws of gravity, Charles Darwin took on the quest to discover the laws operating the Tree of Life through "successive, slight" changes. However, between then and now, science has not discovered a single natural law of evolution, after 150 years. The origin of life remains beyond any known natural law.

References

Preface

1. Anon. 1999. The top minds of the 20th century. *USA Today*, *Life* section, commenting on the book *1,000 people: ranking the men and women who shaped the millennium*, 1998, by Bowers B., Bowers B., and Gottlieb A. H. and H. Kodansha America.

2. Anon. 1999. *The Wall Street Journal*, May 27, p. A24.

3. Zirkle, C. 1959. Evolution, *Marxian Biology, and the Social Scene*. Philadelphia: University of Pennsylvania Press, p. 86.

4. Dewey, J. 1910. *The Influence of Darwinism on Philosophy*. Holt, pp. 9–19.

5. Carnegie, A. 1920. *Autobiography of Andrew Carnegie*, ed. John C. Van Dyke. 1986, reprint, Boston: Northeastern University Press, p. 327.

6. Hofstadter, R. 1959. *Social Darwinism in American thought*. New York George, Braziller, p. 45.

7. Dobzhansky, T. 1973. Nothing in biology makes sense except in the light of evolution, *The American Biology Teacher*. National Association of Biology Teachers, 35:125–129.

8. Pauling, L. 1958. *No More War!* Dodd Mead, NY, p. 209.

9. Anon. 1998. *Teacing about evolution and the nature of science*. National Academy of Science. National Academies Press, available online at www.http://books.nap.edu/openbook.

10. Samuelsson, B. 2002. *Opening Address*, available online at http://nobelprize.org/award_ceremonies/ceremony_sthlm/speeches/opening-2002.html.

11. Huxley, T. 1880. *The Coming of Age of The Origin of Species (1880);* Collected Essays, Volume 2, available online at http://en.wikiquote.org/wiki/Creation_and_evolution.

Chapter One: The Early Years

1.   Darwin, C. R. 1958. *The autobiography of Charles Darwin 1809–1882.* With the original omissions restored. Edited and with appendix and notes by his granddaughter Nora Barlow. London: Collins, p. 68.

2.   Ibid. p. 23.

3.   Ibid. p. 22.

4.   Ibid. p. 23.

5.   Ibid. p. 23.

6.   Ibid. p. 23.

7.   Ibid. p. 25.

8.   Ibid. p. 43.

9.   Ibid. p. 45.

10. Ibid. p. 25.

11. Ibid. p. 44.

12. Ibid. p. 45.

13. Ibid. p. 46.

14. Ibid. p. 27.

15. Ibid. p. 28.

16. Ibid. p. 28.

17. Ibid. p. 46.

18. Darwin, F., editor. 1887. *The life and letters of Charles Darwin,* including an autobiographical chapter. Volume1. London: John Murray, p. 22.

19. Darwin, E. 1796. *Zoönomia, Vol I: Or, the Laws of Organic Life. London: J. Johnson.* Zoonomia. Available at http://www.gutenberg.org/etext/15707.

20. Darwin, C. R. 1958. *The autobiography of Charles Darwin 1809–1882.* With the original omissions restored. Edited and with appendix and notes by his granddaughter Nora Barlow. London: Collins, p. 48.

21. Ibid. p. 49.

22. Ibid. p. 62.

23. Ibid. p. 47.

24. Ibid. p. 54.

25. Ibid. p. 57.

26. Ibid. p. 86.

27. Ibid. p. 62.
28. Ibid. p. 68.
29. Ibid. p.128.
30. Himmelfarb, G. 1959. *Darwin and the Darwinian Revolution.* Doubleday, pp. 33–34.
31. Darwin, C.R. 1958. *The autobiography of Charles Darwin 1809–1882.* With the original omissions restored. Edited and with appendix and notes by his grand-daughter Nora Barlow. London: Collins, p. 60.
32. Brent, P. 1981. *Charles Darwin: A man of enlarged curiosity.* Harper and Rowe, p. 89.
33. Darwin, C. R. 1958. *The autobiography of Charles Darwin 1809–1882.* With the original omissions restored. Edited and with appendix and notes by his grand-daughter Nora Barlow. London: Collins, p. 61.
34. Ibid. p. 58.
35. Ibid. p. 59.
36. Ibid. p. 59.
37. Ibid. p. 68.
38. Ibid. p. 65.
39. Ibid. p. 64.
40. Adam Sedgwick to Charles Darwin. 24 November 1859. Darwin Correspondence Project. 2008. Letter — 2548, *Sedgwick, Adam to Darwin, C.R.,* November 24, 1859. Available at http://www.darwinproject.ac.uk/darwinletters/calendar/entry-2548.html.
41. Darwin, F., editor. 1887. *The life and letters of Charles Darwin,* including an autobiographical chapter. London: John Murray. Volume 3, p. 177.

Chapter Two: The Voyage

1. Darwin, C. R. 1958. *The autobiography of Charles Darwin 1809–1882.* With the original omissions restored. Edited and with appendix and notes by his grand-daughter Nora Barlow. London: Collins, p. 71.
2. Anon. 2008. *Darwin, Young Naturalist.* American Museum of Natural History. Available at www.amnh.com.

3.   Darwin, C. R. 1958. *The autobiography of Charles Darwin 1809–1882*. With the original omissions restored. Edited and with appendix and notes by his grand-daughter Nora Barlow. London: Collins, p. 70.

4.   Ibid. p. 71.

5.   Ibid. p. 71.

6.   Ibid. p. 72.

7.   Fitzroy, R. 1839. *Narrative of the surveying voyages of His Majesty's ships adventure and Beagle between the years 1826 and 1836, describing their examination of the southern shores of South America, and the Beagle's circumnavigation of the globe.* Appendix to Volume II. London: Henry Colburn, p. 91.

8.   Darwin, F., editor. 1887. *The life and letters of Charles Darwin*, including an autobiographical chapter. London: John Murray. Volume 1, p. 277.

9.   Ibid. p. 235.

10.   Ibid. p. 237.

11.   Darwin, C. R. 1860. *Journal of researches into the natural history and geology of the countries visited during the voyage of HMS Beagle round the world, under the command of Capt. Fitz Roy R.N.* London: John Murray. Tenth thousand. Final text. p. 449.

12.   Darwin, C. R. 1958. *The autobiography of Charles Darwin 1809–1882*. With the original omissions restored. Edited and with appendix and notes by his grand-daughter Nora Barlow. London: Collins, p. 73.

13.   Darwin, F., editor. 1887. *The life and letters of Charles Darwin*, *including an autobiographical chapter.* London: John Murray. Volume 1, p. 308.

14.   Ibid. p. 229.

15.   Ibid. p. 233.

16.   Anon. 2008. *Letter 171 — Darwin, C.R., to Henslow, S., 18 May & 16 June, [1832].* Darwin Correspondence Project. Available at www.darwinproject.ac.uk.

17.   Darwin, C. R. 1860. *Journal of researches into the natural history and geology of the countries visited during the voyage of HMS Beagle round the world,* under the command of Capt. Fitz Roy

R.N. London: John Murray. Tenth thousand. Final text, p. 152.

18. Anon. 2008. *Letter 188 — Darwin, C. R. to Darwin, C. S., 24 Oct and 24 Nov [1832]*. Darwin Correspondence Project. Available at www.darwinproject.ac.uk.

19. Darwin, C. R. 1860. *Journal of researches into the natural history and geology of the countries visited during the voyage of HMS Beagle round the world, under the command of Capt. Fitz Roy R.N.* London: John Murray. Tenth thousand. Final text, p. 48.

20. Ibid. p. 503.

21. Ibid. p. 188.

22. Ibid. p. 69.

23. Ibid. p. 330.

24. Darwin, F., editor. 1887. *The life and letters of Charles Darwin, including an autobiographical chapter*. London: John Murray. Volume 1, p. 251.

25. Darwin, C. R. 1860. *Journal of researches into the natural history and geology of the countries visited during the voyage of HMS Beagle round the world, under the command of Capt. Fitz Roy R.N.* London: John Murray. Tenth thousand. Final text, p. 216.

26. Ibid. p. v.

27. Ibid. p. 305.

28. Anon. 2008. *Letter 275 — Darwin, C. R. to Darwin, S. E., 23 Apr 1835*. Darwin Correspondence Project. Available at www. darwinproject.ac.uk.

29. Darwin, C. R. 1860. *Journal of researches into the natural history and geology of the countries visited during the voyage of HMS Beagle round the world, under the command of Capt. Fitz Roy R.N.* London: John Murray. Tenth thousand. Final text, p. 373.

30. Melville, H. 1854. "The Encantadas or, Enchanted Isles." In *Putnam's Monthly Magazine*, Vol. III (March, April, May issues). New York: G. P. Putnam & Co.

31. Darwin, C. R. 1860. *Journal of researches into the natural history and geology of the countries visited during the voyage of HMS*

Beagle round the world, under the command of Capt. Fitz Roy R.N. London: John Murray. Tenth thousand. Final text, p. 377.

32. Ibid. p. 379.

33. Darwin, C. R. 1839. *Narrative of the surveying voyages of His Majesty's Ships Adventure and Beagle between the years 1826 and 1836, describing their examination of the southern shores of South America, and the Beagle's circumnavigation of the globe. Journal and remarks. 1832–1836.* London: Henry Colburn. p. 474.

34. Darwin, C. R. 1860. *Journal of researches into the natural history and geology of the countries visited during the voyage of HMS Beagle round the world, under the command of Capt. Fitz Roy R.N.* London: John Murray. Tenth thousand. Final text, p. 380.

35. Anon. 2008. *Letter 301 — Darwin, C. R. to Darwin, C. S., 29 Apr 1836.* Darwin Correspondence Project. Available at www.darwinproject.ac.uk.

36. Anon. 2008. *Letter 306 — Darwin, C. R. to Darwin, S. E., 4 Aug [1836],* Darwin Correspondence Project. Available at www.darwinproject.ac.uk.

37. Keynes, R. D. editor, 2001, *Charles Darwin's Beagle Diary.* Cambridge: Cambridge University Press, p. 357.

38. Darwin, C. R. 1860. *Journal of researches into the natural history and geology of the countries visited during the voyage of HMS Beagle round the world, under the command of Capt. Fitz Roy R.N.* London: John Murray. Tenth thousand. Final text, p. 380.

39. Darwin, C. R. 1958. *The autobiography of Charles Darwin 1809–1882.* With the original omissions restored. Edited and with appendix and notes by his grand-daughter Nora Barlow. London: Collins, p. 77.

40. Ibid. p. 78.

41. Anon. 2008. *Letter 729 — Darwin, C. R. to Hooker, J. D., [11 Jan 1844].* Darwin Correspondence Project. Available at www.darwinproject.ac.uk.

42. Anon. 2008. *Letter 307 — Darwin, C. R. to Wedgwood, Josiah, II, [5 Oct 1836]*. Darwin Correspondence Project. Available at www.darwinproject.ac.uk.

43. Darwin, C. R. 1958. *The autobiography of Charles Darwin 1809–1882*. With the original omissions restored. Edited and with appendix and notes by his grand-daughter Nora Barlow. London: Collins, p. 78.

44. Darwin, F. and Seward, A. C., editors. 1903. *More letters of Charles Darwin*. A record of his work in a series of hitherto unpublished letters. London: John Murray. Volume 1, p. 4.

45. Anon. 1882. Charles Darwin, Controversial Scientist, Dies at 73, *London News Obituaries,* 20 April.

46. Darwin, C. R. 1958. *The autobiography of Charles Darwin 1809–1882*. With the original omissions restored. Edited and with appendix and notes by his grand-daughter Nora Barlow. London: Collins, p. 81.

47. Anon. 2008. http://en.wikiquote.org/wiki/Creation_and_evolution.

48. Darwin, C. R. 1958. *The autobiography of Charles Darwin 1809–1882*. With the original omissions restored. Edited and with appendix and notes by his grand-daughter Nora Barlow. London: Collins, p. 76.

Chapter Three: Sketching

1. Darwin, C. R. 1958. *The autobiography of Charles Darwin 1809–1882*. With the original omissions restored. Edited and with appendix and notes by his grand-daughter Nora Barlow. London: Collins, p. 237.

2. Ibid. p. 141.

3. Darwin, F., editor. 1887. *The life and letters of Charles Darwin,* including an autobiographical chapter. London: John Murray. Volume 1, p. 266.

4. Darwin, F. and Seward, A. C., editors. 1903. *More Letters of Charles Darwin* Volume II, Letter 275.

5. Darwin, C. R. 1958. *The autobiography of Charles Darwin 1809–1882*. With the original omissions restored. Edited and

with appendix and notes by his grand-daughter Nora Barlow. London: Collins, p. 83.

6. Anon. 2008. *Letter 3247 — Darwin, C. R. to Jamieson, T. F., 6 Sept [1861].* Darwin Correspondence Project. Available at www.darwinproject.ac.uk.

7. Darwin, C. R. 1958. *The autobiography of Charles Darwin 1809–1882.* With the original omissions restored. Edited and with appendix and notes by his grand-daughter Nora Barlow. London: Collins, p. 85.

8. Sebright, S. J. 1836. *Observations upon the instinct of animals,* London.

9. De Beer, G., editor. 1960. *Darwin's notebooks on transmutation of species. Part II. Second notebook [C]* (February to July 1838). Bulletin of the British Museum of Natural History. Historical Series 2(3):96.

10. Darwin, C. R. 1958. *The autobiography of Charles Darwin 1809–1882.* With the original omissions restored. Edited and with appendix and notes by his grand-daughter Nora Barlow. London: Collins, p. 130.

11. Darwin, C. R. 1860. *Journal of researches into the natural history and geology of the countries visited during the voyage of HMS Beagle round the world, under the command of Capt. Fitz Roy R.N.* London: John Murray. Tenth thousand. Final text, p. 380.

12. Darwin, C. R. 1958. *The autobiography of Charles Darwin 1809–1882.* With the original omissions restored. Edited and with appendix and notes by his grand-daughter Nora Barlow. London: Collins, p. 85–86.

13. Ibid. p. 86.

14. Ibid. p. 87.

15. Eldredge, N. 2007. *Darwin.* American Museum of Natural History. Available at www.amnh.org/exhibitions/darwin. And, *Inception of Darwin's Theory.*

16. Darwin, F. and Seward, A. C., editors. 1903. *More letters of Charles Darwin. A record of his work in a series of hitherto unpublished letters.* London: John Murray. Volume 1, p. 29.

17. Darwin, F., editor. 1887. *The life and letters of Charles Darwin, including an autobiographical chapter.* London: John Murray. Volume 1, p. 281.
18. Darwin, C. R. 1958. *The autobiography of Charles Darwin 1809–1882.* With the original omissions restored. Edited and with appendix and notes by his grand-daughter Nora Barlow. London: Collins, p. 114.
19. Ibid. p. 120.
20. Darwin, F., editor. 1887. *The life and letters of Charles Darwin,* including an autobiographical chapter. London: John Murray. Volume II, p. 16.
21. Darwin, F., editor. 1909. *The foundations of the origin of species. Two essays written in 1842 and 1844.* Cambridge: Cambridge University Press, p. 52.
22. Ibid. p. 52.
23. Anon. 2008. *Letter 734 — Hooker, J. D. to Darwin, C. R., 29 Jan 1844.* Darwin Correspondence Project. Available at www.darwinproject.ac.uk.
24. Anon. 2008. *Letter 471 — Darwin, Emma to Darwin, C. R. [c. Feb 1839].* Darwin Correspondence Project. Available at www.darwinproject.ac.uk.
25. Darwin, C. R. 1958. *The autobiography of Charles Darwin 1809–1882.* With the original omissions restored. Edited and with appendix and notes by his grand-daughter Nora Barlow. London: Collins, p. 237.
26. Ibid. p. 121.
27. Ibid. p. 121.
28. Wallace, R. 1858. On the Tendency of Varieties to Depart Indefinitely From the Original Type. *Journal of the Proceedings of the Linnean Society, Zoology* 3: 45–62. p. 20.
29. Darwin, F., editor. 1887. *The life and letters of Charles Darwin, including an autobiographical chapter.* London: John Murray. Volume II, p. 117.
30. Ibid. p. 117.
31. Darwin, C. R. 1958. *The autobiography of Charles Darwin 1809–1882.* With the original omissions restored. Edited and

with appendix and notes by his grand-daughter Nora Barlow. London: Collins, p. 122.

32. Ibid. p. 124.

33. Darwin, F., editor. 1887. *The life and letters of Charles Darwin,* including an autobiographical chapter. London: John Murray. Volume III, p. 121.

34. Wallace, A. R. 1903. The dawn of a great discovery "My relations with Darwin in reference to the theory of natural selection." *Black and White,* 25 (17 January), p. 78.

35. Anon. 2008. Letter 2533 — Darwin, C. R. to Fox, W. D. [16 Nov 1859]. Darwin Correspondence Project. Available at www.darwinproject.ac.uk.

36. Darwin, C. R. 1958. *The autobiography of Charles Darwin 1809–1882.* With the original omissions restored. Edited and with appendix and notes by his grand-daughter Nora Barlow. London: Collins, p. 238.

37. Litchfield, H. E., editor. 1915. *Emma Darwin, A century of family letters,* 1702–1896. London: John Murray. Volume 2, p. 175.

38. Darwin, F., editor. 1887. *The life and letters of Charles Darwin,* including an autobiographical chapter. London: John Murray, Volume II, p. 109.

39. Anon. 2008. *Letter 2117 — C. R. to Hooker, J. D., 5 July [1857].* Darwin Correspondence Project. Available at www.darwinproject.ac.uk.

40. Anon. 2008. *Letter 8449 — Darwin, C. R. to Hooker, J. D., 4 Aug [1872].* Darwin Correspondence Project. Available at www.darwinproject.ac.uk.

41. Darwin, C. R. 1958. *The autobiography of Charles Darwin 1809–1882.* With the original omissions restored. Edited and with appendix and notes by his grand-daughter Nora Barlow. London: Collins, p. 130.

42. Darwin, C. R. 1871. *The descent of man, and selection in relation to sex.* London: John Murray. Volume I. 1st edition, p. 153.

43. Winch, D. 2001. Darwin Fallen Among Political Economists. *Proceedings of the American Philosophical Society,* 145(4):415.

44. Darwin, F., editor. 1887. *The life and letters of Charles Darwin,* including an autobiographical chapter. London: John Murray. Volume III, p. 236.

45. Darwin, C. R. 1958. *The autobiography of Charles Darwin 1809–1882.* With the original omissions restored. Edited and with appendix and notes by his grand-daughter Nora Barlow. London: Collins, p. 92.

46. Darwin, F. and Seward, A. C., editors. 1903, *More Letters of Charles Darwin.* Volume II, Letter 236, p. 321.

47. Darwin, C. R. 1958. *The autobiography of Charles Darwin 1809–1882.* With the original omissions restored. Edited and with appendix and notes by his grand-daughter Nora Barlow. London: Collins, p. 94.

48. Ibid. p. 138.

49. Darwin, F. and Seward, A. C., editors. 1903, *More Letters of Charles Darwin* Volume III, p. 146.

50. Darwin, C. R. 1958. *The autobiography of Charles Darwin 1809–1882.* With the original omissions restored. Edited and with appendix and notes by his grand-daughter Nora Barlow. London: Collins, p. 7.

51. Ibid. p. 138.

52. Desmond, A and Moore, J. 1991. *Darwin.* Michael Joseph, London, 1991, p. 46.

53. Darwin, F., editor. 1887. *The life and letters of Charles Darwin,* including an autobiographical chapter. London: John Murray, Volume III, p. 358.

54. Darwin, F. and Seward, A. C., editors. 1903, *More Letters of Charles Darwin.* Volume I, p. 30.

55. Huxley, T. H., 1882, Charles Darwin. April 27, 1882. *Nature* (London); reprinted in Thomas H. Huxley, 1896, *Darwiniana Essays* [1970: New York AMS Reprint], p. 247.

56. Clark, R. W. 1985. *The survival of Charles Darwin: a biography of a man and an idea.* Weidenfeld & Nicholson, p. 199.

57. Moore, J. 1994. *The Darwin Legend. Baker Book House.* MI., p. 21.

58. Litchfield, H. 1922. Charles Darwin's Death-Bed: Story of Conversion Denied. *The Christian,* February 23, p. 12.

59. Clark, R. 1985. *The survival of Charles Darwin: a biography of a man and an idea.* Weidenfeld & Nicholson, p. 199.

60. Peckham, M, editor. 1959, *The Origin Of Species By Charles Darwin: A Variorum Text.*

Chapter Four: The Stage

1. Darwin, C. R. 1872. *The origin of species by means of natural selection, or the preservation of favoured races in the struggle for life.* London: John Murray. 6th edition; with additions and corrections. Eleventh thousand. p. xiii.

2. Curtis, L. P., Jr. 1968. *Anglo-Saxons and Celts: A Study of Anglo-Irish Prejudice in Victorian England.* Connecticut: University of Bridgeport Conference on British Studies, p. 84.

3. Lloyd, G. E. R. 1970. *Early Greek science: Thales to Aristotle.* New York: Norton. pp. 17–18.

4. 4. Ibid. pp. 17–18.

5. Aristotle. 384–322, B.C. *History of Animals,* pp. 608b–609b.

6. Barnes, J., editor. 1995. *The Cambridge companion to Aristotle.* Cambridge: Cambridge UP, p. 239.

7. 7. Anon. 2008. 384–322, B.C. Quoted in *Aristotle,* Wikipedia. Available at www.wikipedia.org.

8. Aristotle. 384–322, B.C. *The History of Animals, Book III,* 350 B.C.E. Translated by D'Arcy Wentworth Thompson, Part 28. Available at www.classics.mit.edu.

9. Ibid. Part 29. Available at www.classics.mit.edu.

10. Darwin, C. R. 1872. *The origin of species by means of natural selection, or the preservation of favoured races in the struggle for life.* London: John Murray, 6th edition, p. xiii.

11. De Beer, G. 1969. Reprint of Robert Chambers, *Vestiges of the Natural History of Creation,* p. 11.

12. De Maupertuis, P. L. M. 1745. *Venus physique/The Earthly Venus,* La Haye. Available at cogweb.ucla.edu/EarlyModern/Maupertuis_1745.html.

13. Ibid. Available at cogweb.ucla.edu/EarlyModern/Maupertuis_1745.html.

14. Ibid. Available at cogweb.ucla.edu/EarlyModern/Maupertuis_1745.html.

15. Darwin, C. R. 1872. *The origin of species by means of natural selection, or the preservation of favoured races in the struggle for life*, London: John Murray, 6th edition, p. 423.
16. Ibid. p. xiii.
17. Buffon, G. L. 1783. *Histoire Naturelle*, Volume IV, p. 382.
18. Darwin, F., editor. 1887. *The life and letters of Charles Darwin*, including an autobiographical chapter. London: John Murray, Volume III, p. 45.
19. Darwin, C. R. 1872. *The origin of species by means of natural selection, or the preservation of favoured races in the struggle for life*, London: John Murray, 6th edition, p. xiii.
20. Quoted in Lamarck, J. B. 1984. *Zoological philosophy: An exposition with regard to the natural history of animals.* The University of Chicago Press, p. 70.
21. Darwin, C. R. 1872. *The origin of species by means of natural selection, or the preservation of favoured races in the struggle for life*, London: John Murray, 6th edition, p. xiv.
22. Ibid. p. 98.
23. Darwin, E. 1794. *Zoönomia; or the laws of organic life.* J. Johnson, London. Republished by The Project Gutenberg EBook, April 25, 2005 [EBook #15707]. Available at www.guterberg.org.
24. Ibid. Available at www.guterberg.org.
25. Darwin, E. 1794. *Zoönomia; or the laws of organic life.* J. Johnson, London. Reprint 1974 by AMS Press, New York, p. 505.
26. Darwin, C. R. 1872. *The origin of species by means of natural selection, or the preservation of favoured races in the struggle for life*. London: John Murray, 6th edition, p. xiv.
27. Darwin, C. R. 1958. *The autobiography of Charles Darwin 1809–1882.* With the original omissions restored. Edited and with appendix and notes by his grand-daughter Nora Barlow. London: Collins, p. 124.
28. Ibid. pp. 119–120.
29. Ibid. p. 120.
30. Anon. 2008. *Letter 2449—Darwin, C. R. to Wallace, A. R., 6 Apr 1859.* Darwin Correspondence Project. Available at www.darwinproject.ac.uk.

31. Darwin, F., editor. 1887. *The life and letters of Charles Darwin,* including an autobiographical chapter. London: John Murray. Volume II, p. 316.

32. Darwin, F., editor. 1887. *The life and letters of Charles Darwin,* including an autobiographical chapter. London: John Murray. Volume 1, p. 69.

33. Copernicus, N. 1543. *De Revolutionibus.*

34. Greene, N. 2008. *Sir Isaac Newton Quotations,* Available at space.about.com/od/astronomerbiographies/a/newtonquotes. htm.

35. Thayer, H. S., editor. 1953. *Principia Mathematica,* Book III; cited in; *Newton's Philosophy of Nature: Selections from his writings.* Hafner Library of Classics, N.Y., p. 42.

36. Hull, D.L. 1983, *Darwin and his critics: the reception of Darwin's theory of evolution by the scientific community.* University of Chicago Press.

37. Smith, W. 1835. Deductions from established facts of geology. *The Literary Gazette,* No. 969 (15 August), p. 522.

38. Richards, R.J. 2001. Commotion over evolution before Darwin. *American Scientist* 89(5):454–456, p. 454.

39. Chambers, W. 1872. *Memoirs,* p. 254. And, Anon. 2008. Quoted in *William Chambers* Wikipedia. Available at www. wikipedia.org.

40. Chambers, R. 1844. *Vestiges of Natural History of Creation,* p. 231. And, Anon. 2008. Quoted in *William Chambers* Wikipedia. Available at www.wikipedia.org.

41. Ibid. p. 388. And, Anon. 2008. Quoted in *William Chambers* Wikipedia. Available at www.wikipedia.org.

42. Matthew, P. 1860. Nature's law of selection. *Gardeners' Chronicle and Agricultural Gazette,* 7 April, pp. 312–313.

43. Darwin, F., editor. 1887. *The life and letters of Charles Darwin,* including an autobiographical chapter. London: John Murray. Volume II, p. 301.

44. Darwin, C. R. 1860. Natural selection. *Gardeners' Chronicle and Agricultural Gazette* No. 16 (21 April), p. 362.

45. Darwin, C. R. 1861. *On the origin of species by means of natural selection, or the preservation of favoured races in the struggle for life.* London: John Murray. 3rd edition, p. xv.

46. Anon. 2008. Quoted in *Patrick Matthew* Wikipedia. Available at www.wikipedia.org.

47. Michael, S. 2002. In Darwin's shadow: the life and science of Alfred Russel Wallace. Oxford University Press. And, Anon. 2008. Quoted in *Edward Blyth* Wikipedia. Available at www. wikipedia.org.

48. Eiseley, L. 1979. *Darwin and the Mysterious Mr. X*. New York: E.P. Dutton, p. 55.

49. Darwin, C. R. 1861. *On the origin of species by means of natural selection, or the preservation of favoured races in the struggle for life*. London: John Murray. 3rd edition, p. 19.

50. Anon. 2008. Wallace's views on the book *Vestiges of the Natural History of Creation*. *The Natural History Museum, London. Available at www.nhm.ac.uk.*

51. Wallace, A. R. 1855. *On the Law Which has Regulated the Introduction of Species*. And, Anon. 2008. Quoted in *Alfred Russel Wallace* Wikipedia. Available at www.wikiquote.org.

52. Darwin, C. R. 1861. *On the origin of species by means of natural selection, or the preservation of favoured races in the struggle for life*. London: John Murray. 3rd edition, p. 75.

53. Ibid. p. xiii.

54. Job 12:7.

55. Anon. 2008. *The Huxley file:* Letters Index, TH Huxley to Henrietta, 1872. Available at www.aleph0.clarku.edu/Huxley.

56. Barton, R. 1998. Huxley, Lubbock, and Half a Dozen Others: Professionals and Gentlemen in the Formation of the X Club, 1851–1864, Isis, 89(3):437–438.

57. MacLeod, R. M. 1970. The X-Club a Social Network of Science in Late-Victorian England. *Notes and Records of the Royal Society of London*, 24(2):310.

58. Teller, J. D. 1943. Huxley's "Evil" Influence. *The Scientific Monthly*, 56(2):177.

59. Dewey, J. 1909. A lecture in a course of public lectures on "Charles Darwin and His influence on Science," given at Columbia University in the winter and spring of 1909. Reprinted from the *Popular Science Monthly* for July.

Chapter Five: Origin of Species

1. Darwin, C. R. 1872. *The origin of species by means of natural selection, or the preservation of favoured races in the struggle for life. London: John Murray. 6th edition, p. 424.*

2. *Darwin, C. R. 1958. The autobiography of Charles Darwin 1809–1882.* With the original omissions restored. Edited and with appendix and notes by his grand-daughter Nora Barlow. London: Collins, p. 123.

3. Ibid. p. 13.

4. Ibid. p. 121.

5. Ibid. p. 235.

6. Ibid. p. 237.

7. Ibid. p. 237.

8. Ibid. p. 122.

9. Cirkle, C. 1959. *Evolution, Marxian Biology, and the Social Scene*, Univ. of Pennsylvania Press, p. 88.

10. Darwin, C. R. 1958. *The autobiography of Charles Darwin 1809–1882.* With the original omissions restored. Edited and with appendix and notes by his grand-daughter Nora Barlow. London: Collins, p. 137.

11. Anon. 2008. Available at http://cs.clarku.edu/huxley/letters/88.html#14feb1888.

12. Darwin, C. R. 1872. *The origin of species by means of natural selection, or the preservation of favoured races in the struggle for life.* London: John Murray. 6th edition, p. 404.

13. Barzun, J. 1958. *Darwin, Marx and Wagner: Critique of a Heritage.* Garden City, New York: Doubleday and Co. pp. 32 and 74.

14. Huxley, T. 1888, Collected Essays, Huxley. Volume II, Chapter X, p. 242. Re-published in Obituary Notices of the *Proceedings of the Royal Society*, 44:1888.

15. Peckham, M. 1959. *The Origin of Species by Charles Darwin: A Variorum Text* Philadelphia: University of Pennsylvania Press.

16. Darwin, C. R. 1859. *The origin of species by means of natural selection, or the preservation of favoured races in the struggle for life.* London: John Murray, 1st edition, p. 81.

17. Darwin, C. R. 1869. *The origin of species by means of natural selection, or the preservation of favoured races in the struggle for life.* London: John Murray, 5th edition, p. 92.

18. Darwin, C. R. 1958. *The autobiography of Charles Darwin 1809–1882.* With the original omissions restored. Edited and with appendix and notes by his grand-daughter Nora Barlow. London: Collins, p. 8.

19. Darwin, F., editor. 1887. *The life and letters of Charles Darwin, including an autobiographical chapter.* London: John Murray. Volume II, p. 312.

20. Darwin, C. R. 1859. *The origin of species by means of natural selection, or the preservation of favoured races in the struggle for life.* London: John Murray, 1st edition, p. ii.

21. Ibid. p. ii.

22. Darwin, C. R. 1872. *The origin of species by means of natural selection, or the preservation of favoured races in the struggle for life.* London: John Murray, 6th edition, p. 5.

23. Ibid. p. 31.

24. Ibid. p. 31.

25. Ibid. p. 47.

26. Ibid. p. 50.

27. Ibid. p. 61.

28. Ibid. p. 62.

29. Ibid. p. 86.

30. Ibid. p. 106.

31. Ibid. p. 131.

32. Ibid. p. 133.

33. Ibid. p. 165.

34. Ibid. p. 203.

35. Ibid. p. 205.

36. Ibid. p. 263.

37. Ibid. p. 265.

38. Ibid. p. 289.

39. Ibid. p. 289.

40. Ibid. p. 312.

41. Ibid. p. 264.

42. Ibid. p. 270.

43. Ibid. p. 312.
44. Ibid. p. 396.
45. Ibid. p. 396.
46. Ibid. pp. 395–396.
47. Ibid. p. 403.
48. Ibid. p. 430.
49. Ibid. p. 428.
50. Darwin, C. R. 1871. *The descent of man, and selection in relation to sex.* London: John Murray. Volume 1. 1st edition, p. 34.
51. Darwin, F., editor. 1887. *The life and letters of Charles Darwin, including an autobiographical chapter.* London: John Murray. Volume II, pp. 223–224.
52. Huxley, T. H. 1891. Letter to Darwin, 23 November 1859, in Huxley L. *Life and Letters of T. H. Huxley,* Macmillan and Co Ltd, Volume 2, p. 176.
53. Darwin, C. R. 1872. *The origin of species by means of natural selection, or the preservation of favoured races in the struggle for life.* London: John Murray, 6th edition, p. 424.
54. Ibid. p. 423.
55. Darwin, F, and Seward, A. C., editors. 1903. *More letters of Charles Darwin. A record of his work in a series of hitherto unpublished letters.* London: John Murray. Volume 1, p. 195.
56. Darwin, C. R. 1872. *The origin of species by means of natural selection, or the preservation of favoured races in the struggle for life.* London: John Murray, 6th edition, p. 120.
57. Ibid. p. 403.
58. Anon. 2008. *Letter 2545—Darwin, E. A. to Darwin, C. R., 23 Nov [1859].* Darwin Correspondence Project. Available online at www.darwinproject.ac.uk.
59. Darwin, C. R. 1872. *The origin of species by means of natural selection, or the preservation of favoured races in the struggle for life.* London: John Murray, 6th edition, p. 404.
60. Ibid. p. 121.
61. Darwin, C. R. 1958. *The autobiography of Charles Darwin 1809–1882.* With the original omissions restored. Edited and with appendix and notes by his grand-daughter Nora Barlow. London: Collins, p. 141.

62. Ibid. p. 119.

63. Ibid. p. 158.

64. Ibid. p. 159.

65. Darwin, F. and Seward, A. C., editors. 1903. *More letters of Charles Darwin. A record of his work in a series of hitherto unpublished letters.* London: John Murray. Volume 1, p. 39.

66. Darwin, F., editor. 1887. *The life and letters of Charles Darwin, including an autobiographical chapter.* London: John Murray. Volume 1, p. 149.

67. Darwin, C. R. 1958. *The autobiography of Charles Darwin 1809–1882.* With the original omissions restored. Edited and with appendix and notes by his grand-daughter Nora Barlow. London: Collins, p. 120.

68. Darwin, F and Seward, A. C., editors. 1903. *More letters of Charles Darwin. A record of his work in a series of hitherto unpublished letters.* London: John Murray. Volume 2, p. 133.

69. Darwin, C. R. 1958. *The autobiography of Charles Darwin 1809–1882.* With the original omissions restored. Edited and with appendix and notes by his grand-daughter Nora Barlow. London: Collins, p. 161.

70. Darwin, F. and Seward, A. C., editors. 1903. *More letters of Charles Darwin. A record of his work in a series of hitherto unpublished letters.* London: John Murray. Volume 2, p. 323.

71. Darwin, F., editor. 1887. *The life and letters of Charles Darwin, including an autobiographical chapter.* London: John Murray. Volume 1, p. 70.

72. Shermer, M. 2005. *The Woodstock of Evolution. Scientific America*, quoting Niles Eldredge, June 27.

73. Dennett, D. C. 1995, *Darwin's Dangerous Idea: Evolution And The Meanings Of Life.* Simon & Schuster Adult Publishing Group.

74. Gould, S. J. 1986. Evolution and the Triumph of Homology: Or, Why History Matters. *American Scientist. 74:61.*

75. Darwin, F., editor. 1887. *The life and letters of Charles Darwin, including an autobiographical chapter.* London: John Murray. Volume II, p. 108.

76. Darwin, C. R. 1872. *The origin of species by means of natural selection, or the preservation of favoured races in the struggle for life.* London: John Murray, 6th edition, p. 425.

77. Huxley, T. H. 1907. *Reflection #9, Aphorisms and Reflections, selected by Henrietta A. Huxley,* Macmillan.

78. Darwin, F., editor. 1887. *The life and letters of Charles Darwin, including an autobiographical chapter.* London: John Murray. Volume II, p. 231.

79. Ibid. p. 241.

80. Darwin, F and Seward, A. C., editors. 1903. *More letters of Charles Darwin. A record of his work in a series of hitherto unpublished letters.* London: John Murray. Volume 1, p. 189.

81. Appleman, P. 2001. *Darwin and His Critics: The Reception of Darwin's Theory of Evolution by the Scientific Community.* Cambridge: Harvard University Press; reprinted 1983, Chicago: University of Chicago Press; extract in *Darwin: A Norton Critical Edition,* 3rd edition by, pp. 257–264.

82. Darwin, F. and Seward, A. C., editors. 1903. *More letters of Charles Darwin. A record of his work in a series of hitherto unpublished letters.* London: John Murray. Volume 1, p. 126.

83. Darwin, C. R. 1872. *The origin of species by means of natural selection, or the preservation of favoured races in the struggle for life.* London: John Murray, 6th edition, p. 2.

84. Ibid. p. 233.

85. Ibid. p. 404.

86. Darwin, F. and Seward, A. C., editors. 1903. *More letters of Charles Darwin. A record of his work in a series of hitherto unpublished letters.* London: John Murray. Volume 1, p. 321.

87. Darwin, C. R. 1958. *The autobiography of Charles Darwin 1809–1882.* With the original omissions restored. Edited and with appendix and notes by his grand-daughter Nora Barlow. London: Collins, pp. 92–93.

88. Huxley, T. H. 1907. *Reflection #219,* Aphorisms *and Reflections,* selected by Henrietta A. Huxley, Macmillan. London.

89. Huxley, J. 1939. *Julian Huxley Presents the Living Thoughts of Darwin.* David McKay Company, Philadelphia, p. 15.

90. Mirsky, S. 2006. The Evolution of Ernst: Interview with Ernst Mayr. *Scientific American,* editor, July 2006. Available at www.sciam.com.

91. Darwin, C. R. 1958. *The autobiography of Charles Darwin 1809–1882.* With the original omissions restored. Edited and with appendix and notes by his grand-daughter Nora Barlow. London: Collins, p. 8.

92. Darwin, C. R. 1872. The origin of species by means of natural selection, or the preservation of favoured races in the struggle for life. London: John Murray. 6th edition; with additions and corrections, p. 68.

Chapter Six: Species

1. Darwin, C. R. 1872. *The origin of species by means of natural selection, or the preservation of favoured races in the struggle for life.* London: John Murray. 6th edition; p. 33.

2. Ibid. p. 37.

3. Nicholson, H. A. 1872. *A manual of zoology.* Appleton and Company, New York, p. 20.

4. Darwin, C. R. 1872. *The origin of species by means of natural selection, or the preservation of favoured races in the struggle for life.* London: John Murray. 6th edition, p. 313.

5. Ibid. p. 38.

6. Ibid. p. 278.

7. Ibid. p. 38.

8. Ibid. p. 38.

9. Ibid. p. 42.

10. Ibid. p. 47.

11. Ibid. p. 42.

12. Ibid. p. 86.

13. Ibid. p. 48.

14. Ibid. pp. 48–49.

15. Ibid. p. 49.

16. Ibid. p. 120.

17. Ibid. p. 48.

18. Ibid. p. 235.

19. Ibid. p. 241.

20. Ibid. p. 251.

21. Ibid. p. 364.

22. Ibid. pp. 364–365.

23. Ibid. p. 372.

24. Ibid. p. 368.

25. Ibid. p. 426.

26. Ibid. p. 383.

27. Ibid. p. 404.

28. Ibid. p. 45.

29. Dobzhansky T. 1937. Genetics and the origin of species. Columbia University Press, New York, p. 310.

30. Hey J. 2001. The mind of the species problem. *Trends in Ecology and Evolution* 16:326–329.

Chapter Seven: Natural Selection

1. Darwin, C. R. 1958. *The autobiography of Charles Darwin 1809–1882*. With the original omissions restored. Edited and with appendix and notes by his grand-daughter Nora Barlow. London: Collins, p. 87.

2. Eldredge, N. 2005. *Darwin, Discovering the Tree of Life*. WW Norton & Company, Inc, p. 16.

3. Darwin, C. R. 1859. *On the origin of species by means of natural selection, or the preservation of favoured races in the struggle for life*. London: John Murray. 1st edition, p. 18.

4. Gould, S.J. 1982. Darwinism and the expansion of evolutionary theory. *Science*, 216(4544):380.

5. Anon. 2008. *Darwin*. American Museum of Natural History. Available at www.amnh.org/exhibitions/darwin.

6. Anon. 2008. *Darwin*. American Museum of Natural History. Available at www.nmnh.si.edu/paleo/geotime/main/foundation_life3.html.

7. Shapiro, K. 2006. Misplaced Sympathies, *The Wall Street Journal*, Friday, April 21, 2006. Available at www.opinionjournal.com/taste/?id=110008270.

8. Darwin, C. R. 1872. *The origin of species by means of natural selection, or the preservation of favoured races in the struggle for*

life. London: John Murray. 6th edition; with additions and corrections, pp. 295–296.

9. Ibid. p. 233.
10. Ibid. p. 282.
11. Ibid. p. 133.
12. Ibid. p. 63.
13. Darwin, C. R. 1859. *On the origin of species by means of natural selection, or the preservation of favoured races in the struggle for life*. London: John Murray. 1st edition, p. 127.
14. Darwin, C. R. 1872. *The origin of species by means of natural selection, or the preservation of favoured races in the struggle for life*. London: John Murray. 6th edition; with additions and corrections, p. 63.
15. Hitching, F. 1982. *The Neck of the Giraffe: Where Darwin Went Wrong*. Ticknor & Fields: New York, p. 104.
16. Waddington, C. H. 1967. Quoted by Moorhead P. S. and M. M. Kaplan, M. M., editors. *Mathematical Challenges to the neo-Darwinian Interpretation of Evolution*, Monograph #5. Philadelphia, PA: Wistar University Press. p. 14 – and 1976 quoted by Bethell, T. in Darwin's Mistake. *Harper's Magazine*, February, p. 75.
17. Hitching, F. 1982. *The Neck of the Giraffe*. New York: Ticknor and Fields, p. 84.
18. Darwin, C. R. 1872. *The origin of species by means of natural selection, or the preservation of favoured races in the struggle for life*. London: John Murray. 6th edition; with additions and corrections, p. 96.
19. Ibid. p. 222.
20. Ibid. p. 70.
21. Ibid. p. 87.
22. Darwin, F., editor. 1887. *The life and letters of Charles Darwin, including an autobiographical chapter*. London: John Murray. Volume 1, p. 20.
23. Darwin, C. R. 1872. *The origin of species by means of natural selection, or the preservation of favoured races in the struggle for life*. London: John Murray. 6th edition; with additions and corrections, p. 49.

24. Ibid. p. 63.
25. Ibid. p. 146.
26. Ibid. p. 412.
27. Ibid. pp. 68–69.
28. Anon. 2008. *Evolution.* University of California, Berkley. Available at evolution.berkeley.edu/evosite/evo101.
29. Darwin, C. R. 1872. *The origin of species by means of natural selection, or the preservation of favoured races in the struggle for life.* London: John Murray. 6th edition; with additions and corrections, p. 152.
30. Ibid. p. 117.
31. Ibid. p. 176.
32. Ibid. p. 307.
33. Ibid. p. 67.
34. Ibid. p. 67.
35. Ibid. p. 143.
36. Ibid. p. 103.
37. Ibid. p. 156.
38. Ibid. p. 99.
39. Ibid. p. 429.
40. Anon. 2008. *Darwin.* American Museum of Natural History. Available at www.amnh.org/exhibitions/darwin/
41. Darwin, C. R. 1872. *The origin of species by means of natural selection, or the preservation of favoured races in the struggle for life.* London: John Murray. 6th edition; with additions and corrections, p. 175.
42. Ibid. p. 146.
43. Ibid. p. 98.
44. Ibid. p. 85.
45. Ibid. p. 90.
46. Ibid. pp. 65–66.
47. Ibid. p. 291.
48. Anon. 2008. *Darwin.* American Museum of Natural History. Available at www.amnh.org/exhibitions/darwin.
49. Darwin, C. R. 1872. *The origin of species by means of natural selection, or the preservation of favoured races in the struggle for*

life. London: John Murray. 6th edition; with additions and corrections, p. 408.

50. Ibid. p. 422.
51. Ibid. p. 129.
52. Ibid. p. 154.
53. Ibid. p. 106.
54. Ibid. p. 153.
55. Ibid. p. 120.
56. Ibid. p. 201.
57. Ibid. p. 64.
58. Ibid. p. 188.
59. Ibid. p. 146.
60. De Vries, H. 1905, *Species and Varieties: Their Origin by Mutation.*
61. Patterson, C. 1982. *Cladistics,* Interview by Brian Leek, interviewer Peter Franz, March 4, BBC.
62. Darwin, C. R. 1872. *The origin of species by means of natural selection, or the preservation of favoured races in the struggle for life*. London: John Murray. 6th edition; with additions and corrections, p. 159.
63. Ibid. p. 131.
64. Ibid. p. 107.
65. Ibid. p. 31.
66. Ibid. p. 107.
67. Ibid. p. 63.
68. Ibid. p 164.
69. Sermonti, G. 2005, *Why a Fly is Not a Horse*, Seattle, Washington, Institute, p. 13.
70. Ibid. pp. 13–14.
71. Darwin, C. R. 1872. *The origin of species by means of natural selection, or the preservation of favoured races in the struggle for life*. London: John Murray. 6th edition; with additions and corrections, p. 313.
72. Ibid. p. 121.
73. Ibid. p. 103.
74. Ibid. p. 67.
75. Ibid. p. 114.

76. Ibid. p. 116.
77. Ibid. p. 152.
78. Ibid. p. 425.
79. Ibid. p. 103.
80. Ibid. p. 67.
81. Ibid. p. 292.
82. Ibid. p. 68.
83. Ibid. p. 143.
84. Ibid. p. 421.
85. Ibid. p. 175.
86. Ibid. p. 307.
87. Ibid. p. 404.
88. Anon. 2008. *Darwin*. American Museum of Natural History. Available at www.amnh.org/exhibitions/darwin.
89. Darwin, C. R. 1872. *The origin of species by means of natural selection, or the preservation of favoured races in the struggle for life*. London: John Murray. 6th edition; with additions and corrections, p. 3.
90. Ibid. p. 75.
91. Ibid. p. 350.
92. Ibid. p. 102.
93. Ibid. p. 104.
94. Ibid. p. 206.
95. Ibid. p. 98.
96. Ibid. p. 132.
97. Ibid. p. 121.
98. Ibid. p. 415.
99. Ibid. p. 10.
100. Ibid. p. 62.
101. Ibid. p. 85.
102. Ibid. p. 156.
103. Ibid. p. 67.
104. Darwin, F., editor. 1887. *The life and letters of Charles Darwin, including an autobiographical chapter*. London: John Murray. Volume 2, p. 23.
105. Ibid. p. 29.
106. Ibid. p. 215.

107. Darwin, C. R. 1872. *The origin of species by means of natural selection, or the preservation of favoured races in the struggle for life.* London: John Murray. 6th edition; with additions and corrections, p. xiv.

108. Morgan, T. H. 1903. *Evolution and Adaptation.* New York: Macmillan.

109. Osborn, H. F. 1929. *From the Greeks to Darwin.* Charles Scribner's Sons, New York.

110. Darwin, F. and Seward, A. C., editors. 1903. *More letters of Charles Darwin. A record of his work in a series of hitherto unpublished letters.* London: John Murray. Volume 1, p. 360.

111. Darwin, C. R. 1872. *The origin of species by means of natural selection, or the preservation of favoured races in the struggle for life.* London: John Murray. 6th edition; with additions and corrections, p. 421.

112. Ibid. p. 206.

113. Ibid. p. 188.

114. Ibid. p. 178.

115. Ibid. p. 188.

116. Ibid. p. 190.

117. Darwin, C. R. 1858. *The autobiography of Charles Darwin 1809–1882. With the original omissions restored. Edited and with appendix and notes by his grand-daughter Nora Barlow.* London: Collins, p. 139.

118. Darwin, F. and Seward, A. C., editors. 1903. *More letters of Charles Darwin. A record of his work in a series of hitherto unpublished letters.* London: John Murray. Volume 2, p. 84.

119. Darwin, C. R. 1872. *The origin of species by means of natural selection, or the preservation of favoured races in the struggle for life.* London: John Murray. 6th edition; with additions and corrections, p. 69.

120. Ibid. p. 69.

121. Ibid. p. 70.

122. Ibid. p. 158.

123. Ibid. p. 69.

124. Ibid. p. 124.

125. Darwin, C. R. 1871. *The descent of man, and selection in relation to sex*. London: John Murray. Volume 2. 1st edition. p. 327.

126. Darwin, C. R. 1872. *The origin of species by means of natural selection, or the preservation of favoured races in the struggle for life*. London: John Murray. 6th edition; with additions and corrections, p. 84–85.

127. Ibid. pp. 413–414.

128. Ibid. p. 166.

129. Ibid. p. 156.

130. Ibid. p. 266.

131. Ibid. p. 282.

132. Ibid. p. 408.

133. Ibid. p. 145.

134. Ibid. p. 289.

135. Anon. 2008. *Darwin*. American Museum of Natural History. Available at www.amnh.org/exhibitions/darwin/evolution/take.

136. Darwin, C. R. 1872. *The origin of species by means of natural selection, or the preservation of favoured races in the struggle for life*. London: John Murray. 6th edition; with additions and corrections, p. 156.

137. Anon. 2008. *Darwin*. American Museum of Natural History. Available at www.amnh.org/exhibitions/darwin/evolution/time.

138. Darwin, C. R. 1872. *The origin of species by means of natural selection, or the preservation of favoured races in the struggle for life*. London: John Murray. 6th edition; with additions and corrections, p. 158.

139. Ibid. p. 134.

140. Ibid. p. 3.

141. Ibid. pp. 295–296.

142. Anon. 2008. *Darwin*. American Museum of Natural History. Available at www.amnh.org/exhibitions/darwin/evolution/opportunity.

143. Darwin, C. R. 1872. *The origin of species by means of natural selection, or the preservation of favoured races in the struggle for*

life. London: John Murray. 6th edition; with additions and corrections, p. 178.

144. Ibid. p. 3.
145. Ibid. p. 81.
146. Ibid. p. 134.
147. Ibid. p. 146.
148. Anon. 2008. *Darwin*. American Museum of Natural History. Available at www.amnh.org/exhibitions/darwin/evolution/vestigial.php - 25k.
149. Williams, P.L. and Warwick, R. 1980. Gray's Anatomy, Churchill Livingstone, 36th edition.
150. Darwin, C. R. 1872. *The origin of species by means of natural selection, or the preservation of favoured races in the struggle for life*. London: John Murray. 6th edition; with additions and corrections, p. 397.
151. Ibid. p. 420.
152. Ibid. p. 397.
153. Ibid. p. 118.
154. Ibid. p. 401.
155. Ibid. p. 401.
156. Ibid. p. 420.
157. Ibid. p. 401.
158. Ibid. p. 131.
159. Ibid. p. 63.
160. Ibid. p. 119.
161. Ibid. p. 166.
162. Ibid. p. 398.
163. Ibid. p. 165.
164. Ibid. p. 165.
165. Ibid. p. 412.
166. Darwin, F., editor. 1887. *The life and letters of Charles Darwin, including an autobiographical chapter*. London: John Murray. Volume 2. p. 371.
167. Darwin, C. R. 1872. *The origin of species by means of natural selection, or the preservation of favoured races in the struggle for life*. London: John Murray. 6th edition; with additions and corrections, p. 146.

168. Ibid. p. 231.

169. Ibid. p. 143.

170. Eldredge, N. 2005. *Darwin, Discovering the Tree of Life*. WW Norton & Company, Inc, p. 17.

171. Sermonti, G. 2005. *Why is a Fly Not a Horse*. Discovery Institute Press. Seattle, Washington, p. 13.

172. Darwin, F. and Seward, A. C., editors. 1903. *More letters of Charles Darwin. A record of his work in a series of hitherto unpublished letters*. London: John Murray. Volume 1, p. 209.

173. Eldredge, N. 2005. *Darwin, Discovering the Tree of Life*, 2005, WW Norton & Company, Inc, p. 17.

174. Darwin, C. R. 1872. *The origin of species by means of natural selection, or the preservation of favoured races in the struggle for life*. London: John Murray. 6th edition; with additions and corrections, p. 426.

175. Ibid. p. 63.

176. Sermonti, G. 2005. *Why a Fly is not a Horse*. Discovery Institute Press, Seattle, Washington, p. 13.

177. Ibid. p. 51.

178. Shermer, M, 2005. The Woodstock of Evolution. *Scientific American,* June 27. Available at www.sciam.com.

179. Darwin, F. and Seward, A. C., editors. 1903. *More letters of Charles Darwin. A record of his work in a series of hitherto unpublished letters*. London: John Murray. Volume 2. p. 72.

180. Darwin, C. R. 1872. *The origin of species by means of natural selection, or the preservation of favoured races in the struggle for life*. London: John Murray. 6th edition; with additions and corrections, p. 163.

181. Ibid. pp. 118–110.

182. Ibid. p. 162.

183. Ibid. p. 85.

184. Ibid. p. 3.

185. Ibid. p. 175.

186. Ibid. pp. 124–125.

187. Ibid. p. 132.

188. Ibid. p. 120.

189. Ibid. p. 138.

190. Ibid. p. 233.
191. Ibid. p. 102.

Chapter Eight: The Challenge

1. Darwin, C. R. 1872. *The origin of species by means of natural selection, or the preservation of favoured races in the struggle for life*. London: John Murray. 6th edition, p. 266.

2. Sedgwick, I. 1898. Grandmother's tales, *Macmillan's Magazine*, LXXVIII, no. 468, A 433–4. Quoted in Lucas, J. R. 1979. Wilberforce and Huxley: A Legendary Encounter. *The Historical Journal*, 22 (2):313–330.

3. Stewart, K. 2000. Thomson, Huxley, Wilberforce and the Oxford Museum, *American Scientist*, 88(3):210.

4. Anon. 1860. *The Athenaeum*. Nos. 1705, 1706 and 1707, 30 June, 7 July and 14 July 1860. Quoted in Lucas, J. R. 1979. Wilberforce and Huxley: A Legendary Encounter. *The Historical Journal*, 22(2):313–330.

5. Anon. 1860. *Jackson's Oxford Journal*, Saturday 7 July, p.2, col. 6. Quoted in Lucas, J. R. 1979. Wilberforce and Huxley: A Legendary Encounter. *The Historical Journal*, 22(2):313–330.

6. Horan, P.G. 1979. The wife of the Bishop of Worcester upon hearing of 'Origins,' *The Origin of Species*. Avenel Books, New York.

7. Darwin, F., editor. 1887. *The life and letters of Charles Darwin, including an autobiographical chapter*. London: John Murray. Volume 2. p. 324.

8. Anon. 1860. *Professor Agassiz on the Origin of Species. American Journal of Science*, WW Norton & Company, p. 150.

9. Huxley, T. H. *Autobiography and Selected Essays*, p. 18. Available www.human-nature.com/darwin/huxley/notes.html.

10. Darwin, C. R. 1872. *The origin of species by means of natural selection, or the preservation of favoured races in the struggle for life*. London: John Murray. 6th edition, p. 84.

11. Ibid. p. 282.
12. Ibid. p. 264.
13. Ibid. p. 404.
14. Ibid. p. 286.

15. Ibid. p. 287.
16. Ibid. p. 313.
17. Ibid. p. 270.
18. Ibid. p. 289.
19. Ibid. p. 134.
20. Ibid. p. 134.
21. Ibid. p .134.
22. Ibid. p. 302.
23. Ibid. p. 302.
24. Ibid. p. 408.
25. Wheen, F. 2000. *Karl Marx, a life.* Fourth Estate, ISBN: 1-85702-637-3, as quoted by Friends of Charles Darwin. Available at www.darwin.gruts.com/articles/2000/marx/index. htm.
26. Joseph Matkin, J. 2008. Available at aquarium.ucsd.edu/ Education/Learning_Resources/Challenger/people.php.
27. Marshall N.B. 1954. *Aspects of Deep Sea Biology.* Hutchinson Publishing Co., London, p. 19.

Chapter Nine: Fossils

1. Darwin, C. R. 1872. *The origin of species by means of natural selection, or the preservation of favoured races in the struggle for life.* London: John Murray, 6th edition, p. 265.
2. Ibid. p. 264.
3. Ibid. p. 404.
4. Ibid. p. 146.
5. Denton, M. 1985. *Evolution: A Theory in Crisis.* Adler & Adler Publishers, Chevy Chase, MD, p. 160.
6. von Meyer, H. 1861. Archaeopteryx litographica (Vogel-Feder) und Pterodactylus von Solenhofen. *N. Jhb. Mineralogie, Geognosie, Geologie und Petrefakten-Kunde* 1861: V + 678–679.
7. Feduccia, A. 1999. *The Origin and Evolution of Birds.* New Haven, CT: Yale University Press, p. 29.
8. Huxley T.H. 1868. Remarks upon Archaeopteryx lithographica. *Proc Roy Soc* 16:243–48.

9. Darwin, C. R. 1872. *The origin of species by means of natural selection, or the preservation of favoured races in the struggle for life.* London: John Murray, 6th edition, p. 302.

10. Shipman, P. 1999. *Taking Wing: Archaeopteryx and the Evolution of Bird Flight,* pp. 14–16.

11. Stahl, B.J. 1974. *Vertebrate History: Problems in Evolution.* McGraw-Hill Book Co. New York, p. 349.

12. Feduccia, A and Tardoff, H. 1979. Feathers of Archaeopteryx: Asymmetry Vanes Indicate Aerodynamic Function, *Science,* 203(3484):1022.

13. Ostrom, J. 1979. "Bird Flight: How Did It All Begin?" *American Scientist,* 67(1):46–56.

14. Feduccia, A. and Harrison, B. 1979. Tordoff Feathers of Archaeopteryx: Asymmetric Vanes Indicate Aerodynamic Function. *Science,* 203(4384):1021–1022.

15. Mayr, E. 1982. *The Growth of Biological Thought.* Cambridge, MA; Harvard University Press, p. 430.

16. Martin, L. D. 1985. The relationship of Archaeopteryx to other Birds. *The Beginnings of Birds,* Eichstatt: Freunda des Jura Museums, p. 182.

17. Feduccia, A. 1984. *Es Begann am Jura-Meer, in* German edition. (*The Age of Birds*). Hildesheim, Germany: Gerstenberg Buchuerlag.

18. Dodson, P. 1985. International *Archaeopteryx* Conference. *Journal of Vertebrate Paleontology,* 5:179.

19. Denton, M. 1985. *Evolution: A Theory in Crisis.* Adler & Adler Publishers. Chevy Chase, MD, p. 175.

20. Carroll, R. 1997. *Patterns and Processes of Vertebrate Evolution,* Cambridge University Press, 1997, pp. 8–10.

21. Gee, H. 1999. *In Search of Deep Time.* New York: The Free Press, pp 195–197.

22. Sunderland, L. 1998. *Darwin's Enigma.* Master Books, p. 102.

23. Lingham-Soliar T, et al. 2007. A new Chinese specimen indicates that 'protofeathers' in the Early Cretaceous theropod dinosaur Sinosauropteryx are degraded collagen fibres. *Proc Biol Sci.* 274(1620):1823–9.

24. Lingham-Soliar T. 2003. Evolution of birds: ichthyosaur integumental fibers conform to dromaeosaur protofeathers. *Naturwissenschaften.* 90(9):428–32.

25. Feduccia A, et al. 2005. Do feathered dinosaurs exist? Testing the hypothesis on neontological and paleontological evidence. *J Morphol.* 266(2):125–66.

26. Hoyle, F, et al. 1985. *Archaeopteryx. British Journal of Photography,* June 21, 132:693; Spetner, L. M., et al. 1988. *Archaeopteryx—* More Evidence for a Forgery. *The British Journal of Photography,* 135:14–17; Watkins, R.S., et al. 1985. *Archaeopteryx:* A Photographic Study. *British Journal of Photography,* 132:264–266.

27. Kennedy, E. 2000. Solnhofen Limestone: Home of *Archaeopteryx. Geoscience Reports,* 30:1–4.

28. Darwin, C. R. 1872. *The origin of species by means of natural selection, or the preservation of favoured races in the struggle for life.* London: John Murray, 6th edition, p. 146.

29. Ibid. p. 428.

30. Ibid. p. 270.

31. Ibid. p. 289.

32. Ibid. p. 134.

33. Ibid. p. 313.

34. Ibid. p. 427.

35. Denton, M. 1985. *Evolution: A Theory in Crisis.* Adler & Adler Publishers. Chevy Chase, MD, p. 160.

36. Darwin, C.R. 1872. *The origin of species by means of natural selection, or the preservation of favoured races in the struggle for life.* London: John Murray, 6th edition, p. 76.

37. Ibid. p. 265.

38. Ibid. p. 287.

39. Hirsch, E. D. Jr., et al. 2002. *The New Dictionary of Cultural Literacy.* Boston: Houghton Mifflin Company, New York. Available at www.Bartleby.com.

40. Morris, S. C. and Whittington, H.B. 1979. The Animals of the Burgess Shale. *Scientific America,* 24(1):119.

41. Denton, M, 1985. *Evolution: A Theory in Crisis.* Adler & Adler Publishers. Chevy Chase MD, p. 161.

42. Darwin, C. R. 1872. *The origin of species by means of natural selection, or the preservation of favoured races in the struggle for life.* London: John Murray, 6th edition, p 286.
43. Darwin, F, and Seward, A. C., editors. 1903. *More letters of Charles Darwin. A record of his work in a series of hitherto unpublished letters.* London: John Murray. Volume 2, p. 323.
44. Gould, S. J. 1994. The Evolution of Life on Earth. *Scientific American,* 271:85–91.
45. Darwin, C. R. 1872. *The origin of species by means of natural selection, or the preservation of favoured races in the struggle for life.* London: John Murray, 6th edition, p 286.
46. Conway-Morris, S. 1998. *The Crucible of Creation.* Oxford University Press. Oxford. pp. 2, 28.
47. Glaessner, M. P. 1961. Pre-Cambrian Animals. *Scientific America,* 204(3)72–78.
48. Darwin, C. R. 1872. *The origin of species by means of natural selection, or the preservation of favoured races in the struggle for life.* London: John Murray, 6th edition, p. 282.
49. Schopf, J. W. 1994. The early evolution of life: solution to Darwin's dilemma. *Trends in Ecology and Evolution,* pp. 375–377.
50. Glaessner, M. F. 1959. The oldest fossil faunas of South Australia. *International Journal of Earth Sciences,* 47(2):522–531.
51. Simpson, G. G. 1987. *The Evolution of Life.* University of Chicago Press, p. 149.
52. Barnes, R. D. 1980. Invertebrate Beginnings. *Paleobiology,* 6:365.
53. Whittington, H. 1985. *The Burgess Shale,* p. 131.
54. Denton, M. 1985. *Evolution: A Theory in Crisis.* Adler & Adler Publishers. Chevy Chase, MD, pp. 157–198.
55. Ibid. pp. 157–198.
56. Jeffrey H Schwartz, J. H. 1999. Homeobox genes Fossils, and the Origin of Species. *Anatomical Record (New Anatomist),* 257:15–31.
57. Valentine, J. W. et al. 1991. The Biological Explosion at the Precambrian-Cambrian Boundary. *Evolutionary Biology,* 25:279.

58. Ibid. p. 281.

59. Gould, S. J. 1980. *The Panda's Thumb.* WW Norton and Co, Inc, New York and London.

60. Anon. 2008. Marjorie Eileen Doris Courtenay-Latimer. Wikipedia. Available at www.wikipedia.com.

61. Stahl, B. J. 1974. *Vertebrate History, Problems in Evolution.* McGraw-Hill Book Co, New York, p. 146.

62. Denton, M. 1985. *Evolution: A Theory in Crisis.* Adler & Adler Publishers, Chevy Chase, MD, p. 177.

63. Ibid. p. 218.

64. Carroll, R. 1988, *Vertebrate Paleontology and Evolution.* New York: W.H. Freeman and Co., p. 698.

65. Carroll, L. L. 1969. *Problems of the Origin of Reptiles. Biological Reviews of the Cambridge Philosophical Society,* Volume 44.

66. Colbert, E. H. and Morales, M. 1991. *Evolution of the Vertebrates.* New York: John Wiley and Sons, p. 510.

67. Sikes, S. K. 1971. *The Natural History of the African Elephant.* Weidenfeld and Nicolson, London, pp. 2–4.

68. Darwin, C. R. 1872. *The origin of species by means of natural selection, or the preservation of favoured races in the struggle for life.* London: John Murray, 6th edition, p. 201.

69. Grinnell, G. B. 1910. *Leading American Men of Science.* Jordan, D. S., edition. New York: Henry Holt and Company, p. 294.

70. Simpson, G. G. 1953. *The Major Features of Evolution.* Columbia University Press, New York, p. 125.

71. Heribert, N. 1954. *Synthetische Artbildung.* Lund. Sweden: Vertag CWE Gleenrup, pp. 551–552.

72. Newell, N. D. 1959. The Nature of the Fossil Record. *Proc of the Amer Phil Soc,* 103(2):267.

73. Kerkut, G. A. 1960. *The Implications of Evolution.* London: Pergamon, pp VII–VIII.

74. Raup, D. 1979. Conflicts Between Darwin and Paleontology. *Field Museum of Natural History Bulletin,* 50(1):25.

75. Rensberger, B. 1980. *Houston Chronicle,* November 5, Section 4, p. 15.

76. Barnes, R. D. 1980. Invertebrate Beginnings, *Paleobiology* 6:365.

77. Taylor, G. R. 1984. *The Great Evolution Mystery*. Abacus, Sphere Books, London, p. 230.

78. Gould, S. J. 1997. *The Spread of Excellence From Plato To Darwin*. Full House, pp 67–69.

79. Gould, S. J. 2000. *Abscheulich! (Atrocious)*. *Natural History*, 109(2):45.

80. Darwin, C. R. 1872. *The origin of species by means of natural selection, or the preservation of favoured races in the struggle for life*. London: John Murray, 6th edition, p. 428.

81. Darwin, C. R. 1871. *The Descent of Man and Selection in Relation to Sex*, London: John Murray. Volume 1, 1st edition, p. 10.

82. Ibid. p. 34.

83. Ibid. p. 35.

84. Ibid. p. 105.

85. Ibid. pp. 48–49.

86. Darwin, C. R. 1871. *The Descent of Man and Selection in Relation to Sex*. London: John Murray. Volume 1, 1st edition, p. 146.

87. Huxley, T. H. 1863. *Evidence for Man's place in Nature*. Reprint of 1863 edition, New York: D Appleton, 1886, p. 125.

88. Ibid. p. 132.

89. Anon. 1912. The Earliest Man? Remarkable Discovery in Sussex. A Skull "Millions of Years" Old, *Manchester Guardian* November 21.

90. Anon, 1938. The Piltdown Man Discovery. *Nature,* July 30.

91. Lewin, R.1997. *Bones of Contention*. University of Chicago Press, p. 70.

92. Ibid. p. 73.

93. Maienschein, J. 1997. *The One and the Many: Epistemological Reflections on the Modern Man Origins Debate*, pp. 413–422.

94. Eldredge, N. and Tattersall, I. 1982. *The Myth of Human Evolution*. Columbia University Press, p. 127.

95. Gee, H. 1999. *In Search of Deep Time: Beyond the Fossil Record to a New History of Life*, New York: The Free Press, pp. 116–117.

96. Durant, J. R. 1981. The myth of human evolution. *New Universities Quarterly*, 35:425–428.

97. Howell, F. C. 1996. Thoughts on the Study and Interpretation of the Human Fossil Record, pp 1–39 in Eric Meikle, E., et al., editors, *Contemporary Issues on Human Evolution*, p. 3.

98. Clark, G. A. 1997. Through a Glass Darkly: Conceptual Issues and Modern Human Origins Research, in Clark, G. A. and Willermet, C. M., editors, *Conceptual Issues in Modern Human Origin Research,* pp. 60–62.

99. Anon. 1980. British Museum of Natural History. *Man's Place in Evolution,* p. 20.

100. Gee, H. 2001. Palaeontology: Return to the planet of the apes. *Nature,* 412:131–132, 12 July.

101. Elliott, S. G. 1922. Hesperopithecus: The ape-man of the western world. *The Illustrated London News,* June 24, p. 944.

102. Anon. 1922. Nebraska's 'Ape man of the western world.' *The New York Times*, Sept. 17, Sect. 7, p. 1.

103 Gregory, W. K. 1927. *Hesperopithecus* apparently not an ape nor a man. *Science,* 66:579–81.

104 Anon. 1928. Nebraska ape tooth proved a wild pig's. *The New York Times,* Feb. 20, p. 1.

105 Anon. 1928. Hesperopithecus dethroned. *The Times (London),* Feb. 21, p. 16.

106 Anon. 1928. *Peccavis and peccaries. The New York Times*, Feb. 21, p. 24.

107 Wray, H. 1982. Was Lucy a Climber? Dissenting Views of Ancient Bones. *Science News,* 122:116.

108 Oxnard, C. E. 1984. *The Order of Man.* New Haven: Yale University Press, pp. iii–iv.

109 Gould, S. J. 1986. A Short Way to Big Ends. *Natural History,* 95:28.

110 Gould, S. J. 1987. Empire of Apes. *Natural History,* 96:24.

111 Lewin, R. 1997. *Bones of Contention: Controversies in the Search for Human Origins.* University of Chicago Press.

112 Kelso, A. J. 1974. *Physical Anthropology.* New York: JB Lippincott, p. 151.

113 Martin, R. D. 1990. *Primate Origins and Evolution.* Princeton University Press, p. 82.

114 Gibbons, A. 1996. Homo erectus in Java: a 250,000 Year Anachronism. *Science,* 274:1841–1842.

115 Watson, L. 1982. Water People. *Science Digest,* 90:44.

116 Spoor, F, et al. 2007. *Nature,* 448(7154):688–691.

117 Pearlman, D. 2007, Anthropologists split on what fossil find means about early humans. *San Francisco Chronicle,* August 9.

118 Noonan J. P., et al. 2006. Sequencing and analysis of Neanderthal genomic DNA. Science, 314(5802):1113–1118.

119 Ruvolo, M. 1997. DNA from an Extinct Human. *Science,* 277:176–178, 11 July 1997, p. 177.

120 Yarris, L. 2006. Neanderthal Genome Sequencing Yields Surprising Results and Opens a New Door to Future Studies." *Research News Berkley Lab,* November 15.

121 Wall, J. D. and Hammer, M. F. 2006. Archaic admixture in the human genome. *Curr Opin Genet Dev.,* 16(6):606–10.

122 Evans, P. D., et al. 2006. Evidence that the adaptive allele of the brain size gene microcephalin introgressed into Homo sapiens from an archaic Homo lineage. *Proc Natl Acad Sci U S A.,* 103(48):18178–18183.

123 Handwerk, B. 2007. Odd Skull Boosts Human, Neanderthal Interbreeding Theory. National Geographic News, August 2.

124 Sloan, C. P. 1999. Feathers for *T. Rex? National Geographic,* 196(5):100.

125 Anon. 1999. Story Number 159226. *Associated Press,* October 15.

126 Olson, S. L. 1999. Letter to Dr. Peter Raven, Secretary of the Committee for Research and Exploration, *National Geographic* Society.

127 Olson, S. L. 1999. Letter to Dr. Peter Raven, Secretary of the Committee for Research and Exploration, *National Geographic* Society.

128 Xing, X. 2000. *National Geographic,* 197(3).

129 Simon, L. M. 2000. Archaeoraptor Fossil Trail. *National Geographic,* 197(10).

130 Darwin, F. and Seward, A. C., editors. 1903. *More letters of Charles Darwin. A record of his work in a series of hitherto unpublished letters.* London: John Murray. Volume 2, p. 323.

131 Storrs, L. and Olson, S. L. 2000. *Smithsonian Institution's National Museum of Natural History.*

132 Rowe, T, et al. 2001. Nature, 410(6828):539–540.

133 Ibid. pp. 539–540.

134 Darwin, C. R. 1872. *The origin of species by means of natural selection, or the preservation of favoured races in the struggle for life.* London: John Murray, 6th edition.

135 Darwin, F. 1888. *The Life and Letters of Charles Darwin,* John Murray London, Volume 3, p. 248.

136 Darwin, C. R. 1872. *The origin of species by means of natural selection, or the preservation of favoured races in the struggle for life.* London: John Murray, 6th edition, p. 87.

137 Ibid. p. 289.

138 Benton, M. J. et al. 2000. Quality of the fossil record through time. *Nature* 403:534–536.

139 Romer, A. S. 1966. *Vertebrate Paleontology,* 3rd edition, pp. 347–96.

140 Simpson, G. G. 1953. *The Major Features of Evolution.* Columbia University Press, New York, Table 8.

141 Valentine, J. W. and Erwin, D. H. 1987. Interpreting Great Developmental Experiments: The Fossil Record. pp 71–107 in Raff, R. A. and Raff, E. C. editors., *Development as an Evolutionary Process.* New York: Alan R Liss, pp. 84–85.

142 Benton, M. J. et al. 2000. Quality of the fossil record through time. *Nature* 403:534–536.

143 Valentine, J. W. and Erwin, D. H. 1987. Interpreting Great Developmental Experiments: The Fossil Record, pp 71–107 in Raff, R. A. and Raff, E. C., editors. *Development as an Evolutionary Process.* New York: Alan R Liss, pp. 84–85.

144 Darwin, C. R. 1872. *The origin of species by means of natural selection, or the preservation of favoured races in the struggle for life.* London: John Murray, 6th edition, p. 293.

145 Gaylord, G. G. 1953. *The Major Features of Evolution,* p. 360.

146 Eldredge, N. 1981. Did Darwin Get It Wrong? *Nova,* 22: 6.

147 Newell, N. D. 1984. *Creation and Evolution: Myth or Reality.* Convergence, New York, N.Y. p. 10.

148 Ridley M, 1985, *Problems with Evolution.* Oxford University Press, p. 11.

149 Futuyma, D. 1983. *Science on Trial: The Case for Evolution,* p. 82.

150 Mayr, E. 1991. *Our Long Argument: Charles Darwin and the Genesis of Modern Evolutionary Thought,* p. 138.

151 Bengtson, S. 1990. The Solution to a Jigsaw Puzzle. *Nature,* 345:765–766.

152 Gould, S. J. 1986. A Short Way to Big Ends. *Natural History,* 95:18–28.

153 Sermonti, G. 2005, *Why a Fly is not a Horse.* Discovery Institute Press, Seattle, Washington, p. 153.

154 Darwin, C. R. 1872. *The origin of species by means of natural selection, or the preservation of favoured races in the struggle for life.* London: John Murray, 6th edition, p. 65.

155 Ibid. p. 84.

156 Gould, S. J. 1994. The Evolution of Life on Earth. *Scientific American,* 271:87.

157 Nelson, G. V. 1971, Origin and Diversification of Telostean Fishes. *Annals of the New York Academy of Sciences,* p. 22.

158 Levinton, J. S. 1992, The Big Bang of Animal Evolution. *Scientific American,* 267:84–91.

159 Gould, S. J. 1980. *The Panda's Thumb.* W. W. Norton and Co. New York and London, pp. 181–182.

160 Eldredge, N. 1985. *Time Frames: The Rethinking of Darwinian Evolution and the Theory of Punctuated Equilibria,* p. 188.

161 Bowler, P. J. 1984. *Evolution: The History of an Idea.* University of California Press, p. 187.

162 Gould, S. J. 1985. The Paradox of the First Tier: an Agenda for Paleobiology. *Paleobiology,* p. 3.

163 Gould, S. J. 1985. The Paradox of the First Tier: an Agenda for Paleobiology. *Paleobiology,* p. 3.

164 Eldredge, N. 1986. *Time Frames.* Columbia University, American Museum of Natural History, p. 144.

165 Ager, D.V. 1976. *Proceedings of the Geological Association,* 87(2)131–159.

166 Eldredge, N. 1996. *Reinventing Darwin: The Great Evolutionary Debate.* Weidenfeld Nicholson History, p. 95.

167 Schopf, J. W. 1994. The early evolution of life: solution to Darwin's dilemma. *Trends in Ecology and Evolution,* pp. 375–377.

168 Patterson, C. 1999. *Evolution,* Ithaca, NY: Cornell University Press, second edition.

169 Gee, H. 1999. *In Search of Deep Time.* New York: Free Press.

170 Futuyma, D.1982. *Science on Trial. The Case for Evolution.* NY: Pantheon, p. 191.

171 Carroll, R. 1997. *Patterns and Processes of Vertebrate Evolution.* Cambridge: Cambridge University Press, pp. 8–10.

172 Eldredge, N. 1985. *Time Frames: The Rethinking of Darwinian Evolution and the Theory of Punctuated Equilibria.* Simon & Schuster: New York NY, p. 44.

173 Denton, M. 1985. *Evolution: A Theory in Crisis.* Adler & Adler Publishers. Chevy Chase, MD, p. 157–198.

174 Wesson, R., *Beyond Natural Selection.* Massachusetts Institute of Technology Press, p. 45.

175 Gould, S. J. 1980. *The Panda's Thumb.* W. W. Norton & Company, pp. 181–182.

176 Darwin, C. R. 1872. *The origin of species by means of natural selection, or the preservation of favoured races in the struggle for life.* London: John Murray, 6th edition, p. 289.

177 Ibid. p. 287.

178 Ibid. p. 282.

179 Darwin, F., editor. 1887. *The life and letters of Charles Darwin,* including an autobiographical chapter. Volume 2. London: John Murray, p. 248.

Chapter Ten: Molecular Biology

1. Darwin, C. R. 1872. *On the origin of species by means of natural selection, or the preservation of favoured races in the struggle for life.* London: John Murray. 6th edition, p. 421.

2. Ibid. p. xx.

3. Ibid. p. 423.

4. Ibid. p. 267.

5. Ibid. p. 225.
6. Ibid. p. 429.
7. Ibid. p. 429.
8. Ibid. p. 421.
9. Darwin, C. R. 1859. *On the origin of species by means of natural selection, or the preservation of favoured races in the struggle for life.* London: John Murray. 1st edition, p. 429.
10. Oparin, A. 1924. *The Origin of Life.* Moscow: Moscow Worker publisher. English translation. 1968. Oparin, A. I. *The Origin and Development of Life.* NASA. Washington: D.C.L GPO TTF-488.
11. Haldane, J. B. S. 1928. *Rationalist Annual,* 148:3–10.
12. Urey, H. 1952. On the Early Chemical History of the Earth and the Origin of Life. *Proceedings of the National Academy of Sciences USA,* 38:351–363.
13. Miller, S. 1953. A Production of Amino Acids Under Probable Primitive Earth Conditions. *Science* 117:528–529.
14. Miller S. and Urey, H. Organic Compound Synthesis on the Primitive Earth. *Science,* 130:245–352.
15. Brown, H. 1952. Rare Gases and the Formation of the Earth's Atmosphere, in Kuiper, G. P., editor, *The Atmospheres of the Earth and Planets,* Revised Edition. Chicago. The University of Chicago Press, pp. 258–266.
16. Holland, H. D. 1962. Model for the Evolution of the Earth's Atmosphere, pp 447–477 in Engel, A. E. J. et al., editors. *Petrologic Studies: A Volume in Honor of AF Buddington, Geological Society of America,* pp. 448–449.
17. Abelson, P. H. 1966. Chemical Events on the Primitive Earth. *Proceedings of the National Academy of Sciences USA,* 55:1365–1372.
18. Ibid. p. 1365.
19. Dimroth, E. and Kimberley, M. M. 1976. Precambrian Atmospheric Oxygen: Evidence in the Sedimentary Distributions of Carbon, Sulfur, Uranium, and Iron. *Canadian Journal of Earth Sciences,* 13(9):1161.

20. Clemmey, H and Badlam, N. 1982. Oxygen in the Precambrian atmosphere: An evaluation of the geological evidence. *Geology* 10:141–146.

21. Florkin, M. 1975. Ideas and Experiments in the Field of Prebiological Evolution. *Comprehensive Biochemistry,* 29B:231–260.

22. Lumsden, J. and Hall, D.O. 1975. Superoxide dismutase in photosynthetic organisms provides an evolutionary hypothesis. *Nature* 257:670–672.

23. Fox, S W. and Dose, K. 1977. *Molecular Evolution and the Origin of Life, Revised Edition.* New York: Marcel Dekker, pp. 43, 74–76.

24. Ibid. pp. 43, 74–76.

25. Schlesinger, G. and Miller, S. L. 1983. Prebiotic Synthesis in Atmospheres containing CH_4, CO, and CO_2; I Amino Acids. *Journal of Molecular Evolution,* 19:376–382.

26. Horgan, J. 1991. In the Beginning. *Scientific American.* February, pp. 116–126.

27. Cohen, J. 1995. Novel Center Seeks to Add Spark to Origins of Life. *Science* 270:1925–1926.

28. Fitch, F. W and Anders, E. 1963. Organized Element - Possible Identification in Orgueil Meteorite. *Science* 140(357):1097 and Anders, E., et al. 1964. Contaminated Meteorite. *Science, New Series,* 146(3648):1157–1161.

29. Harold, F. M. 2001. *The Way of the Cell.* (New York: Oxford University Press, p. 235.

30. Monastersky, R. 1998. The Rise of Life on Earth. *National Geographic,* March, pp. 54–81.

31. Anon. 2008. *Harvard University.* Available at www.origins. harvard.edu.

32. Stein, W. H. 1958. Observations on the amino acid composition of human hemoglobins. Conference on Hemoglobin. May 1957. *National Academy of Sciences,* p. 220.

33. Rossi-Fanelli A., et al. 1955 Amino-acid composition of human crystallized myoglobin and haemoglobin. *Nature,* 17:377–381.

34. Denton, M. 1985. *Evolution: A Theory in Crisis.* Adler & Adler Publishers, Chevy Chase, MD, p. 276.

35. Dickerson, R. E. and Geis, I. 1969. *The Structure and Action of Proteins,* Harper and Row, New York.

36. Anon. 2008. Available at www.ncbi.nlm.nih.gov/sites/entrez?db=pmc.

37. Milo, R., et al. 2007. The relationship between evolutionary and physiological variation in hemoglobin. *Proc Natl Acad Sci U S A.,* 104(43):16998–17003.

38. Denton, M. 1985. *Evolution: A Theory in Crisis.* Adler & Adler Publishers, Chevy Chase, MD, p, 276.

39. Dickerson, R. E. 1972, The Structure and History of an Ancient Protein. *Scientific American* 226:58–72.

40. Denton, M. 1985. *Evolution: A Theory in Crisis.* Adler & Adler Publishers, Chevy Chase, MD, pp. 277–278.

41. Dayhoff, M. D. 1972. *Atlas of Protein Sequences and Structure, National Biomedical Research Foundation.* Silver Springs, Maryland, Volume 5 Matrix 1, p D-8.

42. Denton, M. 1985. *Evolution: A Theory in Crisis.* Adler & Adler Publishers, Chevy Chase, MD, pp. 278–279.

43. Ibid. pp. 280–281.

44. Sermonti, G and Sermonti, I. S. 1987. "L'ipotesi nulla nella evoluzione dei Veretrati," *Rivista di Biologia/Biology Form,* 82:348–349.

45. Denton, M. 1985. *Evolution: A Theory in Crisis.* Adler & Adler Publishers, Chevy Chase, MD, p. 284.

46. Wilkinson, T. N. et al. 2005. Evolution of the relaxin-like peptide family. *BMC Evolutionary Biology,* 5:14.

47. Ibid. 5:14.

48. Di Fiore, M. M., et al. 2000. Mammalian and chicken I forms of gonadotropin-releasing hormone in the gonads of a protochordate. *Ciona intestinalis, Proc Natl Acad Sci U S A.,* 97(5):2343–2348.

49. Cardoso, J. C. R., et al. 2006. Evolution of secretin family GPCR members in the metazoan. *BMC Evolutionary Biology,* 6:108.

50. Cavalier-Smith, T. 2006. Rooting the tree of life by transition analyses, *Biol. Direct.,* 1:19.

51. Sermonti, G. 2005, *Why a Fly is not a Horse,* Discovery Institute, Seattle, Washington, p. 155.

52. Denton, M. 1985. *Evolution: A Theory in Crisis.* Adler & Adler Publishers, Chevy Chase, MD, p. 290.

53. Pereira-Leal, J. B., Levy, E. D. and Teichmann S.A. 2006. The origins and evolution of functional modules: lessons from protein complexes. *Philos Trans R Soc Lond B Biol Sci.,* 361(1467):507–517.

54. Sermonti, G. 2005. *Why a Fly is Not a Horse.* Discovery Institute Press, Seattle, Washington, p. 66.

55. Jacob F, 1977. Evolution and Tinkering, *Science* 196:1611–1166.

56. Blum, H. 1968. *Time's Arrow and Evolution,* 3rd Edition, p. 158. Quoted in Bird, W. R. 1991. *The Origin of Species Revisited.* Thomas Nelson Co., Nashville, p. 304.

57. Hoyle, F. and Wickramasinghe, C. 1984. *Evolution from Space.* New York, Simon & Schuster, p. 148.

58. Sermonti, G. 2005. *Why a Horse is Not a Fly,* 2005, Discovery Institute Press. Seattle, Washington, p. 130.

59. Monod, J. L. 1972. *Chance and Necessity.* Collins, London, p. 136.

60. Orgel, L. E. 1994. The Origin of Life on the Earth. *Scientific American,* 271:78.

61. Dawkins, R. 1996. *Climbing Mount Improbable.* W.W. Norton, New York, pp. 282–283.

62. Anon. 2008. Abiogenesis. Wikipedia. Available at wwwwikipedia. com.

63. Zuckerkandl, E. and Pauling, L. B. 1962. Molecular disease, evolution, and genetic heterogeneity," in Kasha, M. and Pullman, B, editors. *Horizons in Biochemistry.* Academic Press, New York, pp. 189–225.

64. Denton, M. 1985. *Evolution: A Theory in Crisis.* Adler & Adler Publishers, Chevy Chase, MD, p. 296.

65. Ibid. p. 296.

66. Douzery, E. J. P., et al. 2003. Local molecular clocks in three nuclear genes: divergence times for rodents and other mammals,

and incompatibility among fossil Calibrations." *Journal of Molecular Evolution,* 57: S201–S213.1.

67. Kumar, S. 2005. Molecular clocks: four decades of evolution. *Nat Rev Gene.,* 6: 654–662.

68. Kumar, S., and S. B. Hedges, 1998. A molecular timescale for vertebrate evolution. *Nature,* 392: 917–920.

69. Takahata, N. 2007. Molecular Clock: An *Anti*-neo-Darwinian Legacy, *Genetics,* 176(1):1–6.

70. Bromham, L., and Penny, D. 2003. The modern molecular clock. *Nat Rev Genet,* 4: 216–224.

71. Lanfear, R. et al. 2007. Metabolic rate does not calibrate the molecular clock. *Proc Natl Acad Sci U S A.,* 104(39):15388–15393.

72. Thomas, J. A, et al. 2006. There is no universal molecular clock for invertebrates, but rate variation does not scale with body size. *Proc Natl Acad Sci U S A,* 103(19):7366–7371.

73. Woolfit, M. and Bromham, L. 2005. Population size and molecular evolution on islands, *Proc Biol Sci,* 272(1578):2277–2282.

74. Takahata, N. 2007. Molecular Clock: An Anti-neo-Darwinian Legacy, *Genetics,* 176(1):1–6.

75. Peterson, K. J. and Butterfield, N. J. 2005. Origin of the Eumetazoa: Testing ecological predictions of molecular clocks against the Proterozoic fossil record. *Proc Natl Acad Sci U S A.,* 102(27):9547–9552.

76. Thomas Cavalier-Smith, Cell evolution and Earth history: stasis and revolution, *Philos Trans R Soc Lond B Biol Sci.* 2006 June 29; 361(1470): 969–1006.

77. Denton, M. 1985. *Evolution: A Theory in Crisis.* Adler & Adler Publishers, Chevy Chase, MD, p. 305.

78. Sermonti, G. 2005, *Why a Fly is not a Horse,* Discovery Institute, Seattle, Washington, p. 77.

79. Darwin, C. R. 1872. *The origin of species by means of natural selection, or the preservation of favoured races in the struggle for life.* London: John Murray. 6th edition; p. 156.

80. Ibid. p 146.

81. Sermonti, G. 2005, *Why a Fly is not a Horse*, Discovery Institute, Seattle, Washington, p. 130.
82. Ibid. p. 131.
83. Ibid. p. 132.
84. Nierhaus, K. H. and Wilson, D. N. 2004. *Protein Synthesis and Ribosome Structure.* Wiley-VCH, 3. ISBN 3527306382.
85. Fiers, W., et al. 1976. Complete nucleotide-sequence of bacteriophage MS2-RNA: primary and secondary structure of replicase gene. *Nature,* 260: 500–507.
86. Gerald F Joyce, RNA evolution and the origins of life. *Nature,* 338 (1989) pp 217–224.
87. Dose, K. 1988. The origin of life: more questions than answers. Interdisciplinary *Science Reviews,* 13:348–356.
88. Leslie E Orgel, L. E. 1998. The origins of life: a review of facts and speculations. *Trends in Biological Sciences,* 23:491–495.
89. Baaske P., et al. 2007. Extreme accumulation of nucleotides in simulated hydrothermal pore systems. *Proc Natl Acad Sci USA,*104:9346–9351.
90. Koonin, E. V. 2007. An RNA-making reactor for the origin of life. *Proc Natl Acad Sci U S A.* 2007 May 29; 104(22):9105–9106.
91. Wade, N. 2000. Lifes Origins Get Murkier and Messier. *The New York Times,* (Tuesday, June 13, pp. D1–D2.
92. Behe, M.J. 1996. *Darwin's Black Box.* New York: The Free Press, p. x.
93. Dose K. 1988. The origin of life: more questions than answers. Interdisciplinary *Science Reviews,* 13: 348.
94. Behe, M. J. 1996. *Darwin's Black Box.* New York: The Free Press, p. 186.
95. Doolittle. W. F. 1999. Phylogenetic Classification and the Universal Tree. *Science,* 284:2124–2128.
96. Woese, C.1998. The Universal Ancestor. *Proc Nat Acad Sci USA,* 95:6854–9859.
97. Schwabe, C. 1986. On the Validity of Molecular Evolution," *Trends in Biochemical Sciences,* Volume 11, July.
98. Denton, M. 1985. *Evolution: A Theory in Crisis.* Adler & Adler Publishers, Chevy Chase, MD, p. 249.

99. Monod, J. L. 1972. *Chance and Necessity*, Collins, London, p. 135.

100. Crick, F. 1981. *Life Itself.* Simon and Schuster, New York, p. 88.

101. Sermonti, G. 2005. *Why a Fly is not a Horse*, Discovery Institute, Seattle, Washington, p. 15.

102. Ibid. p. 63.

103. As quoted by *Henry Allen,* Fantasia, Washington Post, September 30, 1990.

Chapter Eleven - Embryology

1. Darwin, C. R. 1872. *On the origin of species by means of natural selection, or the preservation of favoured races in the struggle for life.* London: John Murray. 6th edition, p. 396.

2. Darwin, F., editor. 1887. *The life and letters of Charles Darwin, including an autobiographical chapter.* London: John Murray, Volume 2. p 339.

3. Ibid. pp. 121–122.

4. Ibid. p. 169.

5. Darwin, C. R. 1958. *The autobiography of Charles Darwin 1809–1882.* With the original omissions restored. Edited and with appendix and notes by his grand-daughter Nora Barlow. London: Collins, p. 125.

6. Darwin, C. R. 1871. *The descent of man, and selection in relation to sex.* London: John Murray. Volume. 1, 1st edition, p. 2.

7. Darwin, C. R. 1872. *On the origin of species by means of natural selection, or the preservation of favoured races in the struggle for life.* London: John Murray. 6th edition, p. 434.

8. Ibid. p. 382.

9. Darwin, C. R. 1871. *The descent of man, and selection in relation to sex.* London: John Murray. Volume, 1. 1st edition. p. 107.

10. Darwin, C. R. 1958. *The autobiography of Charles Darwin 1809–1882.* With the original omissions restored. Edited and with appendix and notes by his grand-daughter Nora Barlow. London: Collins, p. 125.

11. Darwin, C. R. 1872. *On the origin of species by means of natural selection, or the preservation of favoured races in the struggle for life.* London: John Murray. 6th edition, p. 381.

12. Ibid. p. 397.

13. Ibid. p. 386.

14. Ibid. p. 310.

15. Ibid. p. 310.

16. Sedgwick, A. 1894. On the Law of Development commonly known as von Baer's Law; and on the Significance of Ancestral Rudiments in Embryonic Development. *Quarterly Journal of Microscopic Science,* 36:35–52.

17. Ibid. pp. 35–52.

18. Sedgwick, A. 1894. On the Law of Development commonly known as von Baer's Law; and on the Significance of Ancestral Rudiments in Embryonic Development. *Quarterly Journal of Microscopic Science,* 36:35–52.

19. Anon. 2008. *Letter 5500 — Darwin, C. R. to Haeckel, E. P. A., 12 Apr [1867].* Darwin Correspondence Project. Available at www.darwinproject.ac.uk.

20. Gould, S. J. 2000. Abscheulich! - Atrocious! - the precursor to the theory of natural selection. *Natural History.* March, p. 42.

21. Shapiro, F. R., editor. 2006. *The Yale Book of Quotations.* Yale University Press.

22. Darwin, C. R. 1871. *The descent of man, and selection in relation to sex.* London: John Murray. Volume. 1. 1st edition, p. 32.

23. Sedgwick, A. 1894. On the Law of Development commonly known as von Baer's Law; and on the Significance of Ancestral Rudiments in Embryonic Development. *Quarterly Journal of Microscopical Science,* 36:35–52.

24. Sedgwick, A. 1909. *The influence of Darwin on the study of animal embryology,* pp 171–184, quoted in Seward, A. C., editor. *Darwin and Modern Science.* Cambridge University Press, 1909, pp. 174–176.

25. Lillie, F. R. 1908. *The Development of the Chick. An Introduction to Embryology.* New York, Henry Holt, pp. 4–6.

26. Garstang, W. 1922. The theory of recapitulation: a critical restatement of the biogenetic law. *Journal of the Linnean Society (Zoology),* 35:81–101.

27. Waddington, C. H. 1956. *Principles of Embryology.* London: George Allen and Unwin Ltd., p. 10.

28. De Beer, G. 1958. *Embryos and Ancestors, Third Edition.* Oxford: Clarendon Press, p. 10.

29. Ibid. p. 172.

30. De Beer, G. R. 1951. *Embryos and Ancestors, Revised Edition.* London: Oxford University Press, p. 10.

31. Ehrlich, P. R. and Holm, R. W. 1963. *The Process of Evolution.* New York: McGraw-Hill, p. 66.

32. De Beer, G. R. 1964. *An Atlas of Evolution.* New York: Nelson, p. 38.

33. Simpson, G. G. and Beck, W. S. 1965. *Life: An Introduction to Biology.* New York: Harcourt, Brace & World, Inc., p. 241.

34. Bock, W. J. 1969. Evolution by Orderly Law. *Science,* 164:684–685.

35. Blechschmidt, E. 1977. *The Beginnings of Human Life.* Heidelberg Science Library, p. 32.

36. Ballard, W. W. 1976. Problems of gastrulation: real and verbal. *BioScience,* 26:38.

37. Darwin, C. R. 1872. *On the origin of species by means of natural selection, or the preservation of favoured races in the struggle for life.* London: John Murray. 6th edition, p. 381.

38. Ibid. p. 396.

39. Rutimeyer, L. 1868, Referate, *Archiv fur Anthropologie,* p 301–302.

40. Richard P. Elinson, R. P. 1987. Change in developmental patterns: embryos of amphibians with large eggs. pp. 1–21. Quoted in. Raff, R. A, and Raff, E. C., editors. *Development as an Evolutionary Process,* Volume 8. New York: Alan R. Liss, p. 3.

41. Richardson, M. *et al.* 1997. There is no highly conserved embryonic stage in the vertebrates: implications for current theories of evolution and development. *Anatomy and Embryology,* 196(2):91–106.

42. Hawkes, N. 1997. *The Times* (London). 11 August, p. 14.

43. Pennisi, E. 1997. Haeckel's Embryos: Fraud Rediscovered, *Science,* 277(5331):1435.

44. Hawkes, N. 1997. *The Times* (London), 11 August, p. 14.

45. Richardson. M. K. and Keuck, G. 2001. A Question of Intent: When Is a 'Schematic' Illustration a Fraud? *Nature,* 410(3):144.

46. Richardson, M. K. 1997. Haeckel's Embryos: Fraud Rediscovered. Quoted by Pennisi, E. *Science,* 277(9):1435.

47. Anon. 1997. Embryonic fraud lives on. *New Scientist* 155(2098):23.

48. Pennisi, E. 1997. Haeckel's Embryos: Fraud Rediscovered, *Science* 277(5331):1435.

49. Hutchinson, M. L. 1911. Quoted in "The Truth about Haeckel's Confession." *The Bible Investigator and Inquirer.* Melbourne, 11 March, pp. 22–24.

50. Hitching, F.1982. Quoted in *The Neck of the Giraffe: Where Darwin Went Wrong.* Ticknor and Fields, New York, p. 204.

51. Thomson, K. S. 1988. Ontogeny and Phylogeny Recapitulated. *American Scientist, May–June,* 76:273.

52. Darwin, C. R. 1872. *On the origin of species by means of natural selection, or the preservation of favoured races in the struggle for life.* London: John Murray. 6th edition, p. 382.

53. Ibid. p. 203.

54. Ibid. p. 434.

55. Keith, A. 1932. *The Human Body.* London: Thornton and Butterworth, p. 94.

56. Thompson W. R. 1956. "Introduction," *Origin of Species, by Charles Darwin.* London: Dent, Everyman's Library edition, p. 12.

57. Ehrlich, P. R. and Richard W. Holm, R. W. 1963. *The Process of Evolution.* New York: McGraw-Hill, p. 66.

58. Hardy, A. 1965. *The Living Stream.* Collins, London, p. 213.

59. De Beer, G. 1958. *Embryos and Ancestors,* 3rd Edition. Oxford: Clarendon Press.

60. De Beer, G. 1971. *Homology: An Unresolved Problem.* Oxford University Press, London, p. 8.

61. Ibid. p. 8.
62. Ibid. p. 8.
63. Ibid. p. 8.
64. Simpson, G. G. and Beck, W. S. 1965. *Life: An Introduction to Biology.* New York: Harcourt, Brace & World, p. 240.
65. Simpson, G. G. and Beck, W. S. 1965. *Life: An Introduction to Biology.* New York: Harcourt, Brace & World, p. 241.
66. Alberch, P. 1985. Problems with the Interpretation of Developmental Sequences. *Systematic Zoology* 34(1):51.
67. Denton, M. 1985. *Evolution: A Theory in Crisis.* Adler & Adler Publishers, Chevy Chase, MD, p. 151.
68. Hinchliffe, R. 1990. Towards a Homology of Process: Evolutionary Implications of Experimental Studies on the Generation of Skeletal Pattern in Avian Limb Development, p. 121. Quoted in Smith J. M. and Vida, G, editors. *Organizational Constraints on the Dynamics of Evolution,* pp. 119–131. Manchester, UK: Manchester University Press.
69. McNamara, K. 1999. Embryos and Evolution. *New Scientist,* Volume 12416, 16 October.
70. Laubichler, M. D. and Maienschein, J. 2007. *From Embryology to Evo-Devo. A History of Developmental Evolution.* The MIT Press, p. 3.
71. Hall, B. K. 2005. *Scientific American Magazine.* April. Available at www.sciam.com.
72. Anon. 2007. *ScienceDaily,* March. Available at www.sciencedaily.com.
73. Laubichler M. D. and Maienschein, J. 2007. *From Embryology to Evo-Devo. A History of Developmental Evolution.* The MIT Press, p. 5.
74. Pennisi, E. 2002. Evolutionary Biology: Evo-Devo Enthusiasts Get Down to Details. *Science,* 298 (5595):953–955.
75. Darwin, C. R. 1872. *On the origin of species by means of natural selection, or the preservation of favoured races in the struggle for life.* London: John Murray. 6th edition, p. 204.

Chapter Twelve - The Rise of Genetics

1. Darwin, C. R. 1958. *The autobiography of Charles Darwin 1809–1882.* With the original omissions restored. Edited and with appendix and notes by his grand-daughter Nora Barlow. London: Collins. P. 130.

2. Ibid. p. 57.

3. Ibid. p. 64.

4. Lamarck, Jean-Baptiste. 1809. *Philosophie Zoologique,* p. 113.

5. Ibid. p. 113.

6. Darwin, C. R. 1872. *The origin of species by means of natural selection, or the preservation of favoured races in the struggle for life.* London: John Murray. 6th edition, p. xiii.

7. Ibid. p. 188.

8. Darwin, F. and Seward, A. C., editors. 1903. *More letters of Charles Darwin. A record of his work in a series of hitherto unpublished letters.* London: John Murray. Volume 2. p. 71.

9. Darwin, C. R. 1868. *The variation of animals and plants under domestication.* London: John Murray. First edition, first issue. Volume 2. pp. 357–358.

10. Ibid. p. 377.

11. Darwin, F., editor. 1887. *The life and letters of Charles Darwin, including an autobiographical chapter.* London: John Murray. Volume 3, p. 72.

12. Darwin, F., editor. 1887. *The life and letters of Charles Darwin, including an autobiographical chapter.* London: John Murray. Volume 3, p. 73.

13. Sermonti, G. 2005. *Why a Fly is not a Horse,* Discovery Institute, Seattle, Washington, p. 34.

14. Darwin, F. and Seward, A. C., editors. 1903. *More letters of Charles Darwin. A record of his work in a series of hitherto unpublished letters.* London: John Murray. Volume 2. p. 358.

15. Darwin, F., editor. 1887. *The life and letters of Charles Darwin, including an autobiographical chapter.* London: John Murray. Volume 3, p. 75.

16. Ibid. p. 78.

17. Darwin, F. and Seward, A. C., editors. 1903. *More letters of Charles Darwin. A record of his work in a series of hitherto unpublished letters.* London: John Murray. Volume 1, p. 128.

18. Darwin, C. R. 1872. *The origin of species by means of natural selection, or the preservation of favoured races in the struggle for life.* London: John Murray. 6th edition, p. 380.

19. Ibid. p. 247.

20. Sermonti, G. 2005. *Why a Fly is not a Horse*, Discovery Institute, Seattle, Washington, p. 46.

21. Weismann, A. 1883. *Ueber die Vererbung.* Translated as *Essays upon Heredity and Kindred Biological Problems, Second Edition.* London, in 1891, Volume I, pp. 67–106.

22. De Vries, H. 1905, *Species and Varieties: Their Origin by Mutation.* University of California.

23. Sermonti, G. 2005. *Why a Fly is not a Horse,* Discovery Institute, Seattle, Washington, p. 46.

24. Wallace, A. W. 1908. The Present Position of Darwinism. *Contemporary Review,* August, 1908.

25. Darwin, C. R. 1872. *The origin of species by means of natural selection, or the preservation of favoured races in the struggle for life.* London: John Murray. 6th edition; with additions and corrections, p. 106.

26. Ibid. p. 159.

27. Ibid. p. 131.

28. Ibid. p. 98.

29. Ibid. pp. 413–414.

30. Morgan, T. H. 1934. *Embryology and Genetics.* New York: Columbia University Press, p. 134.

31. Dobzhansky, T. 1937. *Genetics and the Origin of Species.* New York, Columbia University Press, p. 13.

32. Ibid. p. 13.

33. Beadle, G. W. and Tatum, E. L. 1941. Genetic control of biochemical reactions in *Neurospora. Proc Natl Acad Sci,* 27(11):499–506.

34. Avery, O. T. et al. 1944. Studies on the Chemical Nature of the Substance Inducing Transformation of Pneumococcal Types. *Journal of Experimental Medicine.* 79(2)137–158.

35. Simpson, G. G. 1944. *Tempo and Mode in Evolution*. New York: Columbia Univ. Press.

36. Hershey, A. D. and Chase, M. 1952. Independent functions of viral protein and nucleic acid in growth of bacteriophage. *J Gen Physiol.*, 36:39–56.

37. Watson J. D. and Crick, F. H. 1953. Molecular structure of Nucleic Acids. *Nature*, 171:737–738.

38. Simpson, G. G. et al. 1957. *Life: An Introduction to Biology*. NewYork: Harcourt, Brace and World, p. 430.

39. Crick, F. H. C. 1958. On Protein Synthesis. *Symp. Soc. Exp. Biol.* XII, 139–163.

40. Sermonti, G. 2005. *Why a Fly is not a Horse,* Discovery Institute, Seattle, Washington, p. 40.

41. Jacob, F. 1974. *The Logic of Life: A History of Heredity*. Translated by Spillmann, B. E. New York: Pantheon Books.

42. Monod. J. 1971. Chance and Necessity: An Essay on the Natural Philosophy of Modern Biology. New York, Alfred A. Knopf.

43. Crick, F. 1970. Central Dogma of Molecular Biology. *Nature,* 227: 561–563.

44. Futuyma, D. J. 1983. *Science on Trial*. New York: Pantheon Books.

45. Sermonti, G. 2005. *Why a Fly is not a Horse,* Discovery Institute, Seattle, Washington.

46. Lewis, E. B. 1978. A gene complex controlling segmentation in Drosphilia. *Nature,* 276:565–270.

47. Mayr, E. 1970. *Populations, Species, and Evolution,* Belknap Press, Cambridge, p. 235.

48. Sermonti, G. 2005. *Why a Fly is not a Horse,* Discovery Institute, Seattle, Washington, p. 102.

49. Anon. 2008. Quote in *Evolution* Wikipedia. Available at www. wikipedia.org/wiki/evolution.

50. Darwin, C. R. 1872. *The origin of species by means of natural selection, or the preservation of favoured races in the struggle for life*. London: John Murray. 6th edition; with additions and corrections, p. 413–414.

51. Grassé, Pierre-P. *Evolution of Living Organisms*. New York: Acad. Press, p. 130.

52. Sermonti, G. 2005, *Why a Fly is not a Horse*, Discovery Institute, Seattle, Washington, p. 82.
53. Ibid. p. 37.
54. Ibid. p. 39.
55. Ibid. pp. 38, 39.
56. Ibid. p. 82.
57. Ibid. p. 38.
58. Ibid. p. 38.
59. Ibid. p. 43.

Chapter Thirteen - Evidence
1. Darwin, F., editor. 1887. *The life and letters of Charles Darwin, including an autobiographical chapter*. London: John Murray. Volume 3, p. 64.
2. Darwin, C. R. 1872. *The origin of species by means of natural selection, or the preservation of favoured races in the struggle for life*. London: John Murray. 6th edition, p. 4.
3. Ibid. p. 70.
4. Bumpus, H. C. 1899. The elimination of the unfit as illustrated by the introduced House Sparrow, *Passer domesticus. Biol. Lectures, Marine Biol. Lab.*, Woods Hole, pp. 209–226.
5. Kettlewell, H. B. D. 1955. *Selection experiments on industrial melanism in the Lepidoptera, Heredity*, 9:323–342.
6. Ibid. pp. 287–301.
7. Ibid. pp. 287–301.
8. Kettlewell, H. B. D. 1959. Darwin's Missing Evidence. *Scientific America*, 200:48–53.
9. Sheppard, P. M. 1975. *Natural Selection and Heredity, Fourth Edition*. London: Hutchinson University Library, p. 70.
10. Darwin, C. R. 1872. *The origin of species by means of natural selection, or the preservation of favoured races in the struggle for life*. London: John Murray. 6th edition; with additions and corrections, p. 295–296.
11. Bishop, J. A. 1972. An experimental study of the cline of the industrial melanism in *Biston betularia* (L.) (Lepidoptera) between urban Liverpool and rural North Wales. *Journal of Animal Ecology*, 41:209–243.

12. Lees, D. R. and Creed. E. R. 1975. Industrial melanism in Biston betularia: the role of selective predation. *Journal of Animal Ecology.* 44:67–83.
13. Mikkola, K. 1984. On the selective forces acting in the industrial melanism of *Biston* and *Oligia* moths. (Lepidoptera: Geometridae and Noctuidae). *Biological Journal of the Linnean Society* 21:409–421.
14. Sermonti, G and Catastini, P. 1984. On industrial melanism: Kettlewell's missing evidence. *Rivista di Biologia* 77:35–52.
15. Sibatani, A. 1999. Industrial melanism revisited. *Rivista di Biologia* 92:349–356.
16. Sargent, T. et al. 1998. The "Classical" explanation of industrial melanism. Assessing the Evidence. *Evolutionary Biology* 30:299–322.
17. Coyne, J. 1998. Not black and white, a review of Michael Majerus's Melanism: Evolution in action, *Nature* 396:35–36.
18. Ibid. pp. 35–36.
19. Ibid. pp. 35–36.
20. Ibid. pp. 35–36.
21. Ibid. pp. 35–36.
22. Sermonti, G. 2005. *Why a Fly is not a Horse,* Discovery Institute, Seattle, Washington, p. 51.
23. Darwin, C. R. 1860. *Journal of researches into the natural history and geology of the countries visited during the voyage of H.M.S. Beagle round the world, under the command of Capt. Fitz Roy R.N.* London: John Murray. Tenth thousand. Final text, p 373.
24. Steinheimer, F. D. 2004. Charles Darwin's bird collection and ornithological knowledge during the voyage of H.M.S. *Beagle,* 1831–1836. *Journal of Ornithology* 145(4):300–320.
25. Darwin, C. R. 1845. *Journal of researches into the natural history and geology of the countries visited during the voyage of H.M.S. Beagle round the world, under the Command of Capt. Fitz Roy, R.N.* 2nd edition. London: John Murray. p. 395.
26. Steinheimer, F. D. 2004. Charles Darwin's bird collection and ornithological knowledge during the voyage of H.M.S. *Beagle,* 1831–1836. *Journal of Ornithology* 145(4):300–320.

27. Sulloway, F. J. 1982. The *Beagle* collections of Darwin's finches (*Geospizinae*). *Bulletin of the British Museum (Natural History) Historical Series* 43(2):49–94.

28. Darwin, C. R. 1845. *Journal of researches into the natural history and geology of the countries visited during the voyage of H.M.S. Beagle round the world, under the Command of Capt. Fitz Roy, R.N.* 2nd edition. London: John Murray, p.380.

29. Steinheimer, F. D. 2004. Charles Darwin's bird collection and ornithological knowledge during the voyage of H.M.S. *Beagle*, 1831–1836. *Journal of Ornithology* 145(4):300–320.

30. Darwin, C. R. 1845. *Journal of researches into the natural history and geology of the countries visited during the voyage of H.M.S. Beagle round the world, under the Command of Capt. Fitz Roy, R.N.* 2d edition. London: John Murray. p. 380.

31. Sulloway, F. J. 1982. The *Beagle* collections of Darwin's finches (*Geospizinae*). *Bulletin of the British Museum (Natural History) Historical Series* Vol. 43 no. 2: 49–94.

32. Sulloway, F. J. 1982. Darwin and His Finches: The evolution of a Legend, *Journal of the History of Biology* 15:1–53.

33. Weiner J. *The Beak of the Finch,* New York: Vintage Books, 1994, p 9.

34. Gibbs, H. L. and Grant, P. R. 1987. Oscillating Selection on Darwin's Finches. *Nature* 3237:511–513.

35. Grant, P. R. 1991. Natural Selection and Darwin's Finches, *Scientific American* 265: 82–87.

36. Mayr, E. 1942. *Systematics and the Origin of Species.* Columbia Univ. Press. New York, p, 120.

37. Grant B. R and Grant, P.R. 1993. Evolution of Darwin's finches caused by a rare climatic event. *Proceedings of the Royal Society of London Ser B* 251:111–117.

38. Grant, P. R. and Grant, B. R. 1992. *Hybridization of Bird Species. Science,* 256:193–197.

39. Ibid. pp. 193–197.

40. Grant, P. R., et al. 2007. Convergent evolution of Darwin's finches caused by introgressive hybridization and selection, *Evolution* 58(7):1588–1599.

41. Sato, A, et al. 1999. Phylogeny of Darwin's finches as revealed by mtDNA sequences. *Proc Natl Acad Sci U S A* 96(9):5101–5106.

42. Sato, A., et al. 2001. On the Origin of Darwin's Finches. *Molecular Biology and Evolution* 18:299–311.

43. Simpson, G. G. et al. 1957. *Life: An Introduction to Biology*. New York: Harcourt, Brace and World, p. 430.

44. Futuyma, D. 1983. *Science on Trial*. New York: Pantheon Books.

45. Raven P. and Johnson, G. 1999, *Biology*, W C B/McGraw-Hill, 5th Edition.

46. Quammen, Q. 2004. Was Darwin wrong? *National Geographic*, November 2004, p. 21.

47. Luria, S. E. and Delbrück, M. 1943. Mutations of Bacteria from Virus Sensitivity to Virus Resistance. *Genetics* 28:491–511.

48. Zinder, N. D. and Lederberg, J. 1952. *Genetic Exchange in Salmonella. Journal of Bacteriology* 64(5):679–699.

49. Struzik, E. 1990. Ancient bacteria revived. *Sunday Herald*. Calgary, Alberta, Canada. Sept., Section A1, p. 16.

50. Ribeiro R. M. and Bonhoeffer, S. 2000. Production of resistant HIV mutants during antiretroviral therapy, *Proceedings of the National Academy of Sciences* 97:7681–7686.

51. Ibid. pp. 7681–7686.

52. Ayala, F. J. 1978. The Mechanisms of Evolution. *Scientific American* 239(3):56–69. p. 65.

53. Grassé, P. P. 1977. *The Evolution of Living Organisms*. New York: Academic Press, p. 87.

54. Ibid. p. 87.

55. Lovtrup, S. 1987. *Darwinism: The Refutation of a Myth*. London: Croom Helm, p. 422.

56. Darwin, C. R. 1872, *The origin of species by means of natural selection, or the preservation of favoured races in the struggle for life*. London: John Murray. 6th edition; with additions and corrections, p. 129.

57. Ibid. p. 188.

58. Cairns, J. et al. 1988. The origin of mutants, *Nature* 335:142–145.

59. Slechta, E. S. 2002. Evidence That Selected Amplification of a Bacterial lac Frameshift Allele Stimulates Lac⁺ Reversion (Adaptive Mutation) With or Without General Hypermutability. *Genetics* 161:945–956.

60. Sala, M. and Wain-Hobson, S. 2000. *Are RNA Viruses Adapting or Merely Changing?* Journal of Molecular Evolution, 51(1):12–20.

61. Futuyma, D. J. 1983. *Science on Trial: The Case for Evolution.* New York: Pantheon, pp. 137–138.

62. Mayr, E. 1942. *Systematics and the Origin of Species,* p. 296.

63. Muller, H. 1950.Radiation Damage to the Genetic Material. *American Scientist* 38:33–50,126, p. 353.

64. Martin, C. P. 1953. A Non-Geneticist Looks at Evolution. *American Scientist*, January, p. 102.

65. Dobzhansky, T. 1955. *Evolution, Genetics and Man.* New York: John Wiley & Sons, p. 105.

66. Kimura, M. 1976. Population Genetics and Molecular Evolution. *The Johns Hopkins Medical Journal* 138(6):253–261, p. 260.

67. Mayr, E. 2001. *What Evolution Is.* New York: Basic Books, p. 98.

68. Bull J. J. et al. 2007. Theory of Lethal Mutagenesis for Viruses, *J Virology,* 81(6):2930–2939.

69. Baer, C. F. 2008. Does Mutation Rate Depend on Itself? *PLoS Biol.* 6(2): e52.

70. Lodish H, et al. 2004. *Molecular Biology of the Cell.* WH Freeman: New York, NY. 5th ed., p. 963.

71. Lynn M. and Sagan, D. 2002. *Acquiring Genomes: A Theory of the Origins of Species.* New York: Basic Books, pp. 11–12.

72. Ibid. pp. 28–29.

73. Grassé, P. P. 1977. *Evolution of Living Organisms.* New York: Academic Press, p. 103.

74. Sermonti, G. 2005, *Why a Fly is not a Horse*, Discovery Institute, Seattle, Washington, p. 102.

75. Ibid. p. 103.

76. Goodwin, B. C. 1985. What Are the Causes of Morphogenesis? *Bioessays* 3: 32–36, p. 35.

77. Nijhout, H.F. 1990. Metaphors and the Role of Genes in Development. *Bioessays* 12: 441–446, p. 444.

78. Raff, R. A. and Kaufman, T. C. 1983. *Embryos, Genes, and Evolution.* New York: Macmillan. pp. 67.

79. Darwin, C. R. 1872. *The origin of species by means of natural selection, or the preservation of favoured races in the struggle for life.* London: John Murray. 6th edition; with additions and corrections, p. 143.

80. Salisbury, F. 1971. Doubts About the Modern Synthetic Theory of Evolmution. *American Biology Teacher,* p. 338.

81. Sermonti, G. 2005, *Why a Fly is not a Horse,* Discovery Institute, Seattle, Washington, p. 110.

82. De Beer, G. 1971. *Homology: An Unsolved Problem.* London: Oxford University Press, pp. 15–16.

83. Ibid. pp. 15–16.

84. Wray, G. 1999. *Evolutionary dissociations between homologous genes and homologous structures', Homology.* Novartis Symposium 222; Chichester, UK: John Wiley & Sons, pp. 195–196.

85. Sermonti, G. 2005. *Why a Fly is not a Horse,* Discovery Institute, Seattle, Washington, p. 110.

86. Ibid. p. 110.

87. Moyal, A. 2001. *Platypus. Allen and Unwin,* New South Wales, Australia, p. xii.

88. Warren, W.C. et al. 2008. Genome analysis of the platypus reveals unique signatures of evolution. *Nature,* May 8.

89. Warren, W.C. et al. 2008. Genome analysis of the platypus reveals unique signatures of evolution. *Nature,* May 8.

90. Brown, S. 2008. Top billing for platypus at end of evolution tree Monotreme's genome shares features with mammals, birds and reptiles. Published online 7 May 2008 | *Nature* | doi:10.1038/453138a.

91. Murchison, E. P. *et al. Genome Res.* doi:10.1101/gr.73056.107.

92. Archer, M. et al. 1992. Description of the skull and non-vestigial dentition of a miocene platypus. In *Platypus and echidnas.* Augee M. L., editor, Sydney: The Royal Zoological Society of New South Wales, pp. 15–27.

93. Sullivan, R. 2008. Scientists map the genetic makeup of the platypus, *Associated Press*, May 08.

94. Brown. S. 2008. Top billing for platypus at end of evolution tree Monotreme's genome shares features with mammals, birds and reptiles Published online 7 May 2008 | *Nature* | doi:10.1038/453138a.

95. Beadle, G.W., Tatum, E.L. 1941. Genetic control of biochemical reactions in Neurospora. *Proc Natl Acad Sci* 27(11):499–506.

96. Denton, M. 1985. *Evolution: A Theory in Crisis.* Adler & Adler Publishers, Chevy Chase, MD, p. 149.

97. Ibid. p. 149–150.

98. Frederick H. W. et al. 2004. A Cluster of Metabolic Defects Caused by Mutation in a Mitochondrial tRNA. *Science* 306 (5699), 1190.

99. Endler, J. A. and McLellan T. 1988. The Process of Evolution: Toward a Newer Synthesis. *Annual Review of Ecology and Systematics,* 19:397.

100. Orr, H. A. and Coyne, J. A. 1992. The Genetics of Adaptation. A Reassessment. *American Naturalist,* 140:726.

101. Anon. 2008. *Secret of Life: "Accidents of Creation."* 2001 WGBH Educational Foundation and Clear Blue Sky Productions, Inc., at pbs.org/wgbh/evolution/library/01/2/l_012_02.html.

102. Pauling, L. et al. 1949. Sickle Cell Anemia, a Molecular Disease. *Science* 110:543–548.

103. Anon. 2008. Quotation in National Institute of Health. Available at www.nih.gov/about/researchresultsforthepublic/SickleCellDisease.pdf.

104. Anon. 2008. Quotation in National Institute of Health. Available at http://ghr.nlm.nih.gov/handbook/mutationsandd isorders?show=all#mutationscausedisease.

105. Sermonti, G. 2005. *Why a Fly is not a Horse*, Discovery Institute, Seattle, Washington, p. 15.

106. Lipson, H.S. 1980. A Physicist Looks at Evolution. *Physics Bulletin,* 31:138.

107. Sermonti, G. 2005. *Why a Fly is not a Horse*, Discovery Institute, Seattle, Washington, p. 43.

108. Anon. 2008. Quotation in *Evolutionary Development Biology Wikipedia*. Available online at http://en.wikipedia.org/wiki/EvoDevo.

109. Eldredge, N. 2006. Niles Eldredge forecasts the future. Quotation in *NewScientist.com*, November 18.

110. Darwin, C. R. 1872. *The origin of species by means of natural selection, or the preservation of favoured races in the struggle for life*. London: John Murray. 6th edition; with additions and corrections, p. 175.

111. Darwin, C. R. 1872. *The origin of species by means of natural selection, or the preservation of favoured races in the struggle for life*. London: John Murray. 6th edition; with additions and corrections, p. 131.

112. Twain, M. 1883. *Life on the Mississippi*. Boston, MA: J.R. Osgood, p. 156.

113. Mayr, E. 2006. The Evolution of Ernst: Interview with Ernst Mayr. Quotation in *Scientific American*, Mirsky, S, editor, July. Available at www.sciam.com.

114. Darwin, C. R. 1872. *The origin of species by means of natural selection, or the preservation of favoured races in the struggle for life*. London: John Murray. 6th edition, p. 404.

115. Ibid. p. 146.

116. Ibid. p. 134.

Index

Printed in the United States
154587LV00003B/3/P